# GAME THEORY
## A Comprehensive Introduction

# GAME THEORY
## A Comprehensive Introduction

**Hans Keiding**

*University of Copenhagen, Denmark*

**World Scientific**

NEW JERSEY • LONDON • SINGAPORE • BEIJING • SHANGHAI • HONG KONG • TAIPEI • CHENNAI

*Published by*

World Scientific Publishing Co. Pte. Ltd.

5 Toh Tuck Link, Singapore 596224

*USA office:* 27 Warren Street, Suite 401-402, Hackensack, NJ 07601

*UK office:* 57 Shelton Street, Covent Garden, London WC2H 9HE

**British Library Cataloguing-in-Publication Data**
A catalogue record for this book is available from the British Library.

**GAME THEORY**
**A Comprehensive Introduction**

ISBN 978-981-4623-65-0

In-house Editor: Sandhya Venkatesh

Printed in Singapore

# Preface

Game theory has come a long way from its beginning in the 1930s, when a few mathematicians derived the first results, and to its current situation, where its terminology has entered into standard textbooks and where it is taught regularly at universities. There is no longer a need for an initial methodological discussion (as in von Neumann and Morgenstern [1944]) in a book on game theory, since much of what seemed controversial at that time has become accepted and is commonplace today. There are now a large number of textbooks on game theory, ranging from introductory to advanced, and game theory has its own internet sites (e.g. `http://www.gametheory.net`) with textbooks, teaching materials etc.

It appears therefore that proposing yet another textbook must be followed up by some justification. Most authors have at some point aspired to produce a new and contemporary version of the very successful text by Luce and Raiffa [1957], but again most authors – including the present one – have realized that this may be beyond reach, perhaps because that book was written while the discipline was still young, reflecting the freshness of the whole field. As a unifying principle behind the present book I have chosen to emphasize the *diversity* of game theory, its many different aspects and its penetration into other fields of science.

This means that the coverage of topics is somewhat broader than what is usual, and the text is in some respects sitting between two chairs, being introductory in the sense that it does not presuppose any previous acquaintance with game theory, but also being comprehensive in the sense of treating material which is not standard but nevertheless belongs to the fundamental part of the theory, such as strategic stability of equilibria, communication equilibria, values of games without sidepayments to mention a few. In addition, it has been the aim to introduce some other topics not so often seen in textbooks on game theory, such as the combinatorial games, which have had their own development and which seem somewhat overlooked by traditional game theorists. Also, there is a specific chapter on game forms, the basic building blocks of game theory, again a field which hitherto has been reserved to specialists.

A return to the approach of early expositions of game theory is represented by the emphasis on two-person zero-sum games. Without these games and the remarkable minimax theorem, game theory might not have become a scientific discipline in its own right. Modern texts prefer to begin with games having more immediate applications, mainly in economics, but in doing so they lose some of the attractiveness of game theory, the transparent formal structure. The selection of topics treated in this book has been made with the purpose of displaying the structure of the theory rather than its practical applicability.

The book has grown out of lecture notes from different courses and with differing audiences, something which occasionally can be spotted in the text, where some chapters tend to be slightly more formal than others. However, the text is intended to be self-contained, demanding only some acquaintance with standard mathematical notation. Nowadays,there are several books available which can be used as mathematical reference for the reader with no specific mathematical background, and references are given when necessary.

The text has benefited greatly from the suggestions of many generations of students. In its final version, valuable assistance and advice was provided by Bodil O. Hansen, for which I am grateful.

*Hans Keiding*

# Contents

# Chapter 1

# Introduction

## 1 What is game theory?

A convenient way of introducing a theory – and one which actually works reasonably well in our case – is to define a certain object, in this case a game, and then declare that the theory consists of the analysis of this object.

An obvious drawback of this approach is that it gives little or no impression of why the theory is worthwhile; one has to fight one's way through a considerable body of theory before seeing what it can be used for.

Therefore, although we shall indeed use this method, defining first a *game* and then proceeding to analyze games over several chapters before we get to convincing applications, we begin with some more easy-going considerations as a motivation for what comes later.

Broadly speaking, game theory is the analysis of conflict situations. The type of conflict is irrelevant, indeed the attraction of game theory lies partly in the fact that superficially very different phenomena such as parlor games and war has common features, which can be studied with largely the same analytical tools, and probably it was this many-sidedness of game theory, which made it arise and survive as an independent scientific discipline.

On the other hand, not all conflict studies can be characterized as game theory, indeed game theory may constitute only a minor part of the general study of conflicts. It deals with the possibilities and the actual choices of the participants, when the conflict and its consequences are clearly delineated. General research in conflicts may also consider the emergence of conflicts as well as the way in which conflicts influence the minds of the parties.

Game theory is – at least in the version to be presented here – a mathematically formulated discipline, aiming at studying "rational" behavior in conflict situations. The very concept of rationality will occasionally be questioned, so we shall not overdo this program of finding rational behavior in conflicts. As in may other fields, the study of the theory will make us understand its limitations much better.

## 2   Definition of a game

In order to define a game we need some parties to the conflicts, to be called either *agents* or *individuals* or, most often, *players*. These players should have several possible choices, in the following referred to as *strategies*, and the conflict is described in such a way that the choices of a strategy by each player should result in an *outcome* for the players.

DEFINITION 1  *A* game form *is an array $G = (N, (S_i)_{i \in N}, \pi)$, where*

(i) *$N$ is a nonempty set of players,*

(ii) *for each player $i \in N$, $S_i$ is a set (of strategies of $i$),*

(iii) *$\pi : \times_{i \in N} S_i \to X$, where $X$ is a nonempty set (of outcomes), and $\pi$ is a map assigning to every strategy array $s = (s_i)_{i \in N}$ an outcome $\pi(s)$.*

The set $N$ of players can be arbitrarily large (in particular $N$ may be an infinite set, something which turns out to be useful in some contexts although not in this book). In what follows we shall assume that $N = \{1, \ldots, n\}$.

The outcome set $X$ occurring in Definition 1 has been supressed in our notation, where it is implicit in the definition of the outcome map $\pi$. If we need to be explicit about the outcome space, then we say that $G$ is a game form over $X$. If $G = (N, (S_i)_{i \in N}, \pi)$ is a game form and $f : X \to X'$ is a map, then we get another game form $G = (N, (S_i)_{i \in N}, f \circ \pi)$ (over $X'$), where the outcome map is $f \circ \pi$, the composition of $\pi$ with the map $f$. There is a *canonical* game form associated with the player set $N$ and the collection of strategy sets $(S_i)_{i \in N}$, namely the game form over $S = \times_{i \in N} S_i$,

$$G^0 = (N, (S_i)_{i \in N}, \mathrm{id}_S),$$

where $\mathrm{id}_S$ is the identical map on $S$. We also write $G^0$ as $(N, (S_i)_{i \in N})$, without mention of the identity outcome map. Technically, we may consider any game form $G = (N, (S_i)_{i \in N}, \pi)$ as the canonical game form $G^0$ followed by a map $\pi : S \to X$.

While the game form introduced in Definition 1 is the basic concept of the theory, and one to which we shall return at later occasions, it is not yet a game. The game form gives us the *rules of the game*, and to get to the game itself we need also to add some information about what constitutes good or bad outcomes: We need the preferences of the players over the set $X$ of outcomes. Assuming that the latter are given as complete preorders (complete, transitive and reflexive relations) $\succsim_i$ for $i = 1, \ldots, n$, we obtain a game as an array $\Gamma = (N, (S_i)_{i \in N}, \pi, (\succsim_i)_{i \in N})$, where $(N, (S_i)_{i \in N}, \pi)$ is a game form. However, in many cases – as a matter of fact, in all the cases which will be analyzed in this book – it is assumed that the preference relations $\succsim_i$ can be represented by utility functions $v_i$, satisfying

$$x \succsim_i x' \Leftrightarrow v_i(x) \geq v_i(x'),$$

for $i \in N$. If we define the composite map $u_i : \times_{i \in N} S_i \to \mathbb{R}$ by

$$u_i(s) = v_i(\pi(s)),$$

**Example 1.1. Rock-Paper-Scissors.** One of the simplest cases of a game form is the "Rock-Paper-Scissors"-game, where the players simultaneously display one of the three items (using fist, palm or two fingers), and the rules for choosing a winner are given below. The strategies of Player 1 are written as rows, those of Player 2 as columns.

|          | Rock    | Paper   | Scissors |
|----------|---------|---------|----------|
| Rock     | Draw    | 2 wins  | 1 wins   |
| Paper    | 1 wins  | Draw    | 2 wins   |
| Scissors | 2 wins  | 1 wins  | Draw     |

The table defines the outcome function with values in the space {1 wins, 2 wins, Draw}. Assigning for each player some utility values to the three outcomes, the game form is transformed to a game. Traditionally, it is assumed that each player obtains utility 1 from winning, 0 from a draw, and −1 when the other player wins. The resulting game is:

|          | Rock    | Paper    | Scissors |
|----------|---------|----------|----------|
| Rock     | $(0,0)$ | $(-1,1)$ | $(1,-1)$ |
| Paper    | $(1,-1)$| $(0,0)$  | $(-1,1)$ |
| Scissors | $(-1,1)$| $(1,-1)$ | $(0,0)$  |

The game is *zero-sum:* The gain of one player corresponds to a similar loss of the other player.

then reference to the outcome space $X$ becomes redundant. Consequently, the definition of a game will be as follows:

DEFINITION 2 *A game is an array* $\Gamma = (N, (S_i)_{i \in N}, (u_i)_{i \in N})$, *where*

    *(i) $N$ is a nonempty set of players,*

    *(ii) for each player $i$, $S_i$ is a set (of strategies of $i$),*

    *(iii) for each player $i$, $u_i : \bigtimes_{i \in N} S_i \to \mathbb{R}$ is a payoff map assigning a utility value $u_i(s)$ to each strategy array $s = (s_1, \ldots, s_n)$ in $S = \bigtimes_{i \in N} S_i$.*

In the literature a game $\Gamma$ defined in this way (or, to be more precise, the special case of $\Gamma$ where the preference relations $>_i$ are represented by utility functions)

**Example 1.2. Prisoners' dilemma.** This game is well-known to the extent that for many people it represents the only piece of game theory that they have ever encountered.

The story behind the game goes as follows: Two criminals have been caught and are to be put to trial, and the attorney proposes each of them a deal: If he confesses and witnesses against the other one who is not confessing, he will be set free immediately. If both confess, there is no need for a witness and both will be sentenced, and if none of them confesses, the attorney will make sure that they get at least some penalty. This gives us a normal form game as follows, where payoffs are in terms of years in prison, so that e.g. $(-10, 0)$ means that player 1 (choosing rows) gets 10 years of prison, while player 2, the column player, walks out free.

|             | Confess    | Not confess |
|-------------|------------|-------------|
| Confess     | $(-9, -9)$ | $(0, -10)$  |
| Not confess | $(-10, 0)$ | $(-1, -1)$  |

What will happen in this game does not concern us for the moment, we shall deal with this in later chapters, but it is seen at a first glance that confessing is better than not confessing for each player, no matter what the other player will choose. Therefore we might expect that both will confess, an outcome which is quite unpleasant for the two players involved.

is called a game in *normal* or *strategic* form , to distinguish it from the games in extensive form introduced in Section 4 below.

The game $\Gamma = (N, (S_i)_{i \in N}, (u_i)_{i \in N})$ is *constant-sum* if there is a number $K$ (the constant) such that $\sum_{i \in N} u_i(s) = K$ for all $s \in \times_{i \in N} S_i$, and it is *zero-sum* if it is constant sum with constant $K = 0$.

The use of utility payoffs rather than abstract outcomes combined with individual preference relations makes our analysis much simpler, and indeed much of early game theory saw the payoffs just as money gains. There are however problems hidden in the use of utilities; one of them is related to the foundation of utility representations, since we shall need to take expectations over payoffs, so that expected utility should make sense. Another, and perhaps more subtle problem is that the definition of a game as presented in Definition 2 blurs the distinction between *rules of the game* and *player objectives.* We shall return to this only in Chapter 16.

## 3 Games in extensive form

As it could be seen in the examples, many real world conflicts or games cannot immediately be presented in the form given by Definition 2. Such 'games' are often built by moves of one player or another, and there may even be an element of randomness, as is the case when a card is drawn or dice are thrown. In addition, it often happens that when moving, a player will know some but not all relevant details of the play – for example one player will not know the cards of another player.

In order to see how we can make our definition of a game to apply also to such situations we need to take an indirect route, introducing first an alternative way of representing a game. For this, we need some new notions.

A *graph* is a pair $G = (V, E)$, where $V$ is a set (of vertices or nodes) and $E$ a subset of $V \times V$. An element $(v_1, v_2)$ of $E$ is called an *edge* from $v_1$ to $v_2$. A path in $G$ from $v_1$ to $v_k$ is a finite set $C = \{e_1, \ldots, e_{k-1}\}$ of edges such that $e_1 = (v_1, v_2), e_2 = (v_2, v_3), \ldots, e_{k-1} = (v_{k-1}, v_k)$ for some vertices $v_1, v_2, \ldots, v_k$. A cycle in $G$ is a path from some $v \in V$ to itself. The graph $G$ is acyclic if it has no cycles.

A graph $G$ is *connected* if all its vertices can be connected by a path, and $G$ is a *tree* if it is acyclic and connected. If $v_0$ is a vertex such that there is no edge from some vertex to $v_0$, then $v_0$ is called the *root* of $G$ (since $G$ is connected, there can be at most one root). The vertex $v$ is called *terminal* if there is no edge in $G$ beginning at $v$.

**3.1. Extensive form games with perfect information.** We now have the ingredients needed for defining an extensive form game, but we shall proceed in two steps, dealing first with the case of perfect information (meaning that all players can observe what all the other players, including nature, do), and having introduced this case we can move to the more complicated general version.

DEFINITION 3  *A game in extensive form with perfect information is given by*

   *(i)* *a set $N = \{1, \ldots, n\}$ of players,*
   *(ii)* *a finite tree $\mathcal{T} = (V, E)$, with $V' \subset V$ being the non-terminal vertices,*
   *(iii)* *a partition $\mathcal{P}$ of $V'$ into at most $n + 1$ subsets $P_0, P_1, \ldots, P_n$,*
   *(iv)* *for each $v \in V'$, a subset $C(v)$ of $\{v' \in V \mid (v, v') \in E\}$, and if $v \in P_0$, a probability distribution $p_v$ over $C(v)$,*
   *(v)* *a map $u : V \backslash V' \to \mathbb{R}^n$ which to each terminal vertex $v$ assigns a utility payoff $(u_1(v), \ldots, u_n(v))$.*

We may write the extensive form game with perfect information as

$$\Gamma^e = (N, \mathcal{T}, \mathcal{P}, (C(v))_{v \in V}, u).$$

The interpretation of the components is as follows: The vertices of the tree are the *moves* of the game, and the partition $\mathcal{P}$ consists of sets $P_i$, for $i \in N$, where player $i$ is the player who draws, and $P_0$ where 'nature', a fictitious extra player, draws. The

**Example 1.3. Sharing a pie.** Two persons have a pie and must divide it into two pieces before eating it. A well-known method of division consists in letting one person (here Player 1) cut the pie, after which the other person (Player 2) chooses her piece. For simplicity we assume that there are only three possible cuts, namely $(1, 0)$, leaving it in one piece, $(3/4, 1/4)$, cutting into one small and one large piece, and finally $(1/2, 1/2)$, where the pie is cut into two pieces of equal size. Player 2 can then select the largest or the smallest of the two pieces (which of course amounts to the same in the case of equal division). Letting the payoffs be given by the size of the piece obtained, we get a game in extensive form with perfect information:

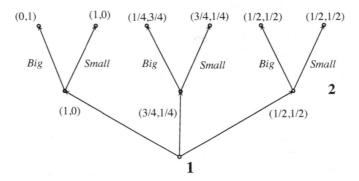

In the associated normal form game, Player 1 has 3 strategies, corresponding to the three ways of cutting the pie, and Player 2 has 2:

|              | Big         | Small       |
|--------------|-------------|-------------|
| $(1, 0)$     | $(0, 1)$    | $(1, 0)$    |
| $(3/4, 1/4)$ | $(1/4, 3/4)$| $(3/4, 1/4)$|
| $(1/2, 1/2)$ | $(1/2, 1/2)$| $(1/2, 1/2)$|

The game is *rectangular* in the sense that for each payoff configuration $(y_1, y_2)$, the set of strategy arrays which gives rise to this outcome is a product; thus $(0, 1)$ comes from $\{(1, 0)\} \times \{Big\}$, $(1/2, 1/2)$ from the set $\{(1/2, 1/2)\} \times \{Big, Small\}$, etc. Rectangular games (or rather, rectangular game forms) are discussed in Gurvič [1982].

draws that a player can make at the move $v$ are specified in $C(v)$ and they amount to selecting an edge from $v$ to some other vertex $v'$. At the terminal moves, each player receives a specified utility payoff.

The fictitious player 'nature' allows for random moves. For $v \in P_0$, 'nature' chooses the outgoing edge from $v$ according to the specified probability distribution. Since we are as yet dealing with perfect information games, the players observe the result of nature's choice. This would be the situation for example when the players choose colors in chess before starting the proper play of the game.

**3.2. Extensive form games with imperfect information.** In many cases – and indeed in many interesting cases – either nature's choices or the choices of some player are not fully revealed to the other players, who therefore have only partial knowledge of the way in which the game has been played before the move. We need a formal representation of such *imperfect information,* and the key to this is the introduction of *information sets,* sets of moves belonging to a player $i$ which she cannot distinguish. Clearly, information sets must be subsets of player sets, and they constitute a subpartition of $\mathcal{P}$, or alternatively, they form a partition which is *finer* than (or rather, as fine as) $\mathcal{P}$. Now the choice sets must also have another nature, since the choices must be the same everywhere in a given information set (otherwise they would reveal some further information), and choosing an element in this abstract choice set must result in selecting an outgoing edge from the actual vertex (although this vertex cannot be identified by the player).

DEFINITION 4 *An n-person game in extensive form is given by*

(i) *a set $N = \{1, \ldots, n\}$ of players,*
(ii) *a finite tree $\mathcal{T} = (V, E)$, with $V' \subset V$ being the non-terminal vertices,*
(iii) *a partition $\mathcal{P}$ of $V'$ into at most $n + 1$ subsets $P_0, P_1, \ldots, P_n$,*
(iv) *for each $v \in V'$, a set $C(v)$ and a map $h_v : C(v) \to \{v' \in V \mid (v, v') \in E\}$,*
(v) *for $P_i \in \mathcal{P}$, $i = 1, \ldots, n$, a partition $\mathcal{I}_i$ of $P_i$ finer than $\mathcal{P}$ and such that $P_0 \in \mathcal{P}$,*
(vi) *for $v \in P_0$, a probability distribution $p_v$ over $C(v)$,*
(vii) *a map $u$ which to each terminal vertex $v$ assigns a utility payoff $(u_1(v), \ldots, u_n(v))$, so that $C(v) = C(v')$ if $v, v' \in I$, some $I \in \mathcal{I}_i$, $i = 1, \ldots, n$.*

We write the general extensive form game as $\Gamma^e = (N, \mathcal{T}, \mathcal{P}, (C(v), h_v)_{v \in V}, (\mathcal{I}_i)_{i \in N}, (p_v)_{v \in P_0}, u)$. Since the writing down of all the details tends to become a lengthy process, one usually skips the parts of the description which are not essential for the reasoning, and much of the latter will rely on the graphical representation of the game.

**3.3. Transforming an extensive form into a normal form game.** There is a standard procedure by which a game in extensive form can be transformed to a game in normal form. For this, we need to define strategy sets for each player (except the fictitious player 'nature') and a payoff function. Intuitively, the method of converting extensive form games consists identifying strategies as full specifications of actions. A player can send a substitute once a strategy has been chosen, since the substitute will know what to do no matter what the other players, including nature, might choose.

DEFINITION 5 *Let $\Gamma^e = (N, \mathcal{T}, \mathcal{P}, (C(v), h_v)_{v \in V}, (\mathcal{I}_i)_{i \in N}, (p_v)_{v \in P_0}, u)$ be an extensive form game, and let $i \in N$ be a player. A strategy for $i$ is a map $s_i$ which to each $v$ in $P_i$, the set of* *i's moves, assigns an element $s_i(v)$ of $C(v)$, subject to the (informational) condition*

$$v, v' \in I \text{ for some } I \in \mathcal{I}_i \implies s_i(v) = s_i(v'), \tag{1}$$

---

**Example 1.4. Paper-Stone-Scissors as extensive form game.** Instead of presenting the game in normal form as in Example 1.1, we could have started with an extensive form game as the following:

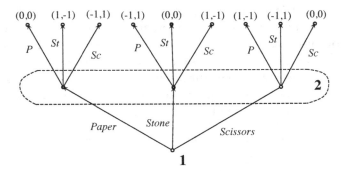

Here player 1 is the first to draw, followed by player 2. Since each player must make the choice without knowing the choice of the other player, we need that all the nodes which are reached after the move by player 1 belong to the same information set, indicated by the area bounded by the dashed curve.

In order to get a normal form game from this extensive form, we must first identify the strategies of the two players, which are the three possible moves at the single instance where the player moves, that is Paper, Stone and Scissors. It is immediate that we get the normal form game from Example 1.1.

If alternatively player 2 had the first move, leaving player 1 with a single information set, the resulting extensive form game would have represented the same conflict, and the resulting normal form game would be the same.

---

In our interpretation, a strategy is a full description of how the player should choose whenever she must make a choice.

To define the outcome functions $u_i$, we need the notion of *play* of the game $\Gamma_E$. Let $s = (s_1, \ldots, s_n)$ be an array of strategies $s_i$ of each player $i = 1, \ldots, n$. Each of the strategies $s_i$ assigns an element of $C(v)$ to each vertex $v \in P_i$, where player $i$ has the move. For each $v \in P_0$, we choose an element $v'$ of $C(v)$ and let $p_v(v')$ be the associated probability. Starting with the root, the collection of strategies for the players and the choice function defined for nature gives rise to a path in $\mathcal{T}$, and by finiteness of paths, it defines a unique terminal vertex $v^*$ and a probability $p(v^*)$ defined as the product of all probabilities $p_v$ for $v$ a vertex in $P_0$ belonging to the path. Repeating the procedure with the same strategies but other choice functions for nature will give a set of terminal vertices and a probability distribution, and taking expectation over the utilities defined in these terminal vertices with respect to the probability distribution gives us the payoff at $(s_1, \ldots, s_n)$.

The above procedure transforms a given extensive form game to a game in normal form. It is easily verified that many different extensive form games may give rise to the same normal form game. Moreover, as illustrated in the examples, not all normal form games can be obtained in this way. This situation will usually not bother us, but there are a few cases where it is of interest to consider all the extensive form games which can be transformed to a given normal form game.

## 4 Utility and payoff

In the derivation of the normal form game belonging to an extensive form game, given in the previous section, we found the payoff of the normal form game only after taking expectation over the utility payoffs of all the outcomes arising from the choice of nature. Implicitly, we have assumed that this procedure is valid, and indeed, we shall rely on it not only here but also in the sequel when dealing with *mixed strategies*, a concept which is central for our theory.

The use of expected utility is by now so commonplace in many other contexts, that it hardly needs a comment, and for a reader feeling comfortable with it, the present section may be skipped. But since expected utility makes sense only if the preferences of the player satisfy suitably properties, it seems appropriate to introduce and discuss these properties, which is what we shall do now.

In the following, we consider a situation where an individual has a choice between different *risky prospects:* We assume that there is a finite number of possible *states of nature* $s_1, \ldots, s_r$; a *risky prospect* is a pair $(x, p)$, where $x : \{s_1, \ldots, s_r\} \to X$ assigns an outcome $x(s) \in X$ to each state $s \in \{s_1, \ldots, s_r\}$, and $p = (p_1, \ldots, p_r)$ is a probability distribution over states. We let $\Xi$ denote the set of all risky prospects.

We shall assume that the information about probabilities is used in a particular way, so that players compute expected utility of the risky prospects and decide according to these expected utilities. Formally, a preference relation $\succsim$ on the set $\Xi$ of risky prospects satisfies the *expected utility hypothesis* if there is a function $u : X \to \mathbb{R}$ such that

$$(x, p) \succsim (y, p') \iff \sum_{h=1}^{r} p_h u(x(s_h)) \geq \sum_{h=1}^{r} p'_h u(x(s_h))$$

for all $(x, p), (y, p') \in \Xi$. The function $u$ is called a *von Neumann-Morgenstern utility* (after von Neumann and Morgenstern [1944], who were the first to consider this kind of utility representation).

As it can be seen from this expression, the expected utility hypothesis amounts to the assumption that there exists a utility function $u$ defined on the "pure" (risk-free) outcomes, such that the utility $U$ of a risky prospect can be found by computing the mean value wrt. the probability distribution involved. Below we give conditions under which the expected utility hypothesis will hold true. In principle, the question of whether or not this type of utility representation is a reasonable one, can be decided on the basis of these conditions.

We shall restrict our discussion to a particularly simple case: We assume that *there are only r different outcomes*, so we may identify states with outcomes. In state $s_h$ the outcome $x(s_h)$ obtains; what can vary is the probability distribution $p_h$ that this state with its associated outcome obtains. The assumption of no more than $r$ different bundles is not important, the main point is there are that only finitely many outcomes.

Thus, the choice problem under consideration in the remainder of this section is that of selecting a probability distribution $(p_1, \ldots, p_r)$ from the set of all such probability distributions, which is the set

$$\Delta = \left\{ (p_1, \ldots, p_r) \in \mathbb{R}_t^r \,\middle|\, \sum_{h=1}^{r} p_h = 1 \right\}.$$

Alternatively, $\Delta$ may be thought of as the set of all lotteries with outcomes from the set $X = \{x(s_1), \ldots, x(s_r)\}$. We assume – as usually – that the player under consideration can order the alternatives in a consistent way, meaning that she has a preference relation $\succsim$ defined on $\Delta$.

AXIOM 1 *The relation $\succsim$ is a continuous total preorder.*

It is wellknown that a continuous total preorder can be represented by a continuous utility function, but this is not enough here; we are looking for a representation with particular properties.

For this we use a further assumption, which in its turn needs some motivating comments. First of all, given two probability distributions $p^0$ and $p^1$ and a number $\alpha \in [0,1]$, we define the *mixture* of $p^0$ and $p^1$ with weights $\alpha$ and $1 - \alpha$ as the probability distribution

$$\alpha p^0 + (1 - \alpha)p^1 = (\alpha p_1^0 + (1 - \alpha)p_1^1, \ldots, \alpha p_r^0 + (1 - \alpha p_r^1)), \tag{2}$$

that is the convex combination of $p^0$ and $p^1$. If we interpret the probability distributions $p \in \Delta$ as lotteries giving the bundle $x(s_h)$ with probability $p_h$, for $h = 1, \ldots, k$, then the mixture corresponds to a lottery, which with probability $\alpha$ gives the right to participate in the lottery $p^0$ and with probability $1 - \alpha$ the right to participate in lottery $p^1$. This mixture lottery can be described in terms of the probabilities of each of the $r$ outcomes, which is exactly what happens in (2).

Consider now two pairs of lotteries, $(p^0, \hat{p}^0)$ and $(p^1, \hat{p}^1)$. We assume that $p^0 > \hat{p}^0$ and $p^1 > \hat{p}^1$. If we mix these lotteries with probabilities $\alpha$ and $1 - \alpha$, we get the new lotteries

$$\gamma = \alpha p^0 + (1 - \alpha)p^1$$

and

$$\hat{\gamma} = \alpha \hat{p}^0 + (1 - \alpha)\hat{p}^1.$$

Here both outcomes of the lottery $\gamma$ (namely, $p^0$ and $p^1$) are better for the player than the corresponding outcomes in the lottery $\hat{\gamma}$, and it is reasonable to assume that the lottery $\gamma$ is preferred to $\hat{\gamma}$; whatever comes out of the lottery, it is better with $\gamma$ that with $\hat{\gamma}$. Technically, it means that the preference relation of the player respects the mixture operation:

AXIOM 2 *Let $p^0, \hat{p}^0, p^1, \hat{p}^1 \in \Delta$ with $p^0 > \hat{p}^0$, $p^1 \succsim \hat{p}^1$, and let $\alpha \in [0, 1]$, $\alpha > 0$. Then $\alpha p^0 + (1 - \alpha)p^1 > \alpha \hat{p}^0 + (1 - \alpha)\hat{p}^1$.*

It may be noticed that we have allowed for indifference in one of the pairs; correspondingly, the other pair must enter into the mixture with a positive weight.

In the following we show that if $\succsim$ satisfies the two Axioms 1 and 2, then the expected utility hypothesis holds for $\succsim$:

PROPOSITION 1 *Let $\succsim$ be a preference relation on $\Xi$, and assume that Axiom 1 and 2 are fulfilled. Then $\succsim$ satisfies the expected utility hypothesis.*

PROOF: For $p, \hat{p} \in \Delta$ with $p > \hat{p}$, let

$$c = p - \hat{p};$$

Since the sum of the coordinates is 1 for both $p$ and $\hat{p}$, it must be 0 for $c$. Now, let $p'$ and $\hat{p}'$ be arbitrary lotteries, and suppose that $p' - \hat{p}' = c'$, see Fig. 1. We shall show that $p' > \hat{p}'$.

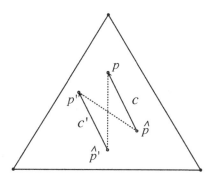

**Fig. 1.** A case with four lotteries, $p, \hat{p}, p', \hat{p}'$ such that $p - \hat{p} = c = c' = p' - \hat{p}'$.

Suppose to the contrary that $\hat{p}' \succsim p'$. We use Axiom 2 on the pairs $(p, \hat{p})$, $(\hat{p}', p')$ with $\alpha = 1/2$ to get that

$$\frac{1}{2}p + \frac{1}{2}\hat{p}' > \frac{1}{2}\hat{p} + \frac{1}{2}p';$$

furthermore, we have that

$$\frac{1}{2}p + \frac{1}{2}\hat{p}' = \frac{1}{2}(\hat{p} + c) + \frac{1}{2}(p' - c) = \frac{1}{2}\hat{p} + \frac{1}{2}p',$$

which tells us that the two mixed lotteries are identical, so that one cannot be preferred to another. From this contradiction we conclude that $p' > \hat{p}'$.

It follows from this that if a vector $c \in \mathbb{R}^r_+$ with $\sum_{h=1}^r c_h = 0$ has a representation

$$c = p - \hat{p} \text{ with } p > \hat{p},$$

then $p' > \hat{p}'$ holds for all pairs $(p', \hat{p}')$ of lotteries with $p' - \hat{p}' = c$. This property will come in useful below:

Define the set

$$C = \{c \in \mathbb{R}^r_+ | \exists p, \hat{p} \in \Delta, p > \hat{p}, p - \hat{p} = c\}.$$

Then $C$ is convex, for if $c$ and $c'$ belong to $C$, then

$$p - \hat{p} = c, \ p > \hat{p},$$

$$p' - \hat{p}' = c', \ p' > \hat{p}',$$

and according to Axiom 2 we must have that $\alpha p + (1 - \alpha) p' > \alpha\hat{p} + (1 - \alpha)\hat{p}'$. But

$$\alpha p + (1 - \alpha)p' - [\alpha\hat{p} + (1 - \alpha)\hat{p}'] = \alpha c + (1 - \alpha)c'$$

and the vector on the right hand side must belong to $C$.

Furthermore, we have that 0 does not belong to $C$ (since $>$ is irreflexive). Consequently we can separate 0 from $C$ by a hyperplane: There exists $u = (u_1, \ldots, u_r)$, $u \neq 0$, such that $u \cdot c > 0$ for all $c \in C$.

Writing this out in detail, we have that

$$\sum_{h=1}^r p_h u_h > \sum_{h=1}^r \hat{p}_h u_h$$

for all $p, \hat{p} \in \Delta$ with $p > \hat{p}$. We leave it to the reader to check that conversely, if $\sum_{h=1}^r p_h u_h > \sum_{h=1}^r \hat{p}_h u_h$ for some pair $(p, \hat{p})$ of lotteries, then $p > \hat{p}$.                                          □

In most applications, the probability distributions $(p_1, \ldots, p_r)$ represent the *beliefs* of the player. This way of treating beliefs is not altogether innocent; it is not apriori obvious that beliefs may be operated with according to the theory of probability. We shall not go further into this subject; the reader is referred to Savage [1954].

## 5  Games and their solutions

**5.1. Types of games.** With the definitions of games in normal and extensive form, the framework for our study has been established, and we may progress towards the study itself. The logically first step would then be to specify the *object* of study,

**Example 1.5. The Allais paradox.** The axioms of expected utility may or may not be satisfied in real world situations. The *Allais paradox* [Allais, 1953] exhibits two different cases of choice between lotteries, namely Case A:

|            | 89% | 1% | 10% |
|------------|-----|----|-----|
| Lottery A1 | 1   | 1  | 1   |
| Lottery A2 | 1   | 0  | 5   |

Here the first lottery represents a sure gain of 1, whereas the second lottery involves an element of gambling. It would seem reasonable – and indeed it is confirmed by many experiments – that lottery A1 is chosen.

Now, consider Case B of choosing between lotteries:

|            | 89% | 1% | 10% |
|------------|-----|----|-----|
| Lottery B1 | 0   | 1  | 1   |
| Lottery B2 | 0   | 0  | 5   |

Now both lotteries involve gambling, and in this situation the lottery B2 might well be chosen. However, if a decision maker satisfies the axioms of expected utility and has the von Neumann-Morgenstern utility function $u$, she cannot choose A1 in Case A and B2 in Case B. Indeed the first choice would imply that

$$u(1) > 0.89u(1) + 0.01u(0) + 0.1u(5)$$

or

$$0.11u(1) > 0.01u(0) + 0.1u(5), \tag{3}$$

whereas the choice of B2 occurs if

$$0.89u(0) + 0.11u(1) < 0.9u(0) + 0.1u(5),$$

which yields that

$$0.11u(1) < 0.01u(0) + 0.1u(5),$$

contradicting (3).

This and other paradoxes have given rise to a voluminous literature on extensions of expected utility theory. We shall not pursue such extensions here, since much of what we shall be doing in the sequel relies rather heavily on expected utility theory.

and a straightforward way of doing so is to define game theory as the study of *solutions* to games.

What is to be understood by a solution to a given game is indeed the main problem that we face throughout this book. We shall approach it from two different angles, giving rise to a split of the material into *non-cooperative* and *cooperative* game theory; in the first part, each player chooses strategies separately, while in the second, players may enter coalitions and make binding agreements as to the way in which the strategies are to be chosen. It should be emphasized that it is not the game as such which is cooperative or non-cooperative, these notions pertain to the solutions selected, intuitively it specifies the way in which the game is played. However, it will facilitate our exposition if we allow ourselves to speak of non-cooperative and cooperative games in the following, as we shall indeed do.

Since the non-cooperative game theory is conceptually simplest, we – and indeed all expositions of game theory – begin there, and we stay within this framework in the Chapters 2 – 9. Historically, the first important results in game theory were derived for games with only two players, indeed for games which are *zero-sum* in the sense that in any outcome, what one player gains is lost by the other player. We follow the tradition (now widely abandoned) and begin with these games, so that we begin our treatment of non-cooperative game theory with the classical results for two-person zero-sum games. Then we proceed to general non-cooperative games and consider solution concepts applying to the general situation as well as specific types of games.

The study of cooperative games, where coalitions matter, calls for other approaches than those used for the analysis of non-cooperative games. To simplify the treatment, attention is transferred from the choice of strategies (by players or coalitions of players) to the strategic possibilities of coalitions acting on behalf of their members. In the simplest approach, the power of a coalition is measured as a total utility or money payoff which the coalition can distribute among its members in any way it would find appropriate; we are then dealing with a *transferable utility* (TU) game. Clearly, this transferability of payoffs may be too restrictive in many cases, so that an extension of the results obtained to *non-transferable utility* (NTU) games is desirable, and we shall be concerned with this as well.

Summing up, our study takes as point of departure the structure of games as shown in the next page.

| Games | | | |
|---|---|---|---|
| Non-cooperative | | Cooperative | |
| Two-person games | *n*-person games | TU games | NTU games |
| zero sum games / non-zero sum games | | | |

For each of these, the theory is concerned with finding a *solution* in the sense of an outcome which will be obtained if the players choose in accordance with their self-interest. What that will be, cannot be summarized briefly, since this is what game theory is all about.

**5.2. Strategic equivalence of games.** In our study of games, we shall often encounter games which are 'the same' even if they distinguish in some respects such as the labeling of strategies or the size of the utility payoffs. To deal with such cases of 'likeness' in a precise way, one introduces the notion of strategic equivalence.

DEFINITION 6 *Let* $\Gamma = (N, (S_i)_{i\in N}, (u_i)_{i\in N})$, $\Gamma' = \left(N, (S'_i)_{i\in N}, (u'_i)_{i\in N}\right)$ *be games. Then* $\Gamma$ *and* $\Gamma'$ *are strategic equivalent if there are bijective maps* $\phi_i : S_i \to S'_i$ *and increasing affine maps* $\psi_i : \mathbb{R} \to \mathbb{R}$, $i = 1, \ldots, N$, *such that the diagram*

$$\begin{array}{ccc} S_1 \times \cdots \times S_n & \xrightarrow{u} & \mathbb{R}^N \\ \downarrow{\phi} & & \downarrow{\psi} \\ S'_1 \times \cdots \times S'_n & \xrightarrow{u'} & \mathbb{R}^N \end{array} \tag{4}$$

*is commutative, where* $\phi$ *is the product map* $\phi_1 \times \cdots \times \phi_n$, *and* $\psi = \psi_1 \times \cdots \times \psi_n$.

It is easily checked that strategic equivalence is indeed an equivalence on the set of games with a given player set: the relation is symmetric, reflexive and transitive. If two games are strategically equivalent, then whatever is known about one of them can be immediately transferred to the other game, so that it should be enough to study the solutions to one representative. Thus, the game

| | confess | not confess |
|---|---|---|
| confess | (1,1) | (10,0) |
| not confess | (0,10) | (9,9) |

is also a Prisoners' Dilemma game (we have added 10 to the payoffs of both players), and

|          | Rock      | Paper     | Scissors  |
|----------|-----------|-----------|-----------|
| Rock     | $(0,0)$   | $(-10,1)$ | $(10,-1)$ |
| Paper    | $(10,-1)$ | $(0,0)$   | $(-10,1)$ |
| Scissors | $(-10,1)$ | $(10,-1)$ | $(0,0)$   |

is strategically equivalent to Paper-Stone-Scissors (Example 1.1), since the only change is that Player 1's payoffs have been multiplied by 10.

We shall use strategical equivalence in the sequel, since it allows us to represent a given game (or rather its equivalence class) in the way which is most convenient for us, but in most cases without mentioning it explicitly.

## 6   Problems

**1.** ("Matching Pennies") Two players simultaneously display one side of a coin. If both players display the same side (heads or tails), Player 1 gets 1 € from Player 2. If they display different sides, Player 1 pays 1 € to Player 2.

Define the relevant strategy spaces and write down the normal form of this game. Consider whether there is a particularly attractive strategy for any of the players.

**2.** Two firms sell the same consumer good in a market. The demand for the good is given by

$$q = 100 - p,$$

where $p$ is the price of the good and $q$ the quantity sold in the market. The production costs are 10 € per unit.

The two firms compete in the sale of the good, each of them proposes a price. If the two prices are different, then all consumers buy their goods from the cheapest firm, if they are equal, the market is split equally between the two.

Describe the competition between firms as a normal form game.

**3.** The following is a simplified discription of a *patent race*, where $n$ firms use resources on research in order to obtain a patent. The firm which uses the largest amount of money on research gets the patent, which has a certain value derived from future sales. For the other firms, the amount used on research is lost.

Formulate this as a normal form game.

**4.** A community consisting of 5 persons must select one out of three projects to be implemented. Each of the persons submits a ranking of the three projects. When the rankings have been submitted, the project which has the largest number of highest rank is selected.

Sketch the extensive form of the game. How may choices are there for each player when she has the move?

**5.** (The St. Petersburg paradox) A gambling house invites customers to participate in a game of chance: A coin is thrown repeatedly until it shows heads for the first time, and then a prize (in \$) is paid out as follows:

$$
\begin{array}{llll}
\text{First heads in} & \text{1st} & \text{throw:} & 2 \\
\text{First heads in} & \text{2nd} & \text{throw:} & 4 \\
\cdots & & & \\
\text{First heads in} & n\text{th} & \text{throw:} & 2^n \\
\cdots & & &
\end{array}
$$

The gambling house demands an entry fee for the participation in this game.

What is the expected gain for the participant given that the entry fee has been paid?

If the entry fee is set so as to cover expected payments of the gambling house (the "fair value" of the game), what should it be? How does this compare to what people would be willing to pay?

Give a suggestion as to how this paradox could be resolved.

**6.** Two firms competing in the market for a product contemplate an increase by \$100,000 of their advertising budget. If only one firm chooses to advertise, its profits are expected to increase by \$120,000 and that of the other firm by \$20,000. If both advertise, the total increase in profits, \$140,000, are shared equally between the firms.

Write down the extensive form of this game and its transformation to a normal form game.

# PART 1
# Non-Cooperative Game Theory

# Chapter 2

# Two Person Zero-Sum Games

## 1 Introduction

The theory of two-person zero-sum games has a special position in game theory: It was from this particular subfield that the discipline as such developed, and in its early stages, it was not only the starting point but also the main research field of game theorists. For reasons that will emerge shortly, the interest in zero-sum games has faded over time, and they tend to become a parenthesis in the exposition of the field.

This is a pity, and we shall devote considerable space to the two-person zero-sum games, not only because of their historical role, but also for pedagogical reasons; the beauty of the main result (known as the minimax-theorem) has played a significant role in the further development of the theory, and it constitutes the most convincing reason for the introduction of mixed strategies (in Section 4 below), which is one of the main ingredients in non-cooperative game theory.

The simplest class of games (apart from one-person games where there is no conflict at all) is that of two person games, where each player has a finite number of strategies, and where the set of possible outcomes consists of pairs $(r_1, r_2)$, interpreted as the gain to player 1 and player 2, respectively, such that $r_1 + r_2 = 0$.

Such a two person zero-sum game may conveniently be represented as follows:

|  |  | Player 2's strategies | | |
|---|---|---|---|---|
|  |  | 1 | $\cdots$ | $n$ |
| Player 1's | 1 | $a_{11}$ | $\cdots$ | $a_{1n}$ |
| strategies | $\vdots$ | $\vdots$ |  | $\vdots$ |
|  | $m$ | $a_{m1}$ | $\cdots$ | $a_{mn}$ |

Here the number $a_{ij}$ in the $i$th row, $j$th column of the table gives the gain or payoff to player 1 when 1 chooses $i$ and 2 chooses $j$, that is $a_{ij} = u_1(i, j)$. The payoff of player 2 is consequently $-a_{ij}$.

The game is uniquely defined by the matrix

$$A = \begin{pmatrix} a_{11} & \cdots & a_{1n} \\ \vdots & \cdots & \vdots \\ a_{m1} & \cdots & a_{mn} \end{pmatrix}$$

so games of this type are often called *matrix games*. The game in Example 1.1 is obviously a matrix game, but that in Example 1.2 is not, since it fails to satisfy the zero-sum condition.

This takes us to a general discussion of the interpretation of the payoffs $a_{ij}$ in the matrix $A$. It is straightforward – and in many situations perfectly satisfactory – to consider $a_{ij}$ as an amount of money to be transferred from player 2 to player 1. But this gives rise to some difficulties:

First of all, in many conflict situations, which are not games in the traditional meaning of the word, it is not possible to reduce the consequences to the change of ownership of a certain amount of money. Presumably the consequences of a conflict will be more complex than that. Therefore it is more natural to interpret $a_{ij}$ as player 1's utility of the result $f(i, j)$ which obtains when player 1 chooses $i$ and player 2 chooses $j$. However, this doesn't quite solve the problem; the zero-sum condition implies some sort of interpersonal comparison of utilities.

Another objection against zero-sum games is that they do not occur very often in applications. In particular, cases taken from economic theory are mostly of the type where some third party having no decision power still carries some of the cost or reaps some of the benefits. Thus the zero-sum games are reduced to a theoretical case – which however does not mean that they are unimportant, as we shall see in the sequel. Indeed, the results obtained can be – and are – used repeatedly in other contexts, game-theoretical or not, as we shall see later in Chapter 3.

## 2  Strategy choices: Domination and maximin

We now turn to the problem of deciding which strategies in a given matrix game can be expected to be chosen by the players – or can be recommended for their choice.

As our point of departure we choose an example, namely the game

$$\begin{pmatrix} 12 & 7 & 3 & 2 \\ 0 & 4 & 3 & 1 \\ 5 & 5 & 4 & 10 \\ 6 & 2 & 1 & 8 \end{pmatrix} \tag{1}$$

Notice (even though it is irrelevant for what follows) that this game is relatively unfavourable for player 2 who will never get a positive payoff and in some cases may loose up to 12 units.

Looking at the four strategies of player 1 in this example, one sees immediately that the gain to be obtained depends on the choices of player 2. However, player 1 has to choose her strategy without any knowledge of the choices of player 2.

For a beginning, player 1 (and 2) may check whether there are strategies that are so unpromising that they may be discarded right away. In the game (1), this is the case for the second strategy of player 1, corresponding to the second row in the matrix: Whatever is chosen by player 2, the first strategy of player 1 gives at least as much as does the second, and in some cases more. It may well be argued that player 1 can just as well forget about strategy 2 in all the following considerations.

In the situation above, we say that strategy 2 is weakly dominated by strategy 1. Notice that for player 2, we similarly have that strategy 2 is weakly dominated by strategy 3. The concept of domination is sufficiently important to warrant further consideration (and we shall have much more to say about it in the chapters to follow), and we will use the occasion to formulate it for general games of the type $\Gamma = (N, (S_i)_{i \in N}, (u_i)_{i \in N} i)$. It is convenient to introduce some specific notation to this purpose:

Let $s_i \in S_i$, $i \in N = \{1, \ldots, n\}$. Then we write $s = (s_1, \ldots, s_n)$ for the strategy array consisting of all the $s_i$, and let

$$s_{-i} = (s_1, \ldots, s_{i-1}, s_{i+1}, \ldots, s_n)$$

for the strategy array with the component corresponding to player $i$ removed. We identify $s$ and the array $(s_i, s_{-i})$. It is convenient to use this notation also for the sets of strategies, so that we write $S = \times_{i \in N} S_i$, $S_{-i} = \times_{j \in N, j \neq i} S_j$.

**DEFINITION 1** *Let* $\Gamma = (N, (S_i)_{i \in N}, (u_i)_{i \in N})$ *be a game, and let* $i \in N$, $s_i, s'_i \in S_i$. *Then* $s_i$ *is dominated by* $s'_i$ *if*

$$u_i(s_i, s_{-i}) < u_i(s'_i, s_{-i}), \text{ all } s_{-i} \in S_{-i},$$

*and* $s_i$ *is weakly dominated by* $s'_i$ *if*

$$u_i(s_i, s_{-i}) \leq u_i(s'_i, s_{-i}), \text{ all } s_{-i} \in S_{-i}$$

*and* $u_i(s_i, s'_{-i}) < u_i(s'_i, s'_{-i})$ *for some* $s'_{-i} \in S_{-i}$.

A very intuitive solution would be a strategy which weakly dominates any other strategy. Unfortunately, there are only few games for which there is such a strategy.

In the example above, leaving out the strategies that were weakly dominated we are left with matrix

$$A = \begin{pmatrix} 12 & 3 & 2 \\ 5 & 4 & 10 \\ 6 & 1 & 8 \end{pmatrix}$$

and we can just as well use this as our basic example, which we will do.

Given the matrix $A$, player 1 can find her best reply on each possible choice by player 2. The resulting *best reply* or *reaction function* is given below:

|                     | Player 2 chooses: | | |
| ------------------- | :-: | :-: | :-: |
|                     | 1 | 2 | 3 |
| Player 1's choice   | 1 | 2 | 2 |

This is of no immediate use to player 1; but if she contemplates the possible choices to be made by player 2, then it would give a similar scheme, seen from the point of view of the other player:

|                     | Player 1 chooses: | | |
| ------------------- | :-: | :-: | :-: |
|                     | 1 | 2 | 3 |
| Player 2's choice   | 3 | 2 | 2 |

The table tells player 1 how player 2 will react on her choices. For each strategy of player 1 the other player will choose so that 1's payoff is as small as possible given this strategy. This results in the so-called *security level* associated with the strategy.

We have thereby reached a proposal as to what player 1 should choose: If player 1 wants to protect herself against the choices of player 2 which in their turn are made with the aim of achieving best possible results, then she should choose the strategy which gives the maximal security level. In the case considered, the security levels associated with strategies 1, 2, and 3 (for player 1) are 2, 4, and 1, respectively, and she should choose strategy 2. What we have found corresponds to minimizing outcome in each row and then choose the row for which this minimum is the largest, and strategy 2 is therefore called a *maximin* strategy.

Exactly the same reasoning works for player 2; the security levels of her strategies 1, 2, and 3 are now $-12$, $-4$, and $-10$, respectively, and the maximin strategy of player 2 is therefore strategy 2. With our convention for writing payoffs in zero-sum games, according to which only the payoff of player 1 figures in the matrix, the maximin strategy of player 2 emerges after maximizing payoff in each column and then minimizing over columns, and it is therefore called a *minimax* strategy.

Summing up the preceding discussion, we have the following

DEFINITION 2 *Let $\Gamma$ be a matrix game with matrix $A$. Let*

$$\text{maximin } A = \max_i \min_j a_{ij},$$

$$\text{minimax } A = \min_j \max_i a_{ij}$$

*A maximin strategy for player 1 is an index $i^0$ such that $a_{i^0 j'} = \text{maximin } A$ for some $j'$, and a maximin strategy for player 2 (also called a minimax strategy) is an index $j^0$ such that $a_{i' j^0} = \text{minimax } A$ for some $i'$.*

**Example 2.1. Security levels in Rock-Paper-Scissors.** Using the method of security levels in the Rock-Paper-Scissor game of Example 1.1, written as a matrix game, we find

|  | Rock | Paper | Scissors | Security level |
|---|---|---|---|---|
| Rock | 0 | −1 | 1 | −1 |
| Paper | 1 | 0 | −1 | −1 |
| Scissors | −1 | 1 | 0 | −1 |
| Security level | 1 | 1 | 1 |  |

The worst case payoff is the same for each of the three strategies, so maximizing worst-case payoffs does not give us a unique strategy choice. This is perhaps not a major problem, since choosing any of the maximizers, that is any strategy, will give the same result (as long as we focus on worst-case scenarios). But it may be noticed that the two players disagree about the payoff that will emerge in this worst case.

This situation suggests that the last word hasn't yet been said about strategy choices, and this is indeed confirmed as we proceed.

The following result is straightforward:

THEOREM 1 *For every matrix A,* maximin $A \leq$ minimax $A$.

PROOF: For every $i'$ and $j$ we have

$$a_{i'j} \leq \max_i a_{ij}.$$

In particular,

$$\min_j a_{i'j} \leq \min_j \max_i a_{ij}.$$

The right-hand side is independent of $i$ and $j$, and $i'$ was arbitrary. Therefore the inequality holds also for the specific $i'$ which maximizes the left-hand side, so that

$$\max_i \min_j a_{ij} \leq \min_j \max_i a_{ij}. \qquad \square$$

In our example we have that minimax $A' =$ maximin $A'$. This is however *not* a general result (meaning that the matrix in (1) was not chosen quite arbitrarily after all). To see this, take for example the matrix $A''$ belonging to the game in Example 2.1; here we obtain that maximin $A'' = -1 < 1 =$ minimax $A''$.

## 3  Equilibria

We stay for a short while with the example considered in the previous section; in our discusion of possible solutions to the game. In the intuitive reasoning leading up to the choice of a maximin strategy, we used as an assumption that the other player was choosing in a rational way. In some cases a player may know more about the other player, and possibly this player is, after all, not so rational as that. In this case maximin strategies may loose the intuitive appeal, since no additional information is used. Thus, maximin strategies represent a reasonable choice – in the sense that catastrophic outcomes are avoided – for as large a class of situations as possible. What was important in the choices of row 2 and column 2 by player 1 and 2 respectively is perhaps not as much that they have the maximin property, but that they have another property: *Given the choice by the other player, the player in question will have no incentive to replace her strategy by any other strategy.*

This property of the strategy choice is quite important, and we say that a strategy pair satisfying the property is an *equilibrium*:

DEFINITION 3 *Let $\Gamma$ be a matrix game with matrix A. A pair $(i^0, j^0)$ is an equilibrium if for all i and j*

$$a_{ij^0} \leq a_{i^0 j^0} \leq a_{i^0 j}$$

*for all i and j.*

Geometrically, the equilibrium property of the strategy pair $(i^0, j^0)$ means that the payoff $a_{ij}$ considered as a function of $i$ and $j$ has a *saddle point* in $(i^0, j^0)$: The payoff decreases in the $i$-direction when moving away from $i^0$ and it increases for movements in the $j$-direction.

Combining the equilibrium property with Theorem 1, we get

THEOREM 2 *Let $\Gamma$ be a matrix game. Then $\Gamma$ has an equilibrium if and only if*

$$\text{minimax } A = \text{maximin } A.$$

PROOF: Let $(i^0, j^0)$ be an equilibrium for $\Gamma$ and let $v = a_{i^0 j^0}$. Then $v \geq a_{ij^0}$ for all $i$ by Definition 1, or, otherwise put,

$$v \geq \max_i a_{ij^0}.$$

Consequently $v \geq \min_j \max_i a_{ij} = \text{minimax } A$. Similarly, we have $v \leq a_{i^0 j}$ for all $j$, from which

$$v \leq \min_j a_{i^0 j},$$

so that maximin $A \geq v \geq$ minimax $A$. Together with Theorem 1, this gives us that maximin $A =$ minimax $A$.

Conversely, suppose that maximin $A$ = minimax $A$ = $v$. Then there must be $(i^0, j^0)$ so that $a_{i^0 j^0} = v$. From $v$ = minimax $A$ we get

$$a_{i^0 j^0} = \min_j \max_i a_{ij},$$

so $j^0$ is the index of a column $j$ for which $\max_i a_{ij}$ is as small as possible, and this number is exactly equal to $a_{i^0 j^0}$, so that the maximum over the numbers in the column $j^0$ is attained at row $i^0$. But this means that $a_{ij^0} \leq a_{i^0 j^0}$ for all $i$, which is the equilibrium condition for player 1. The condition $a_{i^0 j^0} \leq a_{i^0 j}$ for all $j$ is shown in a similar way. □

As an immediate consequence of the above result we get that the following definition is meaningful:

DEFINITION 4 *Let $\Gamma$ be a matrix game with matrix $A$, and assume that $A$ has at least one equilibrium. The value of the game $\Gamma$, denoted $v(\Gamma)$, is the payoff $a_{i^0 j^0}$ of player 1 in any of the equilibria of the game.*

We have the equality $v(\Gamma)$ = maximin $A$ = minimax $A$ whenever the value of the matrix game $\Gamma$ is defined.

## 4  Mixed strategies and mixed extensions of games

By now we have reached the somewhat unsatisfactory result that some matrix games have an equilibrium, but other matrix games may not have one. This is a somewhat unsatisfactory state of affairs, and game theory would probably never have taken off if it hadn't been for the idea, due to John von Neumann, of looking at *mixed strategies*, where the player leaves the choice of strategy to nature, although with preassigned probabilities.

Leaving the choices to chance, the player makes sure that she is never out-guessed by the other player, and this turns out to be quite useful. There is little to be gained for the player herself, who will get the same expected payoff from all the strategies (now to be called *pure* strategies to distinguish them from the new ones) which may be chosen by nature. But mixed strategies offers the kind of protection against the other player that will allow for an equilibrium in the enlarged game, the *mixed extension* of the original game, which we now proceed to define.

Let $\Gamma$ be a matrix game with associated $(m \times n)$-matrix $A$. For every natural number $r \in \mathbb{N}$, the $(r - 1)$-dimensional simplex is defined as

$$\Delta^r = \left\{ x \in \mathbb{R}^r \,\middle|\, x_h \geq 0, h = 1, \ldots, r, \sum_{h=1}^{r} x_h = 1 \right\}.$$

The *mixed extension* mix($\Gamma$) of the game $\Gamma$ has strategy sets $\Sigma_1 = \Delta^{m-1}$, $S\Sigma_2 = \Delta^{n-1}$

and payoff function $u_1 : \triangle^{m-1} \times \triangle^{n-1} \to \mathbb{R}$ defined by

$$u_1(x, y) = \sum_{i=1}^{m} \sum_{j=1}^{n} a_{ij} x_i y_j,$$

so that $u_1$ extends the payoff function of $\Gamma$ in the obvious way.

Our interest in mixed extensions derives mainly from the connection which can be shown to exist between equilibria in the original game and equilibria in the extension. Notice that in Definition 3 we have introduced the concept of an equilibrium only for matrix games. It is however not a problem to extend this definition to general two-person zero-sum games.

DEFINITION 5 *Let $\Gamma$ be a two-person zero-sum game. A pair of strategies $(s_1^0, s_2^0)$ is an equilibrium if*

$$u_1(s_1, s_2^0) \leq u_1(s_1^0, s_2^0) \leq u_1(s_1^0, s_2)$$

*for all $s_1 \in S_1, s_2 \in S_2$.*

It is straightforward that if $(i^0, j^0)$ is an equilibrium in the original matrix game $\Gamma$, then $(\alpha_1(i^0), \alpha_2(j^0))$ is an equilibrium in mix $\Gamma$. Clearly one cannot reason conversely: Given an equilibrium $(s_1^0, s_2^0)$ in mix $\Gamma$ there are not necessarily strategies $i$ and $j$ for players 1 and 2 such that $\alpha_1(i) = s_1^0, \alpha_2(j) = s_2^0$.

Nevertheless, the equilibria in mix $\Gamma$ are of interest. This has to do with the fact that strategies in the mixed extension mix $\Gamma$, known as *mixed strategies*, have an interpretation as probability distributions over the set of original ("pure") strategies in $\Gamma$. We can imagine that the rules of the play give each of the players the possibility of "randomizing" her strategy choice, so that the strategy chosen is found using some random device. This will in many cases be a reasonable way of choosing. Thus, in the game of Example 2.1, if player 1 has a certain preference for one of her three strategies, and player 2 becomes aware of this, then player 2 can force the payoff of the first player down to $-1$. An obvious way of avoiding this would be to make the choice of strategy completely random, corresponding to the mixed strategy $\left(\frac{1}{3}, \frac{1}{3}, \frac{1}{3}\right)$.

It may be noticed that arguments of this type for using mixed strategies are really convincing only when the play is repeated many times, a situation which actually is outside our present scope, since the conflict studied then changes character and becomes a *repeated game*, something to be studied later. However, the ultimate justification of the use of mixed strategies is not so much its intuitive appeal in one or another application, rather it has to do with the nice results that we obtain using the concept.

We shall now formulate the main result for two-person zero-sum games, also known as the *minimax theorem* – or perhaps, more properly, as *a* minimax theorem, since there are many alternative versions.

THEOREM 3 *Let $\Gamma$ be a matrix game. Then mix $\Gamma$ has an equilibrium.*

---

**Box 2.2. A categorical approach to mixed extensions.** A more formalistic approach, following that used in other fields of mathematics (compactification of spaces, completion of number fields, non-standard analysis etc.), would see the extension to mixed strategies as an embedding of the original objects of study into another and larger class of objects where some more general results can be proved.

Let $\Gamma = (N, (S_i)_{i=1}^n, \pi)$ and $\Gamma' = (N, (S_i')_{i=1}^n, \pi')$ be games with the same player set. A *morphism* from $\Gamma$ to $\Gamma'$ is an $(n+1)$-tuple of maps $(\alpha_1, \ldots, \alpha_n, \beta)$ with $\alpha_i : S_i \to S_i'$, $i \in N$, and a map $\beta : \mathbb{R}^N \to \mathbb{R}^N$, such that

(i) $\pi'(\alpha_1(s_1), \ldots, \alpha_n(s_n)) = \beta(\pi(s_1, \ldots, s_n))$ for all $(s_1, \ldots, s_n) \in (S_1, \ldots, S_n)$.

(ii) $u > u' \Rightarrow \beta(u) > \beta(u')$ for all $u, u' \in \mathbb{R}$.

The condition to be satisfied can also be formulated as the demand that the diagram

$$
\begin{array}{ccccccc}
S_1 \times S_2 \times \cdots \times S_n & \xrightarrow{\pi} & \mathbb{R}^N \\
\alpha_1 \downarrow \quad \alpha_2 \downarrow \qquad\qquad \alpha_n \downarrow & & \beta \downarrow \\
S_1' \times S_2' \times \cdots \times S_n' & \xrightarrow{\pi'} & \mathbb{R}^N
\end{array}
$$

should be commutative: Starting in the upper left corner and following the arrows to the lower right corner, the result should be the same whether one goes to the right and then down or down and then to the right.

In the special case of two-person zero-sum games, this simplifies somewhat, since we may restrict attention to the payoff of player 1. Condition (ii) then says that $\beta$ is a monotonic function; if we add that $\beta = id$ (the identical mapping), then condition (i) means that the diagram

$$
\begin{array}{ccc}
S_1 \times S_2 & \xrightarrow{\pi} & \mathbb{R} \\
\alpha_1 \downarrow \quad \alpha_2 \downarrow & & \| \\
S_1' \times S_2' & \xrightarrow{\pi'} & \mathbb{R}
\end{array}
$$

is commutative. The morphism will here be denoted by $\alpha$.

If the maps $\alpha_1$ and $\alpha_2$ are injective, then $\Gamma$ may be considered as embedded in $\Gamma'$. A strategy array $(s_1, s_2)$ induces a strategy array $(\alpha_1(s_1), \alpha_2(s_2))$ in $\Gamma'$. If $\Gamma' = \text{mix}\,\Gamma$, then an equilibrium in $\Gamma$ is also an equilibrium in $\Gamma'$.

---

The theorem may alternatively be formulated as stating that every marix game has an equilibrium in mixed strategies. There are several ways of proving the result. We give an elementary proof here; the result follows also from more general results about existence of Nash equilibria to be established later.

The proof of Theorem 3 will be given in the next section. Here we consider some concepts which will be useful in this proof.

A matrix game $\Gamma$ with matrix $A$ is *symmetric* if the matrix $A$ is skewsymmetric, i.e. if $a_{ij} = -a_{ji}$ for all $i$ and $j$. The game in Example 2.1 is symmetric. If a matrix game $\Gamma$ is symmetric, then $\text{maximin}A = \text{minimax}A$, and if $\Gamma$ has an equilibrium, we must then have that $v(\Gamma) = \text{maximin}A = \text{minimax}A = 0$.

In general a two-person game $\Gamma = (\{1,2\}, S_1, S_2, u_1, u_2)$ which is zero-sum game (not necessarily a matrix game), so that $u_2 = -u_1$, is symmetric if the two players' strategy sets are identical, $S_1 = S_2$, and

$$u_1(s_1, s_2) = u_2(s_2, s_1) = -u_1(s_2, s_1)$$

for all $s_1 \in S_1$, $s_2 \in S_2$. We define the *symmetric extension* of $\Gamma$, $\text{sym}\Gamma$, as the game with identical strategy sets $S_1 \cup S_2 \cup \{*\}$ (disjoint union) for each player, and payoff function $u_1'$ given as

$$u_1'(s_1', s_2') = \begin{cases} 0 & \text{if } s_1', s_2' \in S_1 \text{ or } s_1', s_2' \in S_2, \\ u_1(s_1, s_2) & \text{if } s_1' = s_1 \in S_1, \ s_2' = s_2 \in S_2, \\ -u_1(s_1, s_2) & \text{if } s_1' = s_2 \in S_2, \ s_2' = s_1 \in S_1, \\ 1 & \text{if } s_1' \in S_2, \ s_2' = * \text{ or } s_2' \in S_1, \ s_1' = *, \\ -1 & \text{if } s_1' \in S_1, \ s_2' = * \text{ or } s_2' \in S_2, \ s_1' = *, \end{cases}$$

and $u_1'(*, *) = 0$. It is left to the reader to check that $\text{sym}\Gamma$ is actually an extension of $\Gamma$, and that it is symmetric.

When $\Gamma$ is a matrix game, then $\text{sym}\Gamma$ is also a matrix game, and the matrix of $\text{sym}\Gamma$, which takes the form

$$
\begin{pmatrix}
 & & -a_{11} \cdots -a_{m1} & 1 \\
 & 0 & \vdots \quad\ \vdots & \vdots \\
 & & -a_{1n} \cdots -a_{mn} & 1 \\
a_{11} \cdots a_{1n} & & & -1 \\
\vdots \quad\ \vdots & & 0 & \vdots \\
a_{m1} \cdots a_{mn} & & & -1 \\
-1 \cdots -1 & 1 \ \cdots \ 1 & & 0
\end{pmatrix}
\tag{2}
$$

As was the case for the mixed extension, the symmetric extension can be given an interpretation, this time based on the fact that this game treats the players equally if they start by choosing roles, decide whether they want to be player 1 or player 2. Assume namely that a game with matrix $A$ is given, where players are called "white" and "black", respectively. The symmetric extension then has the following rules: Player 1 and 2 choose independently either to play white, that is one of the strategies of player "white", to play black, or to pass. If players have chosen different colors, the play proceeds as in the original game. If they have chosen the same color or both have chosen "pass", then outcome is 0. Finally, if one player has chosen "pass", she will win 1 if the other player chooses white and lose 1 if she chooses black.

We shall need the following intermediary result.

LEMMA 1 *Let $\Gamma$ be a matrix game with matrix $A$. Let $(x, y)$ be an equilibrium in* mix(sym($\Gamma$)), *where $x = (x^1, x^2, x^3)$, $y = (y^1, y^2, y^3)$ with block partition corresponding to the matrix in (2). Assume that $x^2$, $y^1 \neq 0$. Then $(\hat{x}, \hat{y})$ defined by*

$$\hat{x} = \left( \sum_{i=1}^{m} x_i^2 \right)^{-1} x^2, \quad \hat{y} = \left( \sum_{i=1}^{n} y_i^1 \right)^{-1} y^1$$

*is an equilibrium in* mix($\Gamma$).

PROOF: Suppose that $(\hat{x}, \hat{y})$ is not an equilibrium in mix($\Gamma$). We may assume that for fixed $y$ there is $x' \in \Delta^{m-1}$ such that

$$\sum_{i=1}^{m} \sum_{j=1}^{n} a_{ij} x_i' \hat{y}_j > \sum_{i=1}^{m} \sum_{j=1}^{n} a_{ij} \hat{x}_i \hat{y}_j.$$

Define a new strategy $\bar{x}$ in mix(sym($\Gamma$)) by $\bar{x}^1 = x^1$, $\bar{x}^2 = (\sum_{i=1}^{m} x_i^2) x'$, $\bar{x}^3 = x^3$. It is easily shown that

$$\sum_{i,j} a'_{ij} \bar{x}_i y_j > \sum_{i,j} a'_{ij} x_i y_j,$$

where $a'_{ij}$ are the elements in the matrix in (2), contradicting that $(x, y)$ is an equilibrium. □

In the above lemma we assumed that $x^2, y^1 \neq 0$. If instead $x^1, y^2 \neq 0$, we could have defined equilibrium strategies in mix($\Gamma$) using these. And one of these alternatives must hold, as it can be seen from the following reasoning:

If there is an equilibrium in mix(sym($\Gamma$)), it cannot be $x = (0, 0, 1)$, $y = (0, 0, 1)$, where both players choose the pure strategy "pass", since this is not an equilibrium. Assume now that $x^2 = 0$, $y^1 = 0$. If also $x^1 = 0$, then 1 chooses the pure strategy "pass", and player 2 can get the payoff 1 by choosing a mixed strategy $\bar{y}$ such that $\bar{y}^1 \neq 0$, $\bar{y}^2 = 0$, $\bar{y}^3 = 0$. Since player 2 does not choose this strategy (we assumed $y^1 = 0$), the payoff to player 2 in the equilibrium must be as big as that obtained by using $\bar{y}$, so that the payoff to player 1 must be $\leq -1$. Moreover, according to the above reasoning, $y$ differs from $(0, 0, 1)$. It then follows that player 1, given the choice $y$ of player 2, can obtain a payoff $\geq -1$ by choosing $\bar{x}$ such that $\bar{x}^1 = 0$, $\bar{x}^2 \neq 0$, $\bar{x}^3 = 0$, contradicting our assumption that $x^2 = 0$. We conclude that $x^1 \neq 0$. By symmetry, we have correspondingly that $y^2 \neq 0$. □

## 5  Proof of the main theorem

We now have all the ingredients for the proof of the main result about two-person zero-sum games, Theorem 3. The proof relies on the following version of the separation theorem for convex sets. Recall that a set $C$ is convex if for each pair $c^1, c^2$ of points in $C$ and each $\lambda \in [0, 1]$, the point $\lambda c^1 + (1 - \lambda)c^2$ belongs to $C$. For a proof of the this theorem, see found e.g. [Rockafellar [1970]], Thm. 11.3.

**Example 2.3. Graphical method for finding equilibria.** This method can be applied to matrix games where one of the players, say Player 2, has *only two pure strategies* L and R. In this case a mixed strategy of Player 2 can be characterized by the probability weight $a$ assigned to the R, and the payoff of a given pure strategy $i$ for Player 1 becomes a linear function of $p$, which can be shown as a straight line between the points $(0, a_{i1})$ and $(1, a_{i2})$. Once all the lines are available, the value of $p$ for which the minimax over the strategies of Player 2 is attained can be found.

In the game

|   | L | R |
|---|---|---|
| T | −2 | 6 |
| M | 2 | −1 |
| B | 0 | 4 |

there are only two pure strategies available to Player 2, so the method can be used, and we get the diagram.

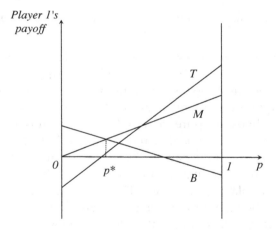

For each value of $p$, that is for each mixed strategy of Player 2, we find the maximal payoff using the upper envelope of the straight lines; this gives us the security levels for each mixed strategy, and Player 2 must then choose $p^*$ such that this security level gets as small as possible.

The minimax is realized in the point where the lines corresponding to pure strategies M and B intersect, and we get that

$$p^* \cdot 4 + (1 - p^*) \cdot 0 = p^* \cdot (-1) + (1 - p^*) \cdot 2$$

or $p^* = 0.286$.

THEOREM 4 *Let C be a convex subset of* $\mathbb{R}^m$, *and let* $N = \{x \in \mathbb{R}^m \mid x_h \le 0, h = 1, \ldots, m\}$. *If* $C \cap N = \emptyset$, *then there is* $p \in \mathbb{R}^m$, $p_h \ge 0, h = 1, \ldots, m, p \ne 0$, *so that* $p \cdot x = \sum_{h=1}^m p_h x_h \ge 0$ *for all C.*

Let $\Gamma$ be a matrix game with matrix $A$. It is no restriction to assume that the game $\Gamma$ is symmetric (so that the matrix $A$ is skewsymmetric), since if this is not the case, we consider instead the game $\mathrm{sym}(\Gamma)$ and use Lemma 1.

We may consider the columns $a^1, \ldots, a^m$ of $A$ as vectors in $\mathbb{R}^m$, where $m$ is the number of pure strategies of each player, which again is equal to the number of rows in $A$ or the number of columns in $A$, and we let $C$ be the convex hull of these vectors, that is

$$C = \left\{ a = y_1 a^1 + \cdots + y_m a^m \,\middle|\, y_i \ge 0, \ i = 1, \ldots, m, \sum_{i=1}^m y_i = 1 \right\}.$$

Notice that the vectors $y = (y_1, \ldots, y_m)$ used in the above description can be considered as mixed strategies of player 2, and the $i$th coordinate of the vector $a$ is the payoff to 1 when she plays the pure strategy $i$ and player 2 plays $y$.

Suppose that $C \cap N \ne \emptyset$, so that there is $y \in \Delta^{m-1}$ with the property that $\sum_{j=1}^m a_{ij} y_j < 0$ for all $i$. This means that player 2 by choosing the mixed strategy $y$ can keep 1's payoff below 0 in all pure strategies. Put $x = y$, so that player 1 replies by using the same strategy. Then $\sum_{i=1}^m a_{ij} x_i = -\sum_{j=1}^m a_{ij} y_j$, since $A$ is skewsymmetric, so that

$$0 > \sum_{i=1}^m \left( \sum_{j=1}^m a_{ij} y_j \right) x_i = \sum_{i=1}^m \sum_{j=1}^m a_{ij} x_i y_j = \sum_{j=1}^m \left( \sum_{i=1}^m a_{ij} x_i \right) y_j > 0,$$

a contradiction. We conclude that $C \cap N = \emptyset$.

By Theorem 4 we now get the existence of an $x \in \Delta^{m-1}$ such that $x \cdot a \ge 0$ for all $a \in C$. Using again that $A$ is skewsymmetric, we get that $x \cdot a = 0$, where $a = x_1 a^1 + \cdots + x_m a^m$. Consequently, we have that

$$0 = \sum_{i=1}^m \sum_{j=1}^m a_{ij} x_i x_j \le \sum_{i=1}^m \sum_{i=1}^m a_{ij} x_i y_j$$

for all $y \in \Delta^{m-1}$. Using the skewsymmetry of $A$ once more, we have also that

$$\sum_{i=1}^m \sum_{j=1}^m a_{ij} z_i x_j \le \sum_{i=1}^m \sum_{i=1}^m a_{ij} x_i x_j = 0$$

for all $z \in \Delta^{m-1}$, and we conclude that the pair $(x, x)$ is an equilibrium strategy in $\mathrm{mix}(\Gamma)$. $\square$

Let $\Gamma$ be a matrix game with (not necessarily skewsymmetric) matrix $A$, and let $(\sigma_1^0, \sigma_2^0)$ be an equilibrium in $\mathrm{mix}(\Gamma)$. Writing the strategy sets of player 1 and 2 as $S_1 = \{1, \ldots, m\}$ and $S_2 = \{1, \ldots, n\}$, respectively, and letting the payoff function of the mixed extension be given by $u_1(\sigma_1, \sigma_2) = \sum_{i,j} a_{ij}\sigma_1(i), \sigma_2(j)$ for $\sigma_1 \in \Delta(S_1)$, $\sigma_2 \in \Delta(S_2)$, we have

$$u_1(\sigma_1^0, \sigma_2^0) \leq u_1(\sigma_1^0, \sigma_2), \text{ all } \sigma_2 \in \Delta(S_2),$$

so that

$$u_1(\sigma_1^0, \sigma_2^0) \leq \min_{\sigma_2 \in \Delta(S_2)} u_1(\sigma_1^0, \sigma_2) \leq \min_{j \in S_2} u_1(\sigma_1^0, j),$$

where the last inequality follows since we take minimum over a smaller set. From this we deduce that

$$u_1(\sigma_1^0, \sigma_2^0) \leq \max_{\sigma_1 \in \Delta(S_1)} \min_{\sigma_2 \in \Delta(S_2)} u_1(\sigma_1, \sigma_2) \leq \max_{\sigma_1 \in \Delta(S_1)} \min_{j \in N} u_1(\sigma_1, j).$$

By similar reasoning, it can be shown that

$$\min_{\sigma_2 \in \Delta(S_2)} \max_{i \in S_1} u_1(i, \sigma_2) \leq \min_{\sigma_2 \in \Delta(S_2)} \max_{\sigma_1 \in \Delta(S_1)} \leq u_1(\sigma_1, \sigma_2) \leq u_1(\sigma_1^0, \sigma_2^0).$$

To proceed from the above two expressions, we notice that we have

$$\max_{\sigma_1 \in \Delta(S_1)} \min_{\sigma_2 \in \Delta(S_2)} u_1(\sigma_1, \sigma_2) = \max_{\sigma_1 \in \Delta(S_1)} \min_{j \in S_2} u_1(\sigma_1, j),$$

since on the left hand side, we have that $\sigma_2$ minimizes $u_1(\sigma_1, \cdot)$ for the given choice of $\sigma_1$, which means that

$$\sum_{i,j} a_{ij}\sigma_1(i)\sigma_2(j) = \sum_j \left(\sum_i a_{ij}\sigma_1(i)\right)\sigma_2(j)$$

is minimal for the given $\sigma_1$. But then we must have $\sigma_2(j) = 0$ for all $j$ not among those for which $\sum_i a_{ij}\sigma_1(i)$ is minimal. It follows that we need only minimize among the pure strategies $j \in S_2$.

Combining this with the corresponding expression

$$\min_{\sigma_2 \in \Delta(S_2)} \max_{\sigma_1 \in \Delta(S_1)} u_1\sigma_1, \sigma_2) = \min_{\sigma_2 \in \Delta(S_2)} \max_{i \in S_1} u_1(i, \sigma_2)$$

and the inequality

$$\max_{\sigma_1 \in \Delta(S_1)} \min_{\sigma_2 \in \Delta(S_2)} u_1(\sigma_1, \sigma_2) \leq \min_{\sigma_2 \in \Delta(S_2)} \max)_{\sigma_1 \in \Delta(S_1)} u_1(\sigma_1, \sigma_2),$$

which is easily shown, we get the following useful result.

THEOREM 5 *Let $\Gamma$ be a matrix game and $(\sigma_1^0, \sigma_2^0)$ an equilibrium. Then*

$$\min_{\sigma_2 \in \Delta(S_2)} \max_{i \in S_1} u_1(\sigma_1, \sigma_2) = \min_{\sigma_2 \in \Delta(S_2)} \max_{\sigma_1 \in \Delta(S_1)} u_1(\sigma_1, \sigma_2)$$

$$= \max_{\sigma_1 \in \Delta(S_1)} \min_{\sigma_2 \in \Delta(S_2)} u_1(\sigma_1, \sigma_2) = \max_{\sigma_1 \in \Delta(S_1)} \min_{j \in S_2} u_1(\sigma_1, j).$$

It follows from Theorem 5 that the main theorem can be interpreted as a statement that maximim = minimax for an arbitrary matrix $A$, when the sets, over which the maximum or minimum is taken, are extended in the proper way. The main theorem is also known as the *minimax* theorem.

It can easily be shown that also the converse of the theorem is true: If a pair $(\sigma_1^0, \sigma_2^0)$ of mixed strategies satisfies

$$\min_{\sigma_2 \in \Delta(S_2)} \max_{i \in S_1} u_1(\sigma_1, \sigma_2) = u_1(\sigma_1^0, \sigma_2^0) = \max_{\sigma_1 \in \Delta^{m-1}} \min_{\sigma_2 \in \Delta(S_2)} u_1(\sigma_1, \sigma_2),$$

then $(\sigma_1^0, \sigma_2^0)$ is an equilibrium.

Based on this we may extend Definition 4 to cover all matrix games:

DEFINITION 6 *Let $\Gamma$ be a matrix game with matrix $A$. The value of $\Gamma$, $v(\Gamma)$, is the payoff to player 1 in an arbitrary equilibrium of* mix($\Gamma$).

Summing up, we have established that every matrix game has an equilibrium, possibly in mixed strategies. The equilibria are interchangable in the sense that if $(\sigma_1, \sigma_2)$ and $(\sigma_1', \sigma_2')$ are equilibria, then so are $(\sigma_1, \sigma_2')$ and $(\sigma_1', \sigma_2)$. And in addition, all equilibria yield the same average payoff to player 1, namely $v(\Gamma)$, the value of the game.

This is a very satisfactory result, in particular in view of the difficulties that we shall encounter when trying to extend it beyond the framework of two-person zero-sum games. The existence of a well-defined value means that one may substitute this value for the underlying game, something which may be useful when considering more complex conflicts; we shall see examples of this in later chapters.

## 6   Problems

**1.** Sherlock Holmes takes the train from London to Dover in order to escape from Professor Moriarty. Moriarty can take an express train and catch Holmes at Dover. However, there is an intermediate station at Canterbury at which Holmes may stop in order to avoid him. Moriarty is aware of this too and may himself stop instead at Canterbury. von Neumann and Morgenstern [1944] estimate the value to Moriarty of these four possibilities to be given by the following matrix:

|  |  | Holmes | |
|  |  | Canterbury | Dover |
| --- | --- | --- | --- |
| Moriarty | Canterbury | 100 | −50 |
|  | Dover | 0 | 100 |

Find optimal strategies for Holmes and Moriarty. What is the value of the game?

**2.** Suppose that a game $\Gamma'$ is obtained from another game $\Gamma$ by removing a pure strategy $s$ for Player 1 which is dominated by a mixed strategy $\sigma$, so that

$$\pi(\sigma, t) > \pi(s, t)$$

for all pure strategies $t$ of Player 2.

Show that the game $\Gamma$ has the same set of equilibria as the game $\Gamma'$.

**3.** [Gale, 1960] The following describes a simplified version of *Poker:* Player 1 draws a card $H$ or $L$ with equal probability and may then choose either "bet" or "pass". If Player 1 bets, then Player 2 may "fold" and lose $a$ or "call" and win or lose $b$ depending on whether Player 1 holds $L$ or $H$. If Player 1 passes, then Player 2 may also pass, giving payoff 0, or Player 2 may bet, in which case she wins $a$ if Player 1 holds $L$ and loses $b$ if Player 1 holds $H$.

Find the payoff matrix for this game.

**4.** Let $\Gamma$ be a matrix game where each player has two pure strategies. Assume that the payoffs in the matrix

$$\begin{pmatrix} a_{11} & a_{12} \\ a_{21} & a_{22} \end{pmatrix}$$

are assigned by independent drawing of four numbers from the uniform distribution on $[0, 1]$. Find the probability that the game has an equilibrium in pure strategies?

**5.** Consider a matrix game where the players have the number of strategies, and where each row and each column has exactly one nonzero element; assume that they are all positive. Find the value of such a game.

**6.** An army $A$ must defend 6 targets against an enemy $B$. The targets are located on a row and any one target together with its immediate neighbors can be defended. $B$ can attack only one target.

If $B$ attacks a defended target there is no loss to both sides but if $B$ attacks an undefended target, $A$ loses it to $B$. The two armies value the targets equally.

Find equilibrium strategies and the value of the game.

**7.** Consider the two-person zero-sum game where the pure strategies of both players consist in choosing one of the numbers 2 or 3. Player 1 wins if the sum of the numbers chosen is odd and Player 2 wins if it is even. The loser pays the winner the product of the two numbers.

Write down the payoff matrix, and find an equilibrium and the value of the game.

# Chapter 3

# Applications of Minimax Theory

## 1  Introduction

Having dealt with the theory of two-person zero-sum games in the previous chapter, it would be natural to proceed to more general games, adding players and – in particular – abandoning the restrictive zero-sum condition. This is indeed what we shall do, but before we get there we digress briefly into some other topics which are closely related to the main results of Chapter 2, either by elaborating on the consequences of the equilibrium existence result – as with linear programming – or by using it as a tool for proving results in other models, gametheoretical or not.

Linear programming, a method for finding maxima or minima of linear functions under constraints given by linear inequalities, was developed roughly at the same time as the theory of two-person zero-sum games and is indeed very closely related to the latter. Indeed, the main result about existence of optimal solutions in linear programming is essentially a reformulation of the minimax theorem. With its emphasis on duality – to each linear programming problem there is a dual problem, and the solutions are closely interrelated – it points towards the many fruitful applications of duality, not only in optimization theory, but in numerous other contexts.

The two other applications are connected with the name of David Blackwell, a mathematical statistician who has contributed significantly to the development of decision theory. In the first of them, we are concerned with *the value of information*, the increase in expected payoff which may be obtained by acquiring additional observations on some relevant phenomena. When there is a choice among different methods of information acquisition, it would be useful to have a ranking of such methods, but the Blackwell infomation theorem tells us that apart from trivial rankings (more is better than less) this will depend on the utility function of the decision maker. There is no utility-independent way of ranking information methods.

In the last section we are moving back to game theory, considering the Blackwell approachability theorem. We consider here zero-sum games with vector payoffs, something which may look strange at first glance. However, as also indicated in the text, the result has applications to more usual games, and it is important for the theory of algorithms and learning. Also, we introduce here the idea of sequences of plays, which shall concern us later in the chapter on repeated games.

## 2  Linear programming

Linear programming deals with solving maximization or minimization problems in the context of linearity, so the objective function is linear, the constraints are formulated as linear inequalities, and in addition, the variables are subject to non-negativity conditions. Since we shall use our acquired knowledge of matrix games in the sequel, we approach the topic from this angle.

Let $\Gamma$ be a matrix game with matrix $A$. We assume that all elements in $A$ are positive (which means that $v(\Gamma) > 0$, so that the game is rather unfavorable for player 2). This assumption is not restrictive in our context, where we are interested in finding equilibria: If $A'$ is a given $(m \times n)$ matrix with both positive and negative elements, then we can add another $(m \times n)$ matrix $B$ to $A'$, where all elements of $B$ are identical, $b_{ij} = b$ for all $i, j$, positive and greater than $-\min_{i,j} a_{ij}$. The new game with matrix $A = A' + B$ has exactly the same equilibria as $\Gamma$ (but the value of the new game differs from that of the old). Intuitively, all that has happened is that player 2 independent of strategy choice has to pay an extra amount $b$ to player 1, and this does not influence whether a strategy choice is the best possible or not.

If $(\sigma_1^0, \sigma_2^0)$ is an equilibrium in mix($\Gamma$), then we have according to Theorem 5 of Chapter 2 that

$$\sum_{i,j} a_{ij}\sigma_1^0(i)\sigma_2^0(j) = v(\Gamma) = \max_{x \in \Delta^{m-1}} \min_j \sum_{i=1}^m a_{ij}x_i,$$

so that the vector $x^0$ with $x_i^0 = \sigma_1^0(i)$, $i = 1, \ldots, m$, and $\sum_{i=1}^m x_i^0 = 1$ satisfies the system of linear inequalities

$$\sum_{i=1}^m a_{ij}x_i \geq v, \ j = 1, \ldots, n,$$

where $v = v(\Gamma) > 0$. Setting $u^0 = \dfrac{1}{v}x^0$, we have that

$$\sum_{i=1}^m u_i \geq \sum_{i=1}^m u_i^0 = \frac{1}{v}\sum_{i=1}^m x_i = \frac{1}{v}$$

for all $u \in \mathbb{R}^m$ satisfying

$$\sum_{i=1}^m a_{ij}u_i = \sum_{i=1}^m a_{ij}\frac{s_1^i}{v} \geq 1, \ j = 1, \ldots, n,$$

and $u_i \geq 0$, $i = 1, \ldots, m$.

An optimization problem of this type, written in matrix notation, whereby $\geq$ for vectors means coordinatewise $\geq$, as

$$\min \; c \cdot u$$
under the constraints
$$uA \geq b \tag{1}$$
$$u \geq 0,$$

(in our case above the vectors $c$ and $b$ are vectors with 1 in all coordinates) is called a *Linear Programming* (LP) problem. For each such LP problem there is a *dual* LP problem, in our case

$$\max \; b \cdot w$$
under the constraints
$$Aw \leq c \tag{2}$$
$$u \geq 0.$$

Upon a closer look, the dual of the problem in (1) (where $b$ and $c$ are vectors of 1's) corresponds to player 2's minimax problem in the same way as the first ("primal") problem corresponded to the maximin problem of player 1.

The following is the main theorem of linear programming.

THEOREM 1 *If the primal and the dual LP problem both have a feasible solution (that is a vector $u'$ ($w'$) which satisfies the constraints $u'A \geq b$, $u' \geq 0$ ($Aw' \leq c$, $w' \geq 0$), then both have an optimal solution. If one of the problems has no feasible solution, then none of them have optimal solutions.*

We could have used this result on the LP problems derived from two-person zero-sum games to give another proof of the existence of equilibria in mix($\Gamma$): $A$ was a matrix with positive entries $a_{ij} > 0$, so for any vector $u \geq 0$, $u \neq 0$, we have $uA > 0$, and multiplying by some $K > 0$ we can obtain that $(Ku)A > e$ (where $e$ is a vector of 1's), so the primal problem has a feasible solution. By similar reasoning it is seen that also the dual problem has feasible solutions.

We now show that Theorem 1 is actually a consequence of Theorem 3 of Chapter 2 (of the previous chapter), an indication that game theoretical reasoning may be useful also in contexts which have nothing to do with games.

PROOF OF THEOREM 1: Let the primal and dual LP problem be as in (1) and (2). We construct from these problems a matrix game $\Gamma$, the matrix of which has the following block structure:

$$\begin{pmatrix} 0 & -A^t & b \\ A & 0 & -c \\ -b^t & c^t & 0 \end{pmatrix}$$

where $\cdot^t$ stands for transposal of matrices (and vectors considered as matrices with 1 column). We know from Theorem 3 of Chapter 2 that there is an equilibrium $(s_1^0, s_2^0)$ in mixed strategies. Using the block structure of the matrix of $\Gamma$, we write mixed strategies as $(w, u, s)$, so that $w$ is an $n$-vector, $u$ an $m$-vector and $s$ a real number.

We start by assuming that there is an equilibrium with strategy $(w, u, s)$ for player 1 satisfying $s > 0$. Since the game is symmetric, its value is 0, so that no matter which pure strategy player 2 chooses, the payoff to player 1, who chooses $(w, u, s)$, is non-negative. If for example player 2 chooses the last pure strategy (corresponding to last column in the matrix), then we get that

$$b \cdot w - u \cdot c \geq 0$$

or $b \cdot w \geq u \cdot c$. Checking similarly what happens if player 2 chooses any of the first $n$ columns, we obtain

$$uA - sb \geq 0$$

or

$$\left(\frac{1}{s}u\right)A \geq b.$$

This gives us the desired result (still under the assumption $s > 0$), since $\frac{1}{s}u$ and $\frac{1}{s}w$ are feasible solutions to the primal and the dual problem, and since for any pair $(u', w')$ of feasible solutions to the primal and dual problem we have that

$$b \cdot w \leq (u'A) \cdot w' = u' \cdot (Aw') \leq u' \cdot c, \tag{3}$$

it follows from the above that $\frac{1}{s}u$ and $\frac{1}{s}w$ are minimal (maximal) among the feasible solutions and therefore solve the respective LP problems.

If $s = 0$, but there is a player, say Player 2, who has an equilibrium strategy $(\bar{w}, \bar{u}, \bar{s})$ with $\bar{s} \neq 0$, then we may reason as above, since the game is symmetric. Suppose finally that $((w, u, s), (\bar{w}, \bar{u}, \bar{s}))$ is an equilibrium with $s = 0$, $\bar{s} = 0$. We assume first that there are feasible solutions $u'$ and $w'$ to the primal and the dual problem, respectively. It then follows that if Player 1 changes strategy to

$$\left(\hat{w}, \hat{u}, \frac{1}{t}\right), \text{ where } t = 1 + \sum_{h=1}^{m} u'_h + \sum_{k=1}^{n} w'_k,$$

then we get that $\hat{u}A \geq \frac{1}{t}b$ and $\hat{w}(-A^t) \geq -\frac{1}{t}c$, from which it can be seen that the strategy $(\hat{w}, \hat{u}, \hat{s})$ guarantees a payoff $\geq 0$ against all pure strategies of player 2 except the last, which however has probability 0 in the mixed strategy of player 2. In particular, we get that $(\hat{w}, \hat{u}, \hat{s})$ is just as good for player 1 as $(w, u, s)$. It is easily checked that $((\hat{w}, \hat{u}, \hat{s}), (\bar{w}, \bar{u}, \bar{s}))$ is an equilibrium, and since $\hat{s} > 0$, we may proceed as above.

If one of the problems has no feasible solutions, then we may use the last inequality in (3) to infer that the other problem is unbounded and therefore has no optimal solution.                                                                    □

The theorem connects solutions to the primal and the dual problem, but it provides no clues to finding a solution. But solving LP problems is a field which has received considerable attention over the years, resulting in several algorithms, among which the simplex method, for a long time the work horse of mathematical programming, and later in the form of faster inner point algorithms. Consequently, linear programming provides a method for finding equilibrium points of matrix games, applicable to any game in contrast to the geometrical methods mentioned in the previous chapter.

## 3 Comparing information services

The application of the minimax theorem to be considered now has some relation to game theory, at least in the sense that we are dealing with decision making under uncertainty. If an action must be chosen which has consequences depending on some unknown events, it may be useful to collect information, collect observations which may reduce this uncertainty. Since information may be collected in many different ways, it is clearly useful to know which way is the best. To some extent this may depend on the preferences of the decision maker, but it might well be the case that some methods are better than others independently of decision maker preferences. In the following, we look closer at such a ranking of methods for collecting information.

We use the following notation: Let $\Omega = \{\omega_1, \ldots, \omega_n\}$ be a set of states of the world, and let $X = \{x_1, \ldots, x_N\}$ be the set of possible signals (or results of experiments). The information service (or the experiment) is described by a stochastic matrix $P = P_{n \times N} = (p_{ij})_{i=1}^{n}{}_{j=1}^{N}$ with $p_{ij} \geq 0$, $\sum_{j=1}^{N} p_{ij} = 1$, all $i$, where $p_{ij}$ represents the probability of signal $j$ when in state $i$.

Further, $A \subset \mathbb{R}^n$ is an action space, where $a = (a_1, \ldots, a_n)$ represents a payoff obtained in each state of the world, and $f : X \to A$ is a decision function, which assigns an action $f(x_j)$ to each signal $x_j$. The expected payoff from $f$ in state $\omega_i$ is

$$u_i(f) = \sum_{j=1}^{N} p_{ij} f_i(x_j),$$

and we let $B(P, A)$ denote the set of expected payoff vectors $u(f) = (u_1(f), \ldots, u_n(f))$ obtained using some decision function $f$.

Now we come to the main concepts used in Blackwell's information theorem:

- The information service $P_{n \times N_1}$ is *more informative than* the information service $Q_{n \times N_2}$, written $P \supset Q$, if $B(P, A) \supseteq B(Q, A)$, that is any vector attainable with $Q$ is also attainable with $P$, for all convex and compact subsets $A$ of $\mathbb{R}^n$.
- The information service is *sufficient for* $Q$, written $P > Q$, if $PM = Q$ for some stochastic matrix $M$ (so that $q_{ij} = \sum_{k=1}^{N_1} p_{ik} m_{kj}$ for $j = 1, \ldots, N_2, i = 1, \ldots, n$.

**Example 3.2.** The following case illustrates the value of information: Suppose that a decision must be made about the use of a medical drug in another country; there are three brands, I, II and III, all produced locally. Each of these drugs may turn out to be cheap or expensive, effective or ineffective, in our example these properties will be shared by *all* brands, and the consequences of them differ among the brands.

The unknown states are the four combinations of price and quality. It is assumed that the utility function of the decision maker has the following form:

|       | effective, expensive | effective, cheap | ineffective, expensive | ineffective, cheap |
| ----- | -------------------- | ---------------- | ---------------------- | ------------------ |
| I     | 10                   | 10               | 10                     | 10                 |
| II    | 20                   | 5                | 15                     | 0                  |
| III   | 12                   | 14               | 8                      | 6                  |

Each state has the same prior probability 1/4.

Now the expected utility of each decision can be computed. For brand I it is

$$\frac{1}{4} \cdot 10 + \frac{1}{4} \cdot 10 + \frac{1}{4} \cdot 10 + \frac{1}{4} \cdot 10 = 10,$$

and using the same method it is seen that all decisions are equally good or bad at this point.

Now the decision maker is offered the information method which consists in revealing whether the country is a high- or low-price country with regard to drugs. If the signal received is "expensive", then only the first and the third columns, each with conditional probability 1/2, matter, so expected utility of I remains 10, but for II it becomes

$$\frac{1}{2} \cdot 20 + \frac{1}{2} \cdot 15 = 17.5$$

which is the best possible. If the signal is "cheap", expected utilities are 10, 2.5 and 10, and best decisions are I and III. The value of the information method is found as the expected value of the best result at each possible signal,

$$\frac{1}{2} \cdot 17,5 + \frac{1}{2} \cdot 10 = 13.75$$

minus expected utility of the best choice without information, which was 10; we have therefore that the value of this information method is $13.75 - 10 = 3.75$.

In similar way, one may compute the value of other possible information methods, for example the method which exactly reveals whether the quality of medicin produced in the country is effective or ineffective. This value can be computed as 1.5, and the value of complete revelation of state, for which expected utility is 17.25, so that the value of complete information is 7.25.

THEOREM 2 $P > Q \Leftrightarrow P \supset Q$.

PROOF: $\Rightarrow$: Assume that $P > Q$, and let $u(f)$ be a point in $B(Q, A)$, obtained using the decision function $f$. We now use the decision function $g$ defined by

$$g(x_k) = \sum_{j=1}^{N_2} m_{kj} f_i(x_j),$$

and then we can write

$$u_i(f) = \sum_{j=1}^{N_1} q_{ij} f_i(x_j) = \sum_{j=1}^{N_1} f_i(x_j) \sum_{k=1}^{N_1} p_{ik} m_{kj} = \sum_{k=1}^{N_1} p_{ik} \sum_{j=1}^{N_2} m_{kj} f_i(x_j) = \sum_{k=1}^{N_1} p_{ik} g_i(x_k) = u_i(g)$$

for each $i$, so that $u(f)$ belongs to $B(P, A)$.

$\Leftarrow$: We have that $B(P, A) \supseteq B(Q, A)$ for any convex and compact $A \subset \mathbb{R}^n$, in particular if $A$ is the convex hull of the rows of the $(N_2 \times n)$-matrix $D = (f_i(x_j))_{j=1 \ i=1}^{N_2 \ n}$ (the outcome vectors according to $f$ of each signal in $X$), so that expected payoff is

$$u_i(f) = \sum_{j=1}^{N_2} q_{ij} d_{ji} = (QD)_{ii}$$

for each $i$.

Now $P \supset Q$, so there must be a decision function $g$ for $(P, A)$ which gives $u(f)$ as expected payoff in each of the states of the world. This decision function must select some $a^j \in A$ when signal $x_j$ is observed, and

$$u_i(g) = \sum_{j=1}^{N_1} p_{ij} a_i^j = u_i(f),$$

all $i$. There is a stochastic matrix $\widehat{M} = (\hat{m}_{jk})_{j=1 \ k=1}^{n \ N_2}$ such that

$$a_i^j = \sum_{k=1}^{N_2} \hat{m}_{jk} d_{ki};$$

indeed, since $A$ is a convex polytope spanned by the rows of $D$, each vector $a^j \in A$ can be written as a convex combination of these rows. We then have

$$u_i(g) = \sum_{j=1}^{N_1} p_{ij} a_i^j = \sum_{j=1}^{N_1} \sum_{k=1}^{N_2} p_{ij} \hat{m}_{jk} d_{ki} = (P\widehat{M}D)_{ii},$$

so that the two matrices $P\widehat{M}D$ and $QD$ have the same diagonal elements.

Now consider a zero-sum game where nature and the decision maker are the two players. Nature chooses a payoff matrix $D$, and the decision maker chooses a stochastic matrix $M$, and nature receives the payoff

$$\Pi(D, M) = \text{tr}\,[(PM - Q)D],$$

where tr $C$ denotes the trace of the matrix $C$ (the sum of the diagonal elements). We assume that all elements of $D$ are in $[0,1]$ (normalizing if this is not the case). Then strategy sets are compact and convex, and payoff is bilinear, so there is an equilibrium $(D^0, M^0)$ in the game, satisfying

$$\Pi(D, M^0) \leq \Pi(D^0, M^0) \leq \Pi(D^0, M)$$

for all strategies $M$ and $D$ of the two players. Choosing $M = \widehat{M}$, we get that $\Pi(D^0, M^0) \leq \Pi(D^0, \widehat{M}) = 0$, since $(P\widehat{M}D)_{ii} = (QD)_{ii}$ for all $i$, and we conclude that

$$\text{tr}[(PM^0 - Q)D] = \Pi(D, M^0) \leq 0.$$

for all $D$. Since nature can choose $D$ with arbitrary entries in $[0,1]$, this can hold only if the two matrices $PM^0$ and $Q$ are identical, so $P > Q$.                     □

The result tells us that a ranking of information methods which is valid for all risk-averse decision makers must be rather coarse, in the sense that one method is better than another one only if it is "at least as detailed", so that the signals obtained in the second method would also be available in the first one (after an encoding using the matrix $M$).

Blackwell's information theorem has found applications not only in statistics and natural sciences (the information methods may be interpreted as experiments) but also in totally different contexts, such as that considered by Demski [1973], who considers accounting standards, rules for presenting the economic activities of a firm to the public in its accounts. Here the theorem tells us that there is no theoretically correct way of setting up the accounts, the only valid rule is that more information is better than less.

## 4   Approachability

In this section, we consider an extension of the concept of matrix games where payoffs are $k$-dimensional rather than one-dimensional. In the theory of two person zero-sum games, there was no uncertainty about the payoffs (of the other player), since the zero-sum property means that it is known to any player knowing her own payoff. However, in some contexts it may be useful to consider games with vector payoffs, corresponding to the payoffs in different states of nature, and following Blackwell [1956], we look for a counterpart of the minimax theorem for such games.

For games $\Gamma$ with scalar payoff, the minimax theorem tells us that there is a mixed strategy $\sigma_1$ for Player 1, such that $u(\sigma_1, s_2) \geq v(\Gamma)$ for all choices of pure strategies $s_2$ by Player 2. This statement can be transformed to one pertaining to sequences of plays, telling us that Player 1 can choose a sequence of pure strategies $(s_1^t)_{t=1}^{\infty}$ such that for all choices $(s_2^t)_{t=1}^{\infty}$, the sequence of average payoffs

$$\frac{1}{t} \sum_{\tau=1}^{t} u(s_1^\tau, s_2^\tau)$$

converges to a number $\geq v(\Gamma)$.

**4.1. Blackwell's approachability theorem.** The generalization to our present case of vector payoffs is provided by the notion of approachability. Suppose that Player 1 has $m$ and Player 2 $n$ pure strategies, and that a payoff matrix is chosen from among $k$ possible $(m \times n)$ matrices $A^1, \ldots, A^k$. Mixed strategies for Player 1 and 2 are probability vectors $p = (p_1, \ldots, p_m)$ and $q = (q_1, \ldots, q_n)$, respectively. As above, we are interested in repetitions of the game whereby the expected payoff of a mixed strategy can be approached by sequences of pure strategy choices. Let $A : \mathbb{R}^m \times \mathbb{R}^n \to \mathbb{R}^k$ be defined by

$$A(p, q) = \sum_{i=1}^{m} \sum_{j=1}^{n} p_i A_{ij} q_j,$$

where $A_{ij}$ are given vectors in $\mathbb{R}^k$, for $i = 1, \ldots, m$, $j = 1, \ldots, n$. Thus, the $(i, j)$th element of $A$ is the vector of elements $A_{ij}^h$ for $h = 1, \ldots, k$.

For $(p^t, q^t)_{t=1}^{\infty}$ a sequence of mixed strategies, define the sequence $(a_t)_{t=1}^{\infty}$ of associated average payoff vectors be given by $a_t = \frac{1}{t} \sum_{\tau=1}^{t} A(p^\tau, q^\tau)$. A set $C \subseteq \mathbb{R}^k$ is *approachable* if there is sequence $(f_t, g_t)_{t=1}^{\infty}$ such that the sequence associated with the resulting sequence $(p^t, q^t)_{t=1}^{\infty}$ satisfies

$$d(a_t, C) = \inf_{a \in C} \|a_t - a\| \xrightarrow[t \to \infty]{} 0$$

THEOREM 3 (Blackwell's approachability theorem) *Let $C \subset \mathbb{R}^k$ be closed and convex. Then the following are equivalent:*

(a) *$C$ is approachable,*
(b) *For every $q \in \Delta_n$ there is $p \in \Delta_m$ such that $A(p, q) \in C$.*
(c) *Every halfspace containing $C$ is approachable.*

PROOF: We first show that (b) and (c) are equivalent. Suppose that (b) holds, and let $H = \{a \in \mathbb{R}^k \mid v \cdot a \leq v\}$ for some $v \in \mathbb{R}^k$ and $v \in \mathbb{R}$ be a halfspace such that $C \subseteq H$. Applying the minimax theorem to the matrix game with matrix $(v \cdot A_{ij})_{i=1 j=1}^{m \ n}$, we get that

$$\forall q \exists p : v \cdot A(p, q) \leq v \implies \exists p \forall q : v \cdot A(p, q) \leq v,$$

which means that $H$ is approachable. Conversely, given (c), choose a halfspace $H = \{a \in \mathbb{R}^k \mid v \cdot a \leq v\}$ containing $C$. Since $H$ is approachable, choosing a sequence $(p^t, q^t)_{t=1}^{\infty}$ where $q^t = q$ for all $t$, we get that

$$\forall q \exists p : v \cdot A(p, q) \leq v,$$

and it follows that

$$\forall q \exists p : v \cdot A(p, q) \in C \tag{4}$$

by separation of convex sets. Thus, (b) $\Leftrightarrow$ (c).

Since (a) $\Rightarrow$ (c) trivially, it remains only to show that (c) implies (a). Suppose that the halfspace $H = \{a \in \mathbb{R}^k \mid v \cdot a \leq v\}$ is approachable. Then we have that for all $q$ there is $p$ such that $v \cdot A(p, q) \leq v$ (use the constant sequence with $q^t = q$ and the definition of approachability), and by the minimax theorem, this means that there exists $p$ such that for all $q$, $v \cdot A(p, q) \leq v$.

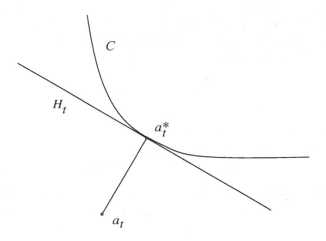

**Fig. 1.** Finding the next value of $p$: At stage $t$, $a_t$ has been obtained. If $a_t$ is not in $C$, then choose the closest point $a_t^*$ in $C$ and a hyperplane containing $C$ which has $a_t^*$ in its boundary. Then the approachability of hyperplans allows us to choose $p_{t+1}$.

Let $(q^t)_{t=1}^\infty$ be given. We construct a sequence $(p^t)_{t=1}^\infty$ inductively as follows: Choose $p^1$ arbitrarily; suppose that $p^\tau$ have been defined for $\tau \leq t$ and let $a_t = \frac{1}{t} \sum_{\tau=1}^t A(p^\tau, q^\tau)$. If $a_t \in C$, then $p^{t+1}$ is chosen arbitrarily. Otherwise, we let $a_t^*$ be the closest point to $a_t$ in $C$ and define the halfspace $H_t$ by

$$H_t = \{a \in \mathbb{R}^k \mid (a_t - a_t^*)) \cdot a \leq (a_t - a_t^*)) \cdot a_t^*\},$$

cf. Fig. 1. The set $C$ is contained in $H_t$, so it is approachable, and from (4) we get that there is $p^{t+1}$ such that $A(p^{t+1}, q) \in H_t$ for all $q$. We then get that

$$\|a_{t+1} - a_{t+1}^*\|^2 \leq \|a_{t+1} - a_t^*\|^2 = \|a_{t+1} - a_t\|^2 + \|a_t - a_t^*\|^2 + 2(a_{t+1} - a_t) \cdot (a_t - a_t^*). \quad (5)$$

If we write

$$a_{t+1} = \frac{1}{t+1} \sum_{\tau=1}^{t+1} A(p^\tau, q^\tau) = \frac{1}{t+1} A(p^{t+1}, q^{t+1}) + \frac{t}{t+1} a_t,$$

then the last term on the right-hand side in (5) can be rewritten as

$$2(a_{t+1} - a_t) \cdot (a_t - a_t^*) = \frac{2}{t+1} A(p^{t+1}, q^{t+1}) \cdot (a_t - a_t^*) - \frac{2}{t+1} a_t \cdot (a_t - a_t^*).$$

Inserting this, we get from (5) that

$$\|a_{t+1} - a_{t+1}^*\|^2 \le \|a_{t+1} - a_t\|^2 + \left(1 - \frac{2}{t+1}\right)\|a_t - a_t^*\|^2 + 2(A(p^{t+1}, q^{t+1}) - a_t^* \cdot (a_t - a_t^*). \quad (6)$$

The first term on the right-hand side satisfies

$$\|a_{t+1} - a_t\|^2 = \frac{\|A(p^{t+1}, q^{t+1}) - a_t\|^2}{(t+1)^2} \le \frac{A_{\max}^2}{(t+1)^2},$$

and the last term is $\le 0$ by our choice of $p^{t+1}$, so (6) can be rewritten as

$$(t+1)^2\|a_{t+1} - a_{t+1}^*\|^2 - t^2\|a_t - a_t^*\|^2 \le 2A_{\max}^2 - \|a_t - a_t^*\|^2. \quad (7)$$

Summing these inequalities from $\tau = 1$ to $\tau = t$ gives

$$t^2\|a_t - a_t^*\|^2 \le 2tA_{\max}^2 - \sum_{\tau=1}^{\infty} \|a_\tau - a_\tau^*\|^2 \le 2tA_{\max}^2,$$

so that

$$\|a_t - a_t^*\| \le A_{\max}\sqrt{\frac{2}{t}} \xrightarrow[t\to\infty]{} 0,$$

showing that $C$ is indeed approachable. $\qquad\square$

**4.2. An application to regret in games.** We return here to the general framework of game theory as introduced in Chapter 1. Consider a game $\Gamma = (N, (S_i)_{i\in N}, (u_i)_{i\in N})$ with finite sets of (pure) strategies $S_i$ for each player. We assume that the game is played repeatedly and let $h^t = (s^1, ..., s^{t-1})$ denote the history up to time $t$ (before the strategies are chosen). The *external regret* of player $i$ against action $\sigma_i$ after history $h^t$ is

$$\mathrm{ER}_i(h^t, \sigma_i) = \sum_{\tau=1}^{t-1} \left[u_i(\sigma_i, s_{-i}) - u_i(s_i^t, s_{-i})\right].$$

It shows the loss in terms of foregone payoff that the player would have experienced if the mixed strategy $\sigma_i$ had been used throughout. A reasonable approach towards an optimal strategy choice for player $i$ would be to minimize regrets, and since the choices of the other players are unknown, the player might minimize worst case regrets.

In the setting of two-person games, a strategy $s_1$ of player 1 is *external regret minimizing*, also known as *Hannan-consistent*, if for all history-dependent strategies of player 2,

$$\limsup_{t\to\infty} \max_{s_1\in S_1} \frac{1}{t}\mathrm{ER}_1(h^t, s_1) = 0.$$

The situation is transformed to one of Blackwell approachability if we define a game with vector-valued payoffs, taken to be the negative regrets. Formally, for each pair $(s_1, s_2) \in S_1 \times S_2$, let

$$A(s_1, s_2)(s_1') = u_1(s_1, s_2) - u_1(s_1', s_2), \quad (8)$$

so that $A(s_1, s_2)$ is an $m$-vector, $m = |S_1|$. We have

$$\mathrm{ER}_1(h^t, s_1) = -\sum_{\tau=0}^{t-1} A(s_1^t, s_2^t)(s_1).$$

We then have

$$a_t = \frac{1}{t} \sum_{\tau=1}^{\infty} A(s_1^\tau, s_2^\tau) = -\frac{1}{t} \mathrm{ER}_1(h^t, \cdot),$$

so that the existence of Hannan-consistent sequences for player 1 amounts to the condition that for all pure strategies $s_1 \in S_1$, the relevant coordinate $a_t(s_1)$ of $a_t$ satisfies

$$\liminf_{t \to \infty} a_t(s_1) \geq 0;$$

in other words, the positive orthant $\mathbb{R}_+^m$ should be approachable in the zero-sum game with payoffs $A$.

We state the findings as a theorem:

THEOREM 4 (*Hannan-Gaddum*). *Let $(S_1, S_2, u)$ be a two-person game, played sequentially. The Player 1 has an external regret minimizing strategy if and only if $\mathbb{R}_+^m$ is approachable in the zero-sum game with vector payoffs given by (8).*

Looking back to the proof of Blackwell's theorem, we notice that it was constructive – the approaching sequence was constructed in a specific way depending on the previous choices. Therefore, Theorem 4 goes beyond establishing that a regret minimizing sequence exists, it also provides an algorithm which can be used to find it.

## 5   Problems

**1.** Consider the LP problem of maximizing

$$2x + 3y$$

under the constraints

$$x - y \geq -11$$
$$x + y \leq 27$$
$$2x + 5y \leq 90$$

together with the nonnegativity constraints $x \geq 0$, $y \geq 0$.

Find the set of feasible solutions, and show that it is the convex hull of intersection points of the boundary lines given by each of the inequalities. Show that the maximum must be attained at such an intersection point.

What is the dual program?

**2.** An LP problem is in *canonical form* if it is written as

$$\max c \cdot x$$
$$Ax \le b \qquad\qquad (9)$$
$$x \ge 0$$

and it is in *standard form* if all constraints are given as equalities, so that it can be written as

$$\max c \cdot x$$
$$Ax = b \qquad\qquad (10)$$
$$x \ge 0$$

Show that an LP in canonical form can be transformed to one in standard form. If the primal LP is in standard form, will the dual also be?

**3.** A country considers the introduction of a mandatory drug testing policy for all professional athletes. The test used gives either a "positive" or "negative" indication. From previous experience with the test it is known that if the person tested is a drug user, there is a probability of 0.93 that the test returns "positive", and if the person is not at drug user, then there is a probability of 0.96 that the test result is "negative". It is expected that around 10% of the athletes to be tested are drug users.

Find the probability that a randomly selected athlete will test positive.

Given that a randomly selected athlete is tested positive, find the probability that she is actually a drug user.

Similarly, given that a randomly selected athlete is tested negative, find the probability that she is actually a drug user.

What are the advantages and disadvantages of the drug testing program?

**4.** Show by an example that the intersection of two sets, which are approachable for a given two-person zero-sum game with vector payoffs, may not be approachable.

# Chapter 4

# Solutions for General Non-Cooperative Games

## 1 Introduction

We now return to our discussion of games, moving from the well-behaved world of two-person zero-sum games to general games, with arbitrary numbers of players and with payoffs not satisfying any zero- or constant-sum condition. Once this latter property is abandoned, the number of players makes little difference in the complications that we encounter. For simplicity, we shall often restrict our treatment to two-person games, since the notation becomes simpler. Indeed, a two-person game with a finite number of (pure) strategies for each player may be presented much in the same way as a matrix game, only the elements of the matrix are pairs of payoff, one for each player:

| | $s_2^1$ | $\cdots$ | $s_2^n$ |
|---|---|---|---|
| $s_1^1$ | $(u_1(s_1^1, s_2^1), u_2(s_1^1, s_2^1))$ | $\cdots$ | $(u_1(s_1^1, s_2^n), u_2(s_1^1, s_2^n))$ |
| $\vdots$ | $\vdots$ | | $\vdots$ |
| $s_1^m$ | $(u_1(s_1^m, s_2^1), u_2(s_1^m, s_2^1))$ | $\cdots$ | $(u_1(s_1^m, s_2^1), u_2(s_1^m, s_2^n))$ |

Much of our reasoning from zero-sum games loses relevance. Thus, selecting maximin strategies is no longer an appealing strategy, as can be seen from the following example, known in the literature as the *Battle of the Sexes:* There are two players (a married couple), and each has two pure strategies $O$ and $S$ (attending either an Opera or a Soccer game). The payoffs are

| | $O$ | $S$ |
|---|---|---|
| $O$ | $(2, 1)$ | $(-1, -1)$ |
| $F$ | $(-1, -1)$ | $(1, 2)$ |

showing that the couple prefers going together but as individuals have different preferences.

Trying out a maximin strategy gives little clue to the choice of optimal strategies, since each of the pure strategies have the worst-case payoff $-1$. If also mixed strategies are taken into consideration, then the row player may keep the worst-case payoff at $1/5$ by choosing $O$ with probability $2/5$ and $S$ with probability $3/5$. The column player may similarly choose the mixed strategy $(3/5, 2/5)$ and achieve the same payoff $1/5$. This solution is not particularly attractive from the players' point of view, since each of the coordinated choices (where both choose the same pure strategy) is better for both, but in our context of searching for a well-behaved solution concept, it at least has the advantage of giving a definite result.

But there is a further objection against this strategy array: Given that the column player chooses $(3/5, 2/5)$, the row player can switch to the pure strategy $O$ and get the payoff $4/5$ instead of the $1/5$ that would follow from the mixed strategy $(2/5, 2/5)$. Thus, the considered array of mixed strategies lacks the stability property which made the maximin-minimax solution so attractive in the context of zero-sum games.

A slightly changed version of the above game (known as Battle of the Sexes 2) further exhibits the weaknesses of maximin:

|   | $O$ | $S$ |
|---|-----|-----|
| $O$ | $(2, 1)$ | $(-1, -1)$ |
| $S$ | $(0, 0)$ | $(1, 2)$ |

Restricting attention to pure strategies, the maximin principle gives a unique choice, namely $(S.O)$ (row player goes to the soccer game, column player to opera). Again this solution is inefficient (in the sense that both players could obtain something better at another array of pure strategies), but furthermore, each of the players could improve by switching to the other pure strategy.

Once it is realized that the approach to solving non-cooperative games using maximin is not a feasible one in general, the logical next task is to look for strategy arrays which retain other properties which made them convincing in the context of zero-games. The obvious candidate for such a property is that of stability against strategy changes, already present in our discussion so far. Formulating this equilibrium property for general $n$-person games, we get the concept of a *Nash equilibrium*, which will be used intensively in the chapters to follow.

DEFINITION 1  *Let $\Gamma = (N, (S_i)_{i \in N}, (u_i)_{i \in N})$ be a game. A Nash equilibrium in $\Gamma$ is an array $s^0 = (s_1^0, \ldots, s_n^0) \in S = \bigtimes_{i \in N} S_i$ such that*

$$u_i(s_i^0, s_{-i}^0) \geq u_i(s_i, s_{-i}^0)$$

*for all $i \in N$ and $s_i \in S_i$.*

We have introduced the notion of a Nash equilibrium without mentioning whether or not mixed strategies are taken into account. The distinction between pure and mixed strategy Nash equilibria is actually not one between different equilibria in the same game, but between equilibria in the game with only pure strategies and equilibria in the mixed extension.

Going back to the Battle of the Sexes (in the first version), it is seen that there are two Nash equilibria in pure strategies, $(S, O)$ and also $(O, S)$. We may also look for mixed strategy Nash equilibria: If the row player chooses first row with probability $x$ and second row with probability $y$, then the two rows must give the same expected payoff given the mixed strategy $(y, 1 - y)$, that is

$$2y + (1 - y)(-1) = y(-1) + (1 - y),$$

which gives $y = 2/5$. Similar reasoning yields the value $x = 3/5$. Since the two pure strategies are equally good for each player given the mixed strategy chosen by the other player, there can be no better choice, and we have found another Nash equilibrum.

It may be noticed that the mixed strategy equilibrium is inferior to the two pure strategy equilibria, since the expected payoff is smaller. This shows that we have not solved all the problems connected with selecting a reasonable solution to a given game if we decide for the Nash equilibrium.

## 2   Dominance and elimination of dominated strategies

Before we look at Nash equilibria in more detail, we consider some other candidates for solutions to a given game. We have already introduced the concept of domination, and it generalizes straightforwardly to non-zero sums and indeed to general $n$-person non-cooperative games.

**2.1. (Strict) domination.** Let $\Gamma = (N, (S_i)_{i \in N}, u)$ be a game and $i \in N$ an arbitrary player. A strategy $s_i$ of player $i$ *dominates* another strategy $s_i'$ if $u_i(s_i, xs_{-i}) > u_i(s_i', s_{-i})$ for all $x_{-i} \in S_{-i}$. The notion of dominance is quite strong, since the strategy $s_i$ gives a strictly better result than the strategy $s_i'$ no matter what the remaining players do.

If $s_i$ dominates $s_i'$, then it would not seem rational for player $i$ to choose $s_i'$. In the case where $s_i$ dominates all other strategies $s_i'$ of player $i$, then it is the only reasonable choice by this player. For two-person games this will give us a complete solution, since the other player must choose the strategy which gives the best payoff given that player $i$ chooses $s_i$. However, the situation does not happen very often (however, a notable case where it does happen is the Prisoners' Dilemma game considered in Example 1.2).

If a dominated strategy $s_i'$ is never used by player $i$, it may be deleted when looking for a solution to the game. Once the dominated strategies of player $i$ have been eliminated, we may proceed to another player $j$, who may now delete some strategies, etc. In this process of *iterated elimination of dominated strategies*, we reduce the number of strategies which may reasonably be considered as possible choices of the candidates, possibly ending up with a single one for each player. In this case, we have a *solution in dominated strategies*.

---

**Example 4.1. Pure strategies may be dominated by mixed strategies.** Consider the 2-person game with payoffs given in the table below:

|   | L | R |
|---|---|---|
| T | $(2,1)$ | $(-1,-1)$ |
| M | $\left(\frac{3}{4},1\right)$ | $(0,0)$ |
| B | $(0,0)$ | $(1,2)$ |

If the row player chooses a mixed strategy with probability 7/16 of playing $T$ and probability 9/16 of playing $B$, then the payoffs in the two pure strategies are 7/8 when the column player chooses $L$ and 1/8 when $R$ is chosen. Thus the strategy $M$, which is not dominated by either $T$ or $B$, can nevertheless be eliminated if mixed strategies are allowed.

---

Some objections may come in at this point: It may not be possible to eliminate all except one strategy for each player, indeed in many cases no strategies of any player are dominated, and then we are left with the original problem of finding a solution. And if successive elimination of dominated strategies does produce a solution, is it then unique? Here we can provide an answer, so we consider the problem somewhat closer.

We consider a game with a finite number of (pure) strategies. Since pure strategies may be dominated by mixed strategies, we must take the latter into consideration as well. So in the following, when we say that a strategy $s_i \in S_i$ is dominated, it means that there is some $\sigma_i \in \Delta(S_i)$ such that

$$u_i(\sigma_i, s_{-i}) > u_i(s_i, s_{-i}) \text{ for all } s_{-i} \in S_{-i}. \tag{1}$$

It is easily seen that if $s_i$ is dominated by a mixed strategy, then there is a mixed strategy $\sigma_i$ dominating $s_i$ such that $\mathrm{supp}\,\sigma_i$, the *support* of $\sigma_i$, i.e. the set of pure strategies which have nonzero probability of being chosen, consists of strategies which are undominated.

The following lemma is a straightforward consequence of the definitions.

LEMMA 1 *Let* $\Gamma = (N, (S_i)_{i \in N}, (u_i)_{i \in N})$ *and* $\Gamma' = (N, (S'_i)_{i \in N}, (u'_i)_{i \in N})$ *be such that* $S'_i \subseteq S_i$ *and* $u'_i = u_i|_{\times_{i \in N} S'_i}$ *for each* $i \in N$. *If* $\sigma_i \in \Delta(S_i)$ *dominates* $s'_i \in S'_i$ *in* $\Gamma$, *then* $\sigma_i$ *dominates* $s'_i$ *in* $\Gamma'$ *as well.*                                                  □

For each $i$, let $S_i^1 \subseteq S_i$ be the set of undominated strategies for player $i$, and let

$$\Gamma^1 = (N, (S_i^2)_{i \in N}, (u_i|_{S^1})_{i \in N})$$

be the game where player $i$ has strategy set $S_i^1$ for $i \in N$, and the payoff functions are the restrictions of $u_i$ to $S^1 = \times_{i \in N} S_i^1$. In general, suppose that $\Gamma^k$ has been defined, $k \geq 1$, let $S^{k+1}$ be the set of strategies that are undominated in $\Gamma^k$, and define

$$\Gamma^{k+1} = (N, (S^{k+1})_{i \in N}, (u_i|_{S^{k+1}})_{i \in N}).$$

Since the sets $S_i$ for $i \in N$ are finite, the sequence $(\Gamma^k)_{k=0}^{\infty}$ becomes stationary at some $K \geq 0$.

Next we turn to sequential deletion of dominated strategies: Let $I = (i_k)_{k=1}^{\infty}$ be a sequence from $N$ such that

(i) $i(k+1) \neq i(k)$ for each $k$ (a new player is selected at each stage), and
(ii) for all $i \in N$ and $k$, there is $k' > k$ such that $i_{k'} = i$ (each player is selected infinitely often).

Define a sequence $\Gamma^{I,k}$ as follows: As before $\Gamma^{I,0} = \Gamma$, and at each step $k \geq 1$, $\Gamma^{I,k}$ is defined as the game where all players except $i_k$ have the same strategy sets as in $\Gamma^{I,k-1}$, and where the strategy set of player $i_k$ consists of the strategies which are undominated in $\Gamma^{I,k-1}$.

Since at each step $k$, the strategy set $S_i^{I,k}$ contains $S_i^k$, all $i \in N$, we get from Lemma 1 that a strategy $s_{i_k}'$ for player $i_k$ in $\Gamma^{I,k-1}$ which is dominated and hence deleted in $\Gamma^{I,k}$, is either not in $S_{i_k}^{k-1}$ or dominated, so that it is not in the strategy set of player $i_k$ in $\Gamma^k$.

Conversely, suppose that $i \in I$ and that $s_i'$ is dominated in $\Gamma^k$. Let $I$ be the sequence which the repeatedly runs through the numbers $(1, \ldots, n)$. Then a strategy $s_i'$ which is dominated in $\Gamma^k$ and therefore not in the stategy sets of $\Gamma^{k+1}$, must be dominated in $\Gamma^{I,ki}$. We conclude that the strategy sets $S_i^K$ in $\Gamma^K$ consists of those strategies that cannot be eliminated in any sequence of elimination of dominated strategies.

These considerations gives us the following result.

THEOREM 1 *Let $\Gamma = (N, (S_i)_{i \in N}, (u_i)_{i \in N})$ be a game, and assume that $S_i$ is a finite set, all $i \in N$. If $\Gamma$ has a solution by iterated elimination of dominated strategies, then this solution is unique.*

PROOF: If $(s_1^0, \ldots, s_n^0)$ is a solution, then the strategy sets in $\Gamma^K$ must be singletons, $S_i^K = \{s_i^0\}$ for all $i$, and uniqueness follows. □

**2.2. Weak domination.** A notion closely related to domination is that of *weak dominance*. A strategy $s_i \in S_i$ of player $i$ *weakly dominates* another strategy $s_i' \in S_i$ in the game $\Gamma = (N, (S_i)i \in N, (u_i)_{i \in N})$ if $u(s_i, s_{-i}) \geq u(s_i', s_{-i})$ for all $s_{-i} \in S_{-i}$ with $u(s_i, s_{-i}) > u(s_i', x_{-i})$ for at least one $x_{-i}' \in S_{-i}$. This extends also to domination by mixed strategies in the obvious way.

**Example 4.2. Nash equilibria may be weakly dominated.** A simple example of a game with a Nash equilibrium where both choose weakly dominated pure strategies is the following:

|   | L | R |
|---|---|---|
| T | (1,1) | (0,0) |
| B | (0,0) | (0,0) |

The strategy *B* is weakly dominated for the row player, and *R* is weakly dominated for the column player. But (*B*, *R*) is a Nash equilibrium, given the choice of the other player, no player can benefit from deviating.

Since weakly dominated strategies are inferior to other strategies (even though we cannot preclude that they may show up as Nash equilibrium strategies), we might want to get rid of them at the outset so as to concentrate our attention on more serious candidates for a solution. However, deleting weakly dominated strategies may give rise to some new problems, since the final result may depend on the order of deletion.

In the game below the row player has three strategies $T, M, B$ and the column player has two, $L, R$.

|   | L | R |
|---|---|---|
| T | (2,3) | (x,3) |
| M | (1,0) | (0,1) |
| B | (0,1) | (1,0) |

Here the order of elimination of weakly dominated strategies will matter for the final outcome. Suppose that player 1 first removes $M$ which is weakly dominated. Then in the remaining game player 2 may remove $R$, and we are left with a game where only player 1 has more than one strategy, while player 2 has only $L$. But then $B$ is dominated and we have only the strategy pair $(T, L)$.

If instead player 1 had started with removing $B$, we would have ended up with $(T, R)$, and if player 1 had removed both $M$ and $B$, we would have had the game

|   | L | R |
|---|---|---|
| T | (2,3) | (x,3) |

as our final result.

There are two ways to improve this very unsatisfactory situation. In the first, one may consider subclasses of games satisfying particular conditions as done e.g. in [Rochet [1980]] who considered games satisfying the condition

$$u_i(s) = u_i(s') \Rightarrow u_j(s) = u_j(s'), \text{ all } s, s' \in S,$$

if one player is indifferent between two particular strategy arrays, then all players are indifferent. With this condition it can be shown that the result of removing weakly dominated strategies – provided that each player removes *all* her weakly dominated strategies in each round – is unique.

An alternative approach consists in changing the concept of weak dominance in such a way that the operation of deleting strategies becomes independent of the order of deletion. This is what is done in Marx and Swinkels [1997]:

DEFINITION 2 *A strategy $s_i$ of player i nice weakly dominates a strategy $s'_i$ if*

(i) $u_i(s_i, s_{-i}) \geq u_i(s'_i, s_{-i})$ *for all $s_{-i} \in S_{-i}$ (weak dominance), and*
(ii) *if $u_i(s_i, s_{-i}) = u_i(s_i, s_{-i})$ for some $s_{-i} \in S_{-i}$, then $u_j(s_i, s_{-i}) = u_j(s_i, s_{-i})$ for all $j \in N$.*

Thus, if one strategy nice weakly dominates another, then not only does it weakly dominate the other strategy, but in the case where the two strategies give the same result to this player, it gives the same result to *all* players.

In the example above, the new concept of nice weak domination is such that after deleting either M or B, player two cannot delete any of columns. This means that the reduction to first row (and both columns) is all we get no matter how we delete (one at a time or all at a time), something which is promising for results to be obtained.

To approach such results, we need to formulate the idea of a reduction of a given game. First we introduce a new notation: We let $V = \cup_{i=1}^{n} S_i$ be the union of the sets of pure strategies of the players (assuming that they are all distinct). A *restriction* of $V$ is a subset $W$ such that $W \cup S_i \neq \emptyset$ for each $i$ (there is at least one strategy left for each player).

DEFINITION 3 *Let G be a game and V the set of strategies in the game. A restriction W of V is a reduction of V (by nice weak dominance) if there are disjoint subsets $X_1, \ldots, X_m$ of V such that $W = V \backslash \cup_{i=1}^{m} X_i$, and for all $x \in X_k$ there is $z \in V \backslash \cup_{i=1}^{k-1} X_i$ such that z nicely weakly dominates x. W is a full reduction if no strategy in W is nicely weakly dominated by another strategy in W.*

A reduction of $V$ is obtained by deleting nice weakly dominated strategies in some order, one or several at the time. We would like to obtain a result stating that all full reductions of $V$ by nice weak dominance are equal, but this is not true. However, it holds except for some operations on the set of all pure strategies which do not really change the nature of the game.

Let $W$ and $W'$ be restrictions of $V$. Then $W$ is *equivalent* to a subset of $W'$ if there are injective maps $m_i : W_i \rightarrow W'_i, i \in N$, such that $u(x) = u(m_1(x_1), \ldots, m_n(x_n))$

for all $x \in W$. This means that up to a renumbering or renaming of strategies, we can identify $W$ with a subset of the strategies in $W'$. If $W$ is a restriction of $V$ and $s_i, s_i' \in S_i$, then $s_i$ is *redundant* to $s_i'$ in $W$ if $u(s_i, s_{-i}) = u(s_i', s_{-i})$ for all $s_{-i} \in W_{-i}$.

Now we may state the main result about nice weak dominance.

THEOREM 2 *Let $X$ and $Y$ be full reductions of $V$ by nice weak dominance. Then, possibly after removal or addition of redundant strategies, $Y$ is equivalent to a subset of $X$.*

For a proof of Theorem 2 we refer to Marx and Swinkels [1997]. Notice that reductions do not give exactly the same result but there is some renaming of strategies such that they become identical except for some duplicates of strategies.

## 3   Rationalizable strategies

One of the problems with the (Nash) equilibrium considered as a solution to a given game is that it provides a good *ex post* reason for choice, but it doesn't give much of a clue *ex ante*, before the other players have chosen among their strategies. A possible way of incorporating the considerations of rational choice given the wide variety of possible choices by other players is provided by the concept of *rationalizable strategies*, introduced independently by Bernheim [1984] and Pearce [1984].

As an example, consider the following game:

|         | $s_2^1$  | $s_2^2$  | $s_2^3$   |
|---------|----------|----------|-----------|
| $s_1^1$ | (5,14)   | (4,-2)   | (6,3)     |
| $s_1^2$ | (3,0)    | (5,5)    | (5,1)     |
| $s_1^3$ | (6,1)    | (2,0)    | (2,16)    |
| $s_1^4$ | (4,0)    | (4,2)    | (3,5)     |

This game has a Nash equilibrium (in pure strategies), namely $(s_1^2, s_2^2)$, and it is the only one. But could it be possible that player 1 might choose $s_1^1$ rather than the Nash equilibrium strategy $s_1^2$? Yes, player 1 might do so if she expected player 2 to not to choose $s_2^2$, but instead $s_2^3$. But can we possible expect player 2 to choose this strategy? Yes, this is the best response of player 2 if she expects player 1 to choose $s_1^3$.

Proceeding in this way, we get that $s_1^3$ is best response of player 1 to the strategy $s_2^1$ of player 2, and the strategy $s_1^1$ which was our point of departure, is best response of player 1 to $s_2^1$. Thus, the choice of $s_1^1$ can be rationalized as best response on something which is indeed a possible choice by the other player, also acting in a rational way.

For the generalization to arbitrary games of the features observed in the example, we consider a game $\Gamma = (N, (S_i)_{i \in N}, (u_i)_{i \in N})$, where each player $i$ has a finite set $S_i$ of (pure) strategies. The set $\Delta(S_i)$ of mixed strategies for player $i$ consists of all probability distributions over $S_i$.

We have already made use of the property of being a *best response*:

DEFINITION 4 *A strategy $\sigma_i'$ of player $i$ is a best response in $C \subseteq \Delta(S_i)$ to the strategy array $\sigma \in \bigtimes_{j \in N} \Delta(S_j)$ if $\sigma_i' \in C$ and*

$$\pi_i(\sigma_i', \sigma_{-i}) \geq \pi_i(s\sigma_i'', \sigma_{-i}), \text{ all } \sigma_i'' \in C.$$

Suppose that we have sequences $(\hat{\sigma}_i^t)_{t=0}^{\infty}$ in $C$ and $\sigma^t$ in a closed subset of $\bigtimes_{j \in N} \Delta(S_j)$, converging to $\hat{\sigma}_i^0$ and $\sigma^0$, respectively, with $\hat{\sigma}_i^t$ a best response to $\sigma^t$ for all $t$. Then $\hat{\sigma}_i^0$ is also best response to the limit strategy array $\sigma^0$: Indeed, if there were $\sigma_i'' \in C$ with $\pi_i(\sigma_i'', \sigma_{-i}^0) > \pi_i(\hat{\sigma}_i^0, \sigma_{-i}^0)$, then by continuity of the payoff function we would also have $\pi_i(\hat{\sigma}_i'', \sigma_{-i}^t) > \pi_i(\hat{\sigma}_i^t, \sigma_{-i}^t)$ for large enough $t$, a contradiction.

Let $H^i$ be the set of best responses to some mixed strategy array $\sigma \in \bigtimes_{j \in N} \Delta(S_j)$. One may think of the elements of $H^i$ as the strategies which may be contemplated as a possible choice by player $i$. We now define the sets $H^i(\nu)$ for each $\nu \in \mathbb{N}$ inductively by

$$H^i(0) = H^i,$$

$$H^i(\nu) = \{\sigma_i \in H^i(\nu-1) \mid \sigma_i \text{ is a best response to some } \sigma' \in H^1(\nu-1) \times \cdots \times H^m(\nu-1)\}.$$

As before, the elements of $H^i(\nu)$ are those strategies which are relevant after a chain of reasoning which goes back at least $\nu$ steps. Using the small argument above, we get that each $H^i(\nu)$ is a closed subset of $S_i$.

For each $i$ define $\mathcal{R}_i$ as the intersection of all the sets $H^i(\nu)$ over all $\nu$,

$$\mathcal{R}_i = \cap_{\nu=0}^{\infty} H^i(\nu).$$

Then $\mathcal{R}_i$ is the set of *rationalizable* strategies of player $i$. Following the interpretation of the sets $H^i(0)$, we can see the rationalizable strategies as those that can be explained as best responses no matter how far back one goes in the chain of argumentation.

Although the idea of rationalizing the choice by a (possibly long) of reasoning has a clear intuitive content, it may not be easy to use, and there is an alternative characterization of rationalizable strategies, avoiding the iteration.

To begin with, we notice the following.

LEMMA 2 *The sets $\mathcal{R}_i$, for $i \in N$, have the best response property, that is for each $i$ and $\sigma_i' \in \mathcal{R}_i$, there is some $\sigma \in \bigtimes_{j \in N} \mathcal{R}_j$ such that $\sigma_i'$ is best response to $\sigma$.*

Proof: This follows from the definition; indeed, $\sigma_i' \in \mathcal{R}_i$ is in all the sets $H^i(\nu)$, so for each $\nu$ the set $K(\nu - 1)$ of strategy arrays $\sigma$ in $\bigtimes_{j \in N} H^j(\nu - 1)$ such that $\sigma_i'$ is a best response to $\sigma$, is nonempty. Using our previous reasoning, we have that $K(\nu - 1)$ is closed. Moreover, if $\sigma \in K(\nu)$, then $\sigma \in K_{\nu'}$ for all $\nu' < \nu$, so that the family $(K(\nu))_{\nu=0}^{\infty}$

has the finite intersection property. Consequently, $\cap_{v=0}^{\infty} K(v) \neq \emptyset$, and $\sigma_i'$ is indeed a best response to something in $\times_{j \in N} \mathcal{R}_j$.                                    $\square$

We may now state the characterization result.

THEOREM 3 *For each i, if $E_i$ is given by*

$E_i = \{\sigma_i \in \Delta(S_i) \mid \exists C_j \subset \Delta(S_j), j \in N,$ with the best response property, s.t. $s_i \in C_i\}$,

*then $E_i = \mathcal{R}_i$, each i.*

PROOF: From the lemma, we have that $\mathcal{R}_i \subseteq E_i$ for all $i$. For the converse inclusion, we note first that $(E_j)_{j \in N}$ has the best response property, since for each $\hat{\sigma}_i \in E_i$, there are $C_j$, $j \in N$ with the best response property such that $\hat{\sigma}_i \in C_i$. Now each family $(C_j)_{j \in N}$ with the best response property must satisfy $C_i \subseteq E_i$ for each $i$, so $\sigma_i$ is a best response to a strategy array in $\times_{j \in N} \Delta(E_j)$, meaning that $(E_j)_{jN}$ has the best response property. Now it is easy to check that $E_i \subseteq H_i(v)$ for each $v$, so that $E_i \subset \mathcal{R}_i$.                                    $\square$

It follows from the characterization that the Nash equilibria of the game are rationalizable: If $(\sigma_i)_{i \in N}$ is a Nash equilibrium,, then $(\{\sigma_j\})_{j \in N}$ has the best response property. Clearly, in many cases the set of rationalizable strategies will be much larger than the set of Nash equilibria.

## 4   Evolutionary stability

In our discussions so far, we have considered games with a finite set $N$ of players choosing among strategies, sometimes aided by a randomization scheme, so that the choice is among mixed strategies. There are several ways by which one may justify the use of mixed strategies, ranging from a conscious device for not hiding the true intents to a limit of choices reflecting subjective beliefs about the other players' strategy choices. But mixed strategies may occur in a totally different context, reflecting the composition of a large population with respect to some basic characteristics. This is the case for the applications of game theory to biology, a field which will be briefly surveyed here.

**4.1. Evolutionarily stable states.** In evolutionary game theory, we assume that there is a large population that plays a mixed strategy as a whole, something which may be achieved either by every player in the population plays this strategy, or by choosing a player which selects the pure strategy (in the support of the mixed strategy) with the relevant probability. The players are not rational in the sense of seeking a maximal payoff, instead the population is assumed to maintain a state which corresponds to maximal fitness, as expressed by the payoff.

Assume that there is a fixed finite set of possible actions $S = \{s_1, \ldots, s_m\}$, and define a matrix $A$ of payoffs by

$$a_{ij} = u(s_i, s_j).$$

We let $\Delta(S)$ be the set of probability distributions over $S$, in the terminology of evolutionary game theory, an element $q = (q_1, \ldots, q_m)$ of $\Delta(S)$ is called a *state* of the population. If we consider an ongoing random pairing of individuals in the population in state $q$, then each individual will meet a other individuals in state $s_j$ with probability $q_j$, and the payoff of an individual taking action $s_i$ against a population in state $q$ is therefore

$$U(s_i, q) = \sum_{j=1}^{m} a_{ij} q_j = e_i \cdot Aq,$$

where $e_i$ is the $i$th unit vector in $\mathbb{R}^m$. A population in state $p$ meeting another population in state $q$ will result in payoff

$$U(p, q) = p \cdot Aq.$$

The population in state $p$ may be thought of as an invading sub-population which chooses the actions with other probabilities that the existing one.

In the present context, we may restrict our attention to *symmetric* Nash equilibria of the (symmetric) game with payoffs given by the matrix $A$, since only symmetric equilibria can be interpreted as stable states of the population. A state $q$ is a Nash equilbrium if

$$U(q, q) \geq U(p, q) \tag{2}$$

for all $p \in \Delta(S)$. However, something more is needed for a solution to be meaningful in the context of evolutionary biology.

Assume that the population is in state $q$ and consider a subpopulation with another state $p$, and that the subpopulation has relative frequency $x_t$ at time $t$. We now assume that the growth rates of subpopulations are proportional to their relative fitness, so that

$$\frac{x_{t+1}}{x_t} = \frac{U(p, q_t)}{U(q_t, q_t)}$$

for all $t$, where the state $q_t$ of the population at time $t$ is defined by $q_t = x_t p + (1 - x_t) q$. From this we get that

$$x_t - x_{t+1} = \frac{U(q_t, q_t) - U(p, q_t)}{U(q_t, q_t)} x_t, \tag{3}$$

and it is seen that $x_{t+1} - x_t < 0$, so that the mutant subpopulation is diminishing, if $U(q_t, q_t) - U(p, q_t) > 0$ for each $t$. Since $U(q_t, q_t) = x_t U(p, q_t) + (1 - x_t) U(q, q_t)$, we get that $U(q_t, q_t) - U(p, q_t) = (1 - x_t)[U(q, q_t) - U(p, q_t)]$ should be positive, and using the definition of $q_t$ once more, we obtain the condition

$$U(q, q_t) - U(p, q_t) = x_t [U(q, p) - U(p, p)] + (1 - x_t)[U(q, q) - U(p, q)] > 0. \tag{4}$$

Here the second bracket on the right-hand side will be nonnegative if $q$ satisfies (2), but the first bracket may be negative, in which case the mutant subpopulation will increase its share over time, and the original state could not be considered as stable. These considerations give rise to the following definition of an *evolutionarily stable* state.

**Example 4.3. The Hawk-Dove Game.** It is customary to introduce the concept of evolutionary equilibrium using the example of the Hawk-Dove game: We consider a (large) population which can be characterized by two different types of behavior $H$ and $D$. This behavior may be represented by two different animals of the same species competing for a scarce resource of value $V$, where the behavior of $H$ is aggressive while $D$ is pieceful. If both choose $H$, they must fight, and each has the same probability of winning. There is a cost $C$ to fighting, and this is avoided if one player chooses $D$. If both choose $D$, we assume that there is some procedure which gives the resource to one of them with equal chance to both. We assume that $C > V$.

This gives a payoff matrix of the following form:

|   | $H$ | $D$ |
|---|---|---|
| $H$ | $\left(\frac{1}{2}(V - C), \frac{1}{2}(V - C)\right)$ | $(V, 0)$ |
| $D$ | $(0, V)$ | $\left(\frac{V}{2}, \frac{V}{2}\right)$ |

This game has two asymmetric equilibria in pure strategies, namely $(H, D)$ and $(D, H)$, but they are not particularly meaningful in a biological context, since the animals should know their player number. But there is a symmetric equilibrium in mixed strategies, namely

$$(q_H, q_D) = \left(\frac{V}{C}, \frac{C - V}{C}\right).$$

One may interpret this equilibrium as one where the composition of animals with the behavior expressed by $H$ and $D$ is stable in the sense that the expected payoff, or the *fitness* obtained cannot be increased be deviations from this composition.

DEFINITION 5 *Let $S$ be a set of actions for a population, and let $U : \Delta(S)^2 \to \mathbb{R}$ be the fitness function. A state $q \in \Delta(S)$ is evolutionarily stable if*

(i) *$q$ is a Nas equilibrium, i.e. $U(q, q) \geq U(p, q)$ for all $p \in \Delta(S)$,*
(ii) *$q$ is stable in the sense that if $U(p, q) = U(q, q)$ and $p \neq q$, then $U(q, p) > U(p, p)$.*

In part (ii) the definition, the state $p$ is that of the mutant subpopulation, and it is assumed to be a best reply to $q$ (giving the same payoff as $q$ which has the equilibrium property in part (i)). Given the condition we obtain that (4) is satisfied, so that the fraction of mutants will diminish over time.

**4.2. Replicator dynamics.** The reasoning which was used to motivate the definition of an evolutionarily stable state may be extended so as to constitute a dynamical system which has these states as their equilibria. Reasoning as in the derivation of

(3), we write the change in state over time as

$$q_{t+1} - q_t = \frac{U(p, q_t) - U(q_t, q_t)}{U(q_t, q_t)} q_t.$$

The difference equation may then be approximated by a differential equation

$$\dot{q} = \frac{U(s, q) - U(q, q)}{U(q, q)} q(p),$$

where $\dot{q}$ is the derivative of $q$ with respect to time. It shows how the displacement from $q$ caused by a mutant subpopulation with state $p$ evolves over time. Since the denominator is constant, we get the *replicator dynamics*

$$\dot{q} = [U(p, q) - U(q, q)] q. \tag{5}$$

An equilibrium of a dynamical system is a state $q$ at which $\dot{q}$. The equilibrium $q$ is *stable* if every neigborhood $U$ of $q$ contains another neighborhood $V$ of $q$ such that a trajectory starting in $V$ stays in $U$, and it is *asymptotically stable* for the replicator dynamics (5) if it is stable and there is a neighborhood $W$ of $q$ such that trajectories starting in $W$ converge to $q$. It can be showed that (e.g. Zeeman [1981]) that an evolutionarily stable state is an asymptotically stable equilibrium of the replicator dynamics.

## 5  Correlated equilibria

In this section, we consider a generalization of the mixed strategies, which served so well in the context of zero-sum games. Instead of having each player randomizing her strategy choices, we might introduce a randomness device which selects the (pure) strategy array and thereby the (pure) strategy to be played by each player. This will lead us to the concept of a *correlated equilibrium*, introduced by Aumann [1974].

We consider a game $\Gamma = (N, (S_i)_{i \in N}, (u_i)_{i \in N})$, where for each player $i \in N$, $S_i$ is finite (the set of pure strategies), and $u_i : S \to \mathbb{R}$ is the payoff of player $i$. We let $m_i = |S_i|$ denote the cardinality of the set of pure strategies of $i$, for each $i \in N$, and let $m_{-i} = |S_{-i}| = \prod_{j \neq i} m_j$. For $i \in N$ and $s_i \in S_i$, we use the notation $u_i(s_i, \cdot)$ for the $m_{-i}$-vector $(u_i(s_i, s_{-i})_{s_{-i} \in S_{-i}}$.

A correlated strategy in $\Gamma$ is a probability distribution $p = (p(s))_{s \in S}$ on $S$. The correlated strategy $p$ is a *correlated equilibrium* if for each player $i$ and each pair of strategies $(s_i, s_i')$ from $S_i$, $p$ satisfies

$$\sum_{s_{-i} \in S_{-i}} p(s_i, s_{-i})[u_i(s_i, s_{-i}) - u_i(s_i', s_{-i})] \geq 0, \tag{6}$$

or, equivalently, $p(s_i, \cdot) \cdot [u_i(s_i, \cdot) - u_i(s_i', \cdot)]$, where we use the notation $p(s_i, \cdot) = p(s_i, s_{-i})_{s_{-i} \in S_{-i}}$. The set of correlated equilibria of the game $\Gamma$ is denoted Corr$(\Gamma)$. It is easily seen that Corr$(\Gamma)$ is convex.

**Example 4.4. Correlated equilibrium in the Battle of the Sexes.** Returning to the game of Section 1,

|   | $O$ | $S$ |
|---|---|---|
| $O$ | $(2,1)$ | $(-1,-1)$ |
| $S$ | $(-1,-1)$ | $(1,2)$ |

it may be noticed that although there are two reasonable Nash equilibria, one lacks a method for selecting one of them, and fixing one of them as the final choice is not quite satisfactory, since this will favorize one of the players.

This may be resolved if the players have access to some randomizing device, which selects $(O,S)$ and $(S,O)$ with equal probability. This gives us the correlated equilibrium where these two strategy arrays have probability $1/2$ each and the remaining two have probability $0$. It is easily seen that this gives a correlated equilibrium treating the two players symmetrically, and expected outcome is $3/2$ for each player.

For the interpretation of the correlated equilibrium, we may think of a correlated strategy as a device made available by a neutral agency, according to which the pure strategy to be chosen by each player is drawn by chance and communicated to the player. However, the player retains the option of choosing something else, and (6) asserts that the correlated strategy is an equilibrium only if it is not advantageous to the player to deviate from what is communicated.

The set of correlated equilibria of a game $\Gamma$ is convex (as a subset of $\mathbb{R}^{m_1 m_2 \cdots m_n}$). Indeed, if both $p$ and $p'$ are correlated equilibria, and $0 \le \lambda \le 1$, then for each $i$ and each pair $(s_i, s'_{-i})$ of pure strategies for $i$ we have by (6) that

$$\sum_{s_{-i} \in S_{-i}} (\lambda p(s_i, s_{-i}) + (1 - \lambda)p'(s_i, s'_{-i}))[u_i(s_i, s_{-i}) - u_i(s'_i, s_{-i})] =$$

$$\lambda \sum_{s_{-i} \in S_{-i}} p(s_i, s_{-i})[u_i(s_i, s_{-i}) - u_i(s'_i, s_{-i})] + (1 - \lambda) \sum_{s_{-i} \in S_{-i}} p'(s_i, s_{-i})[u_i(s_i, s_{-i}) - u_i(s'_i, s_{-i})] \ge 0,$$

so that also $\lambda p + (1 - \lambda)p'$ satisfies the conditions for a correlated equilibrium.

A (mixed strategy) Nash equilibrium of $\Gamma$ is a correlated equilibrium. Indeed, if $\sigma_1, \ldots, \sigma_n$ are the equilibrium strategies, then the correlated strategy $\sigma_1 \times \cdots \times \sigma_n$ with

$$[\sigma_1 \times \cdots \times \sigma_n](s) = \sigma_1(s_1)\sigma_2(s_2) \cdots \sigma_n(s_n)$$

satisfies (6) for all $i$ and all pairs $(s_i, s'_i) \in S_i^2$ of pure strategies for $i$. Consequently, a game $\Gamma$ has a correlated equilibrium whenever it has a Nash equilibrum, and the set of correlated equilibria contains the convex hull of all correlated strategies of the form $\sigma_1 \times \cdots \times \sigma_n$, for $\sigma_1, \ldots, \sigma_n$ a Nash equilbrium of $\Gamma$. However, there may be other correlated strategies than these.

Since the correlated equilibrium is defined by a set of linear inequalities, it might be expected that existence can be shown in a straightforward way (in contrast to the case of Nash equilibria, where existence of equilibria for general games innvolves a fixed-point argument, cf. Chapter 5), and this is indeed the case. The proof given below is due to Hart and Schmeidler [1989]. It may be noticed that we exploit the minimax theorem once again in this proof.

THEOREM 4 *Let* $\Gamma = (N, (S_i)_{i \in N}, (u_i)_{i \in N}$ *be a game where* $S_i$ *is finite for each* $i \in N$. *Then* $\Gamma$ *has a correlated equilibrium.*

PROOF: We begin by constructing an auxiliary two-person zero-sum game. Strategies for player 1 are strategy arrays $\hat{s} \in S$ from $\Gamma$, whereas player 2 chooses triples $(i, s_i, s_i')$ with $i$ a player and $\hat{s}_i, \hat{s}_i' \in s_i$. The payoff (to player 1 paid by player 2) is

$$u_i(s_i, \hat{s}_{-i}) - u_i(s_i', \hat{s}_{-i})$$

if $s_i = \hat{s}_i$ and 0 otherwise. Clearly, a mixed strategy for player 1 is a correlated strategy in $\Gamma$, which is a correlated equilibrium if and only if it yields a nonnegative payoff to player 1 for any strategy chosen by player 1. Now we apply the Minimax Theorem, telling us that such a mixed strategy exists if for each mixed strategy of player 2, there is a strategy for player 1 yielding nonnegative payoff.

Let $(q_i(s_i, s_i'))_{i \in N, s_i, s_i' \in s_i}$ be an arbitrary mixed strategy of player 2. We show that for each $i$ and each $\hat{s} \in S$, there is a probability vector $x_i = x_i(s_i)_{s_i \in s_i}$ such that

$$\sum_{s_i \in s_i} x_i(s_i) \sum_{s_i' \in s_i} q_i(s_i, s_i')[u_i(s_i, \hat{s}_{-i}) - u_i(s_i', \hat{s}_{-i})] = 0. \tag{7}$$

Actually, we show something which is slightly stronger, namely that there is a probability vector $x$ with

$$\sum_{s_i \in s_i} x(s_i) \sum_{s_i' \in s_i} q_i(s_i, s_i')[v(s_i) - v(s_i')] = 0$$

for every array $v = v(s_i)_{s_i \in s_i}$ of numbers.

Writing $\Phi(x, v)$ for this expression and noticing that $\Phi(x, -v) = -\Phi(x, v)$, we need only show that there is $x$ such that $\Phi(x, v) \geq 0$ for all $v$, and we may restrict attention to $v$s which are probability vectors, since one may add constants to all coordinates of $v$ and normalize $v$ without changing the sign of $\Phi(x, v)$. Now the Minimax Theorem applies to $\Phi(x, v)$, so that

$$\max_x \min_v \Phi(x, v) = \min_v \max_x \Phi(x, v). \tag{8}$$

For any given $v$, let $a_i^0$ be such that $v(a_i^0) \geq v(s_i)$ for all $s_i \in s_i$. If $x$ is chosen such that $x(a_i^0) = 1$ and $x(s_i) = 0$ otherwise. Then $\Phi(x, v) \geq 0$, and since $v$ was arbitrary, it follows that $\min_v \max_x \Phi(x, v) \geq 0$, and, by (8), $\max_x \min_v \Phi(x, v) \geq 0$, giving us the desired vector $x$.

Thus, for each $i$ we may chose a vector $x_i$ satisfying (7). Define $p(\hat{a})_{\hat{a} \in A}$ by

$$p(\hat{s}_1, \ldots, \hat{s}_n) = x_1(\hat{s}_1) \cdot \cdots \cdot x_n(\hat{s}_n).$$

Then $p$ is a mixed strategy for player 1 in the auxiliary game, and the payoff at $p$ and the strategy $y$ of player 2 is

$$\sum_a \left[ \prod_{i \in N} x_i(s_i) \right] \left( \sum_{i \in N} \sum_{s_i'} y_i(s_i, s_i') \left[ u_i(\hat{s}_i, s_{-i}) - u_i(s_i', s_{-i}) \right] \right),$$

which can be rewritten as

$$\sum_i \sum_{s_{-i}} \left[ \prod_{j \neq i} x_j(s_j) \right] \left( \sum_{s_i} x_i(s_i) \sum_{s_i'} y_i(s_i, s_i') \left[ u_i(\hat{s}_i, s_{-i}) - u_i(s_i', s_{-i}) \right] \right)$$

which is 0 by (7). It follows that $\Gamma$ has a correlated equilibrium. □

Correlated equilibria are easy to work with, they exist for a large class of games, and the set of equilibria has nice properties. However, the intuition behind this equilibrium – the players rely on some outside device or person to prescribe the actions to be taken during the play – may seem to take us outside game theory proper. On the other hand, correlated equilibria do occur in situations where the setup is genuinely one of game theory, as e.g. in the models of adaptive learning in games, cf. Hart and Mas-Colell [2000].

## 6   Problems

**1.** In the *Game of Chicken*, two players are driving at high speed towards each other, and a collision will occur unless one of them "chickens out" at the last minute. If both chicken out, they both gain 1. If one chickens out and the other one does not, then the latter gets the payoff 2 and the chicken gets ?1. If both players display iron nerves, collision occurs and both lose 2.

Write the payoff matrix of this game and find the Nash equilibria.

**2.** Consider the 2-person game with payoffs

|       | L       | R        |
|-------|---------|----------|
| T     | (5, 10) | (8, 12)  |
| T     | (7, 2)  | (9, 1)   |
| B     | (8, 4)  | (2, 2)   |

Find an equilibrium by iterated elimination of dominated strategies. Find also a pure strategy Nash equilibrium.

**3.** Suppose that in the game $\Gamma = (N, (S_i)_{i \in N}, (u_i)_{i \in N})$. Check the following statements:

If $s_i' \in S_i$ dominates some pure strategy $s_i \in S_i$, then for all choices of *mixed* strategies of players $j$, $j \neq i$, expected payoff of player $i$ when choosing $s_i'$ than when choosing $s_i$.

If the strategy $s_i \in S_i$ is dominated by a mixed strategy $\sigma_i \in \Delta S_i$. Show that there is also a pure strategy $s_i' \in S_i$ which dominates $s_i$.

**4.** Consider a 2-person game $(\{1,2\}, S_1, S_2, u_1, u_2)$ where $S_1, S_2$, the sets of pure strategies, are finite.

Show that if $s_1 \in S_1$ is not dominated (by either a pure or a mixed strategy), then $s_1$ is a best response to some mixed strategy $\sigma_2 \in \Delta(S_2)$.

Use this result to show that rationalizability and iterated elimination of dominated strategies give the same result.

**5.** In the following game, the row player has a dominating strategy: We assume that $a > c, b > d$.

|   | L | R |
|---|---|---|
| T | $(a,a)$ | $(b,c)$ |
| B | $(c,b)$ | $(d,d)$ |

Find the Nash equilibria of the game, and find all evolutionarily stable strategies.

**6.** Consider the 2-person games with payoffs

|   | L | R |
|---|---|---|
| T | $(5,1)$ | $(1,5)$ |
| B | $(3,4)$ | $(2,3)$ |

Show that there is only one Nash equilibrium, namely the one where the row player chooses the mixed strategy $\left(\frac{1}{5}, \frac{4}{5}\right)$ and the column player chooses $\left(\frac{1}{3}, \frac{2}{3}\right)$.

Find all the correlated equilbria.

**7.** Construct an example of a 2-person game which has a correlated equilibrium which does not belong to the convex hull of the correlated strategies belonging to Nash equilibra.

# Chapter 5

# More About Nash Equilibria

## 1  Existence of Nash equilibria

In the previous chapters, Nash equilibria have been a recurrent topic, and it is time to consider the Nash equilibria of a general normal form game in some more detail. First of all, we consider the existence problem, which was a point of concern when dealing with other solution concepts in the previous chapter. Then we look at a few of the many applications of Nash equilibria in the context of economics. Following this, we discuss the Nash equilibria in extensive form games, a discussion which eventually takes us to the notion of subgame perfectness, whereby we open up for a new aspect, namely that of selecting from possibly many different Nash equilibria of a game, a topic which will concern us in Chapter 7.

The existence of mixed strategy Nash equilibria in two-person zero-sum games was established in Chapter 2 using a separation theorem. For general $n$-person games we need another approach using fixed-point theory. We begin with the following basic existence result.

THEOREM 1  *Let* $\Gamma = (N, (S_i)_{i \in N}, (u_i)_{i \in N})$ *be a game such that for each* $i$,
   *(i)*  $S_i$ *is a convex and compact subset of some Euclidean space* $\mathbb{R}^{d_i}$,
   *(ii)  for each strategy array* $s = (s_1, \dots, s_n)$, *the set*

$$\{s_i' \in S_i \mid u_i(s_i', s_{-i}) > u_i(s)\}$$

   *is convex,*
   *(iii)*  $u_i$ *is continuous, each* $i \in N$.
*Then* $\Gamma$ *has a Nash equilibrium.*

We shall need two mathematical results in the proof. The first one is known as Urysohn's lemma (see e.g. Aliprantis and Border [1999], Theorem 2.43).

LEMMA 1  *Let* $U \subseteq \mathbb{R}^d$ *be an open set. Then there is a continuous function* $\psi : \mathbb{R}^d \to [0,1]$ *such that* $\psi(x) > 0$ *for* $x \in U$ *and* $\psi(x) = 0$ *for* $x \notin U$.

The second is Brouwer's fixed-point theorem (Aliprantis and Border [1999],

Theorem 16.52).

LEMMA 2 *Let $K \subset \mathbb{R}^d$ be convex and compact, and let $f : K \to K$ be a continuous function. Then there is $x^0 \in K$ such that $f(x^0) = x^0$.*

PROOF OF THEOREM 1: We prove the theorem by contradiction, so suppose that $\Gamma$ has no Nash equilibria. In that case we have that for each $s = (s_1, \ldots, s_n) \in S = S_1 \times \cdots \times S_n$, there must be $i \in N$ and $s_i' \in S_i$ such that

$$u_i(s_i', s_{-i}) > u_i(s_i, s_{-i}). \tag{1}$$

By continuity of $u_i$, there is an open set $U(s) \subseteq S$ containing $s$ such that

$$u_i(s_i', \hat{s}_{-i}) > u_i(s)$$

for all $\hat{s} \in U(s)$. The family of all open sets $(U(s))_{s \in S}$ found in this way cover $S$, and since $S$ is compact, it is by finitely many of the sets $U(s)$, say $U(s^1), \ldots, U(s^k)$. Let $f^j : U(s^j) \to S$ be the function which to each $\hat{s} \in U(s^j)$ assigns either the strategy $s_i'$ satisfying (1), if $i$ is the player for which (1) holds, or the strategy $s_h$ if $h$ is not such a player.

By Lemma 1, for each of the sets $U(s^j)$, there is a continuous function $\psi_j : S \to S$ which is $> 0$ on $U(s^j)$ and 0 otherwise. Now the function $f$ defined by

$$f(s) = \sum_{j=1}^{k} \frac{\psi_j(s)}{\sum_{h=1}^{k} \psi_h(x)} f^j(s)$$

is continuous and maps points in $S$ to itself. By Lemma 2, it has a fixed point. But this contradicts the construction of $f$, since for each $s$ there is at least one $i$ such that $s_i$ is sent to something different from $s_i$. We conclude that the game has a Nash equilibrium. $\square$

The result may be applied to mixed strategy equilibria in games with finite sets of pure strategies.

COROLLARY 1 *Let $\Gamma = (N, (S_i)_{i \in N}, (u_i)_{i \in N})$ be a game where $S_i$ is a finite set for each $i \in N$. Then $\Gamma$ has a Nash equilibrium in mixed strategies.*

PROOF: The set $\triangle(S_i)$ of mixed strategies for player $i$ is convex and compact, and the extension of $u_i$ to $\times_{j \in N} \triangle(S_j)$ is linear, each $i \in N$, so that Theorem 1 applies to the mixed extension of $\Gamma$. $\square$

We have made use of a fixed-point theorem in the proof of Theorem 1, but we have deliberately chosen the simplest version, that of Brouwer stated as Lemma 2 above. In the following, we shall use a less straightforward version, formulated not in terms of a continuous function. Instead, we consider a set-valued function, also called a multimap or a *correspondence* $\Phi$, which to each $x$ in the compact and convex set $K$ assigns a nonempty subset $\Phi(x)$ of $K$. The fixed-point theorem of Kakutani (Aliprantis and Border [1999], Corollary 15.51) states that if the graph of $\Phi$ (the set $\{x, y) \in K \times K \mid y \in \Phi(x)\}$) is closed and $K(x)$ is convex for each $x \in K$, then there is $x^0 \in K$ such that $x^0 \in \Phi(x^0)$.

## 2 Economic applications of Nash equilibrium

In this section, we consider some examples of applications of Nash equilibrium in economics. The first of them, that of Cournot duopoly, is wellknown even to beginners. Almost if not quite as wellknown is the Bertrand model, dealing with the same phenomenon but obtaining a very different result. Finally, we consider a model of trading in a market, formulated as a game.

**2.1. The Cournot duopoly model.** In this classical model, due to Cournot [1838], we have two firms selling a commodity (mineral water) in a market with a given demand relationship

$$p = b - aq, \ a > 0, b > 0,$$

where $p$ is the price at which the quantity $q$ can be sold. The firms have constant unit costs $c$ in producing the commodity.

If there had been only one firm, it would have obtained maximal profits $pq - cq$ at the quantity $q^{mon}$ which satisfies the first order condition $b - 2aq^{mon} = c$ or

$$q^{mon} = \frac{b - c}{2a}, \tag{2}$$

with the associated price $p^{mon} = \frac{b - c}{2}$. However, there is another firm, and if this firm has already supplied the quantity $q_2$ to the market, the demand from the remaining customers is such that

$$p = (b - aq_2) - aq$$

and the best choice in this market is

$$q_1 = \frac{b - aq_2 - c}{2a} \tag{3}$$

in accordance with (2).

In the terminology of game theory, the strategies of the firm $i$, $i = 1, 2$, are quantities $q_i$ in the interval $\left[0, \frac{b}{a}\right]$, and the payoff of firm $i$ given strategies $q_1, q_2$ of the two players is

$$u_i(q_1, q_2) = [b - a(q_1 + q_2)]q_i - cq_i.$$

A symmetric (i.e. both choose identical quantities) Nash equilibrium can be found by setting $q_1 = q_2 = q^*$ in (3), which gives

$$q^* = \frac{b - c}{3a},$$

and since there are two firms, a total of $q^{(2)} = \frac{2}{3}\frac{b - c}{a}$ is sold in the market.

Comparing this quantity with $q^{mon}$, it is seen that more is sold in the duopoly, and the price is lower. If the same reasoning is used for a case with $n$ firms, the quantity sold in the market would be

$$q^{(n)} = \frac{n}{n + 1}\frac{b - c}{a},$$

converging to the case of a perfectly competitive market, where the commodity is sold at its unit cost $c$.

**2.2. Bertrand duopoly.** In the Cournot duopoly model, firms are assumed to choose *quantities* and market prices adapt to quantities supplied. It may be argued, and it was indeed argued by the French economist Bertrand in the late 19th century, that a model with firms choosing prices instead of quantities would be more in correspondence with reality.

In a model with price-setting duopolists, the strategies of the firms are the *prices* $p_1$ and $p_2$ in the interval $[0, b]$ charged by the two firms. Assuming that customers choose the cheapest seller, the quantity sold by firm $i$ is

$$q_i(p_1, p_2) = \begin{cases} \frac{b-p_1}{a} & p_i < p_j, \\ 0 & p_i > p_j, \\ \frac{b-p_1}{2a} & p_i = p_j, \end{cases}$$

with $j \neq i$, where it is assumed that the available customers are shared equally by the two firms if they have quoted the same price. Payoff in the model is then defined as

$$u_i(q_1, q_2) = (p_i - c)q_i(p_1, p_2), \quad i = 1, 2.$$

Here $(c, c)$ is a Nash equilibrium; the firms get a sale of $(b-c)/2a$ each, and payoff is 0. It is straightforward that an array $(p_1, p_2)$ with $\max\{p_1, p_2\} > c$ for some $i$ cannot be a Nash equilibrium: If $\min\{p_1, p_2\} > c$, then each firm may get all the customers by choosing a price below $\min\{p_1, p_2\}$, and if $\min\{p_1, p_2\} = c$ but $\max\{p_1, p_2\} > c$, then the firm quoting $c$ benefits from a slight increase in its price.

The Bertrand duopoly model has the consequence that as soon as there is more than one firm in the market, the final outcome in terms of price and quantity sold is the same as under perfect competition, where the commodity sells at its cost. The seemingly contradictory situation, where a model with realistic assumptions (price-setting rather than quantity setting) gives an unrealistic result (no monopoly gain whatsoever), has given birth to a voluminous literature in industrial organization theory, cf. e.g. Tirole [1988].

**2.3. A model of trading in a market.** In the above models, customer behavior in the market was taken as given by a demand relationship. It might be preferable to let buyers be players as well. We consider such a model, where general buying and selling are carried out by all individuals in the market. There are several so-called market models in the gametheoretical literature, the present one is adapted from Shubik [1984].

We consider a market with $n$ individuals and $l$ commodities, of which the last, $l$th, commodity has the particular property of being accepted as a medium of exchange. Trade takes place in $l - 1$ different markets, where a commodity $h \in \{1, \ldots, l - 1\}$ is exchanged against money, that is commodity $l$, at a certain rate to be determined below.

For each commodity $h$, individual $i$ must determine an *offer* $a_i^h \geq 0$ and a *bid* $b_i^h \geq 0$ (measured in commodity $l$). What individual $i$ gets, will depend on the exchange rate; it seems reasonable to assume that one of the two (either offer or bid) is 0. From the totality of offers and bids, $(a_i^h, b_i^h)_{i=1}^n$, we determine a net trade of individual $i$ in commodity $h$ as

$$z_i^h = \left(\frac{a^h}{b^h}\right) b_i^h - a_i^h, \text{ where } a^h = \sum_{i=1}^n a_i^h, b^h = \sum_{i=1}^n b_i^h, \tag{4}$$

for $h = 1, \ldots, l - 1$, and the net trade in the last commodity (money) as

$$z_i^l = \sum_{h=1}^{l-1} \left(\frac{b^h}{a^h}\right) b_i^h \tag{5}$$

(with the convention $a^h/b^h = 0$ if $b^h = 0$).

The fraction $b^h/a^h$ has an interpretation as the price of commodity $h$ which achieves equality of offers and bids in the market for commodity $h$, since

$$\sum_{i=1}^n z_i^h = \frac{a^h}{b^h} \sum_{i=1}^n b_i^h - \sum_{i=1}^n a_i^h = 0.$$

This way of modelling markets has the advantage of being explicit on the creation of market clearing prices, contrasting with the standard approach in economic theory, where only the final result of market behavior in the form of equilibrium prices and net trades, are subjects of study.

For each individual a strategy is an array $\left(a_i^h, b_i^h\right)_{i=1}^{l-1}$ of offers and bids in each market. We assume that each individual has a concave utility function $u_i$ defined on all of $\mathbb{R}^l$ (the most intuitive way of having such a utility function is to extend a utility defined on $\mathbb{R}_+^l$ by setting $u_i(x) = -\infty$ for $x \notin \mathbb{R}_+^l$. If they have also an initial endowment $\omega_i$ of goods, then we get payoff functions by assigning to each of the individuals the utility $u_i(\omega_i + z_i)$, where $z_i$ is determined in (4)-(5). In other words, the model gives rise to a (market) game, and we may consider the Nash equilibria of this game as a representation of the trades which are observed in the market.

For the special case of two individuals and two commodities, that is $n = l = 2$, the model and its Nash equilibria can be given a graphical treatment, as shown in Fig. 1. Here, the feasible allocations of commodities constitute a rectangle with sides of length $\omega_1$ and $\omega_2$ (this is the famous Edgeworth box of microeconomic analysis), and the initial endowment of commodities is represented by the point $A$.

We assume here that each individual offers all of the first commodity for sale in the market, so that $a_1^1 = \omega_{11}, a_2^1 = \omega_{21}$. In the course of trading some of this commodity is repurchased (there are no transaction costs), and this offer behavior makes the analysis simpler. Now the strategies $(a_1^1, b_1^1), (a_2^1, b_2^1)$ of the two individuals can be represented by the points $S_1$ and $S_2$ in Fig. 1(a), where individual $i$'s endowment

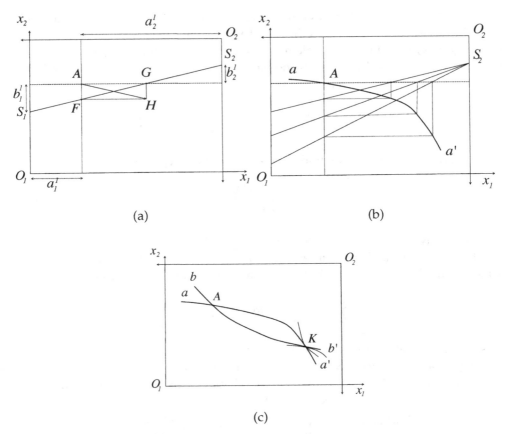

(a)                                              (b)

(c)

**Fig. 1.** Nash equilibria in the market game: In panel (a), the choices $(a_1^1, b_1^1)$ and $(a_2^1, b_2^1)$ correspond to amove from the point $A$ representing initial endowment to the points $S_1$ and $S_2$ respectively. The price defined by these strategies is the slope of the line $S_1 S_2$, and the trade carried out corresponds to a move from $A$ to $H$. Panel (b) illustrates how different strategy choices for individual 1 with fixed strategies of individual 2 give rise to a curve $aa'$. Adding the similar construction for individual 2, one gets that the intersection point (other than $A$) of these curves is a Nash equilibrium outcome if the indifference curves through this point do not intersect the curve $aa'$ (for individual 1) or $bb'$ (for individual 2).

of commodity 1 is reduced by $a_i^1$ and of commodity 2 by $b_i^1$. Connecting the points $S_1$ and $S_2$ we get a line with slope

$$p_1 = \frac{b_1^1 + b_2^1}{\omega_1} = \frac{b_1^1 + b_2^1}{a_1^1 + a_2^1}$$

which then is the market price of commodity 1. Taking the point on the line $F$ for which the first coordinate is $\omega_{11}$, we get that the segment $AF$ has length $p_1 b_1^1$, and similarly, taking $G$ as the point on the line with second coordinate $\omega_{12}$, we get that the length of $AG$ is

$$\frac{1}{p_1}b_1^1 - a_1^1.$$

It is seen from (4) that this is equal to $z_{11}$, the trade of individual 1 in commodity 1, and the trade in commodity 2 is $z_{12} = p_1 b_1^1$, so that the resulting bundle is given by the point $H$. Using $O_2$ as origin, this point also represents the bundle of individual 2.

So far, we have analysed an arbitrary strategy choice by the two individuals. Keeping the strategy of individual 2 fixed, as represented by the point $S_2$ a similar analysis may be carried out for other bid-offer combinations of individual 1, as shown in Fig. 1(b), where the points $H$ found for the different strategies give rise to a curve $aa'$ through $A$. In Fig. 1(c) a similar curve $bb'$ for individual 2 has been added, drawn under the assumption of a fixed strategy choice for individual 1. A Nash equilibrium obtains if the two curves intersect in a point such that for each individual, no other point on the curve yields a higher utility. This is illustrated by the point $K$ in Fig. 1(c).

It may be noticed that the equilibrium allocation is *not* Pareto optimal – both individuals can be made better off by another allocation. This should come as no surprise in view of our previous findings.

## 3  Extensive form games with perfect information

In this section, we return to the *extensive form games* considered in Chapter 1. Since any extensive form game $\Gamma^e$ has a unique associated game in strategic form $\Gamma = \tau(\Gamma^e)$, we may define Nash equilibria of $\Gamma^e$ as Nash equilibria of $\tau(\Gamma^e)$, and in the sequel we skip the distinction between the two.

There are extensive form games for which existence of Nash equilibria is established in a very straightforward way, namely extensive form games with perfect information. Games with perfect information are such that no previous choices (by other players or nature) is hidden from any player. Needless to say, such games occur quite often in applications, and there is an extensive body of theory for particular subclasses of such games (as for example the combinatorial games).

We state and prove the following theorem due to Kuhn (1951).

THEOREM 2  *Let $\Gamma^e = (N, \mathcal{T}, \mathcal{P}, (C(v))_{v \in V}, g)$ be an extensive form game with perfect information. Then $\Gamma^e$ has an equilibrium in pure strategies.*

PROOF: By induction in $h = h(\mathcal{T})$, the height of $\mathcal{T}$, which is defined as the largest number of edges in any path starting at the root and ending in a terminal node. We may assume that $h(\mathcal{T}) \neq 0$ since otherwise the root is the only terminal node and the game is trivial.

Assume that $h(\mathcal{T}) = 1$. Then all the edges starting at the root end in a terminal node, so that there is only one move. If this move is random, none of the players has any influence on the play, and for each player, there is only one strategy for each player, namely "doing nothing", and the choice of this strategy for each player is trivially a Nash equilibrium.

If the move is assigned to one of the players, say player 1, then this player has as many strategies as there are edges pointing away from the root, whereas the other players have only the strategy of doing nothing. If player 1 chooses the edge pointing to a terminal node which maximizes her payoff, then we again have a Nash equilibrium. We conclude that the theorem is true for $h(\mathcal{T}) = 1$.

Assume now that the theorem is true for all games such that $h(\mathcal{T}) = h \geq 1$, and let $\Gamma$ be a game with $h(\mathcal{T}) = h + 1$. Consider any path $(v_0, v_1, \ldots, v_{h+1})$ with $h + 1$ edges; then $v_{h+1}$ is a terminal node, so that $v_h$ must be the last node (of this path) for which some player or nature has a move. Let $v_{h+1}^1, \ldots, v_{h+1}^k$ be the terminal nodes that can be reached from $v_h$ by an edge of the tree; one of these is the edge $v_{h+1}$ encountered already.

Suppose that the move at $v_h$ is random; then we have as before that the result of this move is the mean of the payoffs $g(v_{h+1}^1), \ldots, g(v_{h+1}^k)$, and we can as well consider $v_h$ as terminal with this mean assigned as payoff. If the move at $v_h$ belongs to one of the players, then this player will always choose the edge leading to a node $v_{h+1}^i$ with maximal payoff, and then $v_h$ may again be considered as terminal with $g(v_{h+1}^i)$ assigned as its payoff.

Repeating this procedure for any path of length $h+1$, we obtain a new game form which the game three $\mathcal{T}'$ has height $h(\mathcal{T}') = h$, and by our induction hypothesis, this game has an equilibrium. Translating the equilibrium strategies of this truncated game to strategies of the original game, we obviously obtain an equilibrium of the original game, and we are done. □

It might be noticed that the property of perfect information was only used once in the proof, namely in the reduction of the game tree from $\mathcal{T}$ to $\mathcal{T}'$, but here the property is essential: If a player having the last move does not know at which edge she is actually situated, then the game cannot be reduced.

The approach used in the proof is known as *backwards induction:* We work backwards in the game tree from properties of the terminal nodes and subsequently derive similar properties of the other nodes, ending with the root. This approach can be used also in other contexts, as long as we are dealing with games with perfect information.

## 4   Behavior strategies and perfect recall

In the following we return to the general extensive form game (with possibly imperfect information). The notion of a mixed strategy was introduced in our treatment of equilibria in two-person zero-sum games, but as we have seen in the subsequent chapters, it is no less important when extending the treatment to more general games and other solution concepts.

At this place, we consider mixed strategies in extensive form games. Following the rules established previously for defining strategies in such games, a mixed strategy should be a probability distribution over strategies, which here are maps $s_i$ taking nodes $v$ where $i$ has the move to $C(v)$, the nodes to be reached from $v$. Thus, a mixed strategy corresponds to a randomized choice of full descriptions of the choices to be made whenever necessary. Technically, a mixed strategy for player $i$ in the extensive form game $\Gamma^e = (N, T, \mathcal{P}, (C(v))_{v \in V}, (\mathcal{I}_i)_{i \in N}, (p_v)_{v \in P_0}, u)$ is a probability distribution $\sigma$ over the set $S_i$ of strategies of player $i$ in $\Gamma^e$ (cf. Definition 1 of Chapter 5).

This way of looking at randomization of strategy choices is however not quite intuitive when dealing with extensive form games as a sequence of moves. Here it is more straightforward to think of players randomizing moves rather than randomizing full descriptions. To formalize this notion of randomizing moves whenever necessary, we introduce the concept of a *behavior strategy* introduced by Kuhn [1953].

**DEFINITION 1** *Let $\Gamma^e = (N, \mathcal{T}, \mathcal{P}, (C(v), h_v)_{v \in V}, (\mathcal{I}_i)_{i \in N}, (p_v)_{v \in P_0}, u)$ be an extensive form game, and let $i \in N$ be a player. A behavior strategy for $i$ assigns to each $v$ in $P_i$, the set of moves for $i$, a probability distribution $\beta_v \in \Delta(C(v))$, subject to the (informational) condition*

$$v, v' \in I \text{ for some } I \in \mathcal{I}_i \implies \beta_v = \beta_{v'}. \tag{6}$$

While still representing a randomization, the behavior strategy is a more intuitive, and in some respects simpler, concept than a mixed strategy. Following Hart [1992], a strategy in an extensive form game can be considered as a book of instructions, specifying the choice to be made whenever necessary, and a mixed strategy is a randomized choice of such a book from the library of strategies. A behavior strategy is another book, only here the instructions prescribe a randomized choice instead of a deterministic one.

Since the randomization performed by mixed and behavior strategies is not the same, the question arises whether under suitable conditions one of them corresponds to the other one. We say that the mixed strategy $\sigma_i$ for player $i$ is equivalent to the behavior strategy $\beta_i$ if the probability of visiting any node $v$ in the game tree is the same for any given array $\sigma_{-i}$ of mixed or behavioral strategies of the other players,

$$P\{v \,|\, (\beta_i, \sigma_{-i})\} = P\{v \,|\, (\sigma_i, \sigma_{-i})\},$$

where $P\{v \,|\, (\mu_i, \sigma_{-i})\}$ is the probability that $v$ is visited in a play defined by the strategy array $(\mu_i, \sigma_{-i})$.

To get from a behavior strategy to a mixed strategy is rather straightforward, intuitively one has to perform all randomizations prescribed by the behavior strategy at the outset. Let $s_i$ be a strategy for player $i$. At each node in the set $P_i$ of

$i$'s moves, $s_i$ prescribes a choice $s_i(v)$ of a successor node, which is the same for all $v'$ in the information set $I_i(v)$ from $\mathcal{I}_i$ containing $v$. A behavior strategy defines a probability distribution over all successors of $v$, also the same for all $v' \in I_i(v)$, so that we may consider it as defined on information sets $I_i \in \mathcal{I}_i$ rather than on $P_i$. Defining the mixed strategy $\sigma_i$ by

$$\sigma_i(s_i) = \prod_{I \in \mathcal{I}_i} \beta_i(I_i), \tag{7}$$

we obtain a mixed strategy associated with $\beta$.

THEOREM 3 *Let* $\Gamma^e = (N, \mathcal{T}, \mathcal{P}, (C(v))_{v \in V}, (\mathcal{I}_i)_{i \in N}, (p_v)_{v \in P_0}, u)$ *such that* $|C(v)| \geq 2$ *for all non-terminal nodes* $v$. *Let* $\beta_i$ *be a behavior strategy of player* $i$, *and let* $\sigma_i$ *be given by* (7). *If each information set in* $\mathcal{I}_i$ *intersects any path from the root of* $\mathcal{T}$ *at most once, then* $\sigma_i$ *is equivalent to* $\beta_i$.

PROOF: The probability of visiting a node $v$ during a play can be written as the product $p(v_0) \cdot p(v_n)$ of the probabilities of visiting the nodes before $v$ on the unique path $(v_0, \ldots, v_n, v)$ from the root to $v$, multiplied by the probability of getting from $v_n$ to $v$, and since information sets are visited only once if at all, this probability is the same for $\beta_i$ and for the mixed strategy $\sigma_i$ defined by (7).  $\square$

The condition that information sets intersect paths from the root only once is needed in the proof, and it can be shown that it is also a necessary condition.

The question of whether we can associate an equivalent behavior strategy with any given mixed strategy is less easily answered, and it depends on additional assumptions on the structure of the information sets. What is involved here is the extent to which a player is able to exploit the knowledge obtained at previous moves. More specifically, we shall need the assumption of *perfect recall*. Some additional notation is needed for this.

For $v$ a node in $\mathcal{T}$, we let $\mathcal{T}(v)$ be the subtree with root at $v$; by abuse of notation we shall write $v' \in \mathcal{T}(v)$ when $v'$ is a node in the tree $\mathcal{T}(v)$. For any player $i$ and information set $I_i$, the choice set $C(v)$ is the same for all $v \in I_i$, and for any $c \in C(v)$, let

$$\mathcal{T}(I_i, c) = \cup_{v \in I_i} \mathcal{T}(h_v(c))$$

be the union of all subtrees (or rather: the union of the nodes of all the subtrees) which have their root at a node that can be arrived at following the choice $c$. Thus, $\mathcal{T}(I_i, c)$ can be considered as the remaining game when the choice of $c$ has been made.

DEFINITION 2 *Let* $\Gamma^e = (N, \mathcal{T}, \mathcal{P}, (C(v), h_v)_{v \in V}, (\mathcal{I}_i)_{i \in N}, (p_v)_{v \in P_0}, u)$ *be an extensive form game. A player* $i \in I$ *has perfect recall in* $\Gamma^e$ *if for all* $v, v' \in P_i$ *and* $I_i, I'_i \in \mathcal{I}_i$ *with* $v \in I_i$, $v' \in I'_i$, *if* $v' \in \mathcal{T}(v)$, *then there is a unique* $c \in C(v)$ *such that* $I'_i \subset \mathcal{T}(I_i, c)$.

The condition $v' \in \mathcal{T}(v)$ means that $v'$ comes after $v$ in a play which visits both nodes. Then the inclusion $I'_i \subset \mathcal{T}(I_i, c)$ tells us that at this later stage the information is as detailed as it was at the earlier stage. Indeed, player $i$ (a) recalls what she already knew, in the sense that $I'_i$ is a subset of $\cup_{v \in I_i} \mathcal{T}(v)$. If there had been some $v'' \in I'_i$ which could not be reached from $I_i$, then the player would have forgotten something already known. In addition player $i$ also (b) recalls what she chose, namely the unique $c$ such that $I'_i$ belongs to $\mathcal{T}(I_i, c)$.

We can now state the counterpart of Theorem 3, due to Kuhn [1953].

THEOREM 4 *Let* $\Gamma^e = (N, \mathcal{T}, \mathcal{P}, (C(v), h_v)_{v \in V}, (\mathcal{I}_i)_{i \in N}, (p_v)_{v \in P_0}, u)$ *be an extensive form game, and assume that player $i$ has perfect recall. Then for each mixed strategy $\sigma_i$ of the player $i$ there is an equivalent behavior strategy $\beta_i$.*

PROOF: For $s_i$ a strategy of player $i$, an information set $I_i \in \mathcal{I}_i$ is *reachable under* $s_i$ if there is some play which is consistent with $s_i$ and which intersects $I_i$. For $\sigma_i$ a mixed strategy, we denote by $\sigma_i(I_i)$ the probability according to $\sigma_i$ that $I_i$ is reachable, that is

$$\sigma_i(I_i) = \sum_{\substack{s_i \in S_i: \\ I_i \text{ reachable} \\ \text{under } s_i}} \sigma_i(s_i).$$

Next, for each $c \in C(v)$ with $v \in I_i$, let $\sigma_i(I_i, c)$ be the probability under $\sigma_i$ that $I_i$ is reachable and that $c$ is chosen, found as above but with the summation extended only over $s_i$ for which $I_i$ is reachable and $s_i(v) = c$, $v \in I_i$. Define for each $v \in I_i$ the probability $\beta_i(\cdot, v)$ as

$$\beta_i(c, v) = \frac{\sigma_i(I_i, c)}{\sigma_i(I_i)} \tag{8}$$

if the denominator $\sigma_i(I_i)$ is nonzero, and define $\beta_i(c, v)$ arbitrarily if $\sigma_i(I_i) = 0$, so that $I_i$ is never reached when playing $\sigma_i$.

Let $\sigma_{-i}$ be an arbitrary $(n - 1)$-tuple of mixed strategies of the other players. We show that at each terminal node $v$, we have

$$P\{v \,|\, (\sigma_i, \sigma_{-i})\} = P\{v \,|\, (\beta_i, \sigma_{-i})\}. \tag{9}$$

Let $v$ be an arbitrary terminal node, and let $\mathbf{s} = (v_0, \dots, v)$ be the path from the root $v_0$ to $v$. Let $p^{\sigma_{-i}}$ be the probability that both nature and all players except $i$ choose the edges of $\mathbf{s}$ when choosing at a node from $\mathbf{s}$. If we denote by $p^{\sigma_i}$ and $p^{\beta_i}$ the probability that $i$ chooses from $\mathbf{s}$ when on a node from $\mathbf{s}$ under the strategies $\sigma_i$ and $\beta_i$, respectively, then showing (9) amounts to showing that

$$p^{\sigma_i} p^{\sigma_{-i}} = p^{\beta_i} p^{\sigma_{-i}}. \tag{10}$$

Let $v, v'$ be two consecutive moves of player $i$ on the path $\mathbf{s}$, with associated information sets $I_i$ and $I'_i$, and let $c$ be the choice at $v$ which leads to the edge in $\mathbf{s}$. Since player $i$ has perfect recall, we have that $I'_i \subset T(I_i, c)$, so that $I'_i$ is reachable if and only if $I_i$ is reachable and the choice is $c$. It follows that $\sigma_i(I'_i) = \sigma_i(I_i, c)$, so that the denominator in the expression (8) defining $\beta_i(\cdot, v')$ is equal to $\sigma_i(I_i, c)$, which is the numerator in $\beta_i(\cdot, v)$. To find $p^{\beta_i}$ we need to multiply the probabilities given by $\beta_i$ for all moves of $i$ on the path $\mathbf{s}$, but since the numerator of one probability cancels out against the denominator of the next probability, we get that

$$p^{\beta_i} = \sigma_i(I_i'', c''),$$

where $I_i''$ is the information set associated with the last move of $i$ on $\mathbf{s}$. But $\sigma_i(I_i'', c'')$ is defined as the probability under $\sigma_i$ that $I_i''$ is reachable and the choice is $c''$. Using again perfect recall, this is exactly the probability that all the choices at nodes on $\mathbf{s}$ results in edges of $\mathbf{s}$, which is $p^{\sigma_i}$, so we have shown (9) and therefore also (10). $\square$

We may use the result to show existence of Nash equilibria in behavior strategies for extensive form games with perfect recall in the sense that each player $i \in N$ has perfect recall.

COROLLARY 2 *Let $\Gamma^e$ be an extensive form game with perfect recall. Then $\Gamma^e$ has a Nash equilibrium in behavior strategies.*

PROOF: Using the standard procedure, the game $\Gamma^e$ transforms to a normal form game $\Gamma$ with a finite number of pure strategies, and using the Corollary 1, we get that there is a Nash equilibrium in mixed strategies for $\Gamma$ and hence a Nash equilibrium in mixed strategies $(\sigma_1, \ldots, \sigma_n)$ for $\Gamma^e$. Using now Theorem 4, we can find behavior strategies $\beta_1, \ldots, \beta_n$ such that $\beta_i$ is equivalent to $\sigma_i$ for each $i$, and it is easily seen that $(\beta_1, \ldots, \beta_n)$ is a Nash equilibrium in $\Gamma^e$. $\square$

## 5    Subgame perfectness in extensive form games

In this section, we restrict our attention to games with perfect information, so that information sets contain only a single node. In this context, we have already considered Nash equilibria, and in particular equilibria which were obtained by backward induction (cf. Section 3 of this chapter). Such Nash equilibria have an additional property which is quite important on its own account.

The game in Example 5.1 may be used to introduce the problem. In some cases, Nash equilibrium strategies, which prescribes the choices to be made in cases where the player has to move, may select moves which have very undesirable consequences for the player in question. These choices will never be made if the other players use their Nash equilibrium strategies, but if some other player should deviate, it would place the player in an unpleasant situation.

It may be argued, that this unpleasant situation will not occur, since the equilibrium strategy of the other player prescribes a choice that will prevent the play from reaching the relevant node. However, players might make errors (player 1 might have taken the coin from his purse but it fell from his hands and rolled into the sewer), so that all nodes in the tree remains possible, even though the probability

**Example 5.1. The Eiffel Tower game.** To friends (1 and 2) meet at a cafe in Paris, and 2 says, "Give me 1 € or I shall jump out from the Eiffel Tower. Now 1 has the option of giving or not giving the euro; if he gives it, the game ends here, if not, 2 must either jump, in which case he loses his life and 1 obtains a bad reputation for not helping a friend in trouble, or he may choose not to jump, leaving him in this case with the less than perfect reputation.

The extensive form game may look as follows:

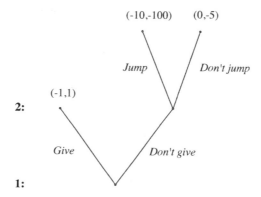

The pair of strategies where 1 chooses "Give" and 2 chooses "Jump" if this choice is relevant (which is not supposed to be, given that the game terminates once the euro has been given), is a Nash equilibrium. However, if player 2 must choose between jumping and not jumping, the latter option is clearly better, so this equilibrium is based on a threat that the player would not like to carry out.

Clearly, the game has another – and more intuitive – Nash equilibrium, where 1 chooses not to give, and where player 2 chooses not to jump in the case that the game gets so far (which it does given 1's choice of strategy).

---

of their occurrence would be very small. This is the *trembling hand hypothesis*, to which we shall return later.

To avoid such equilibria, we need to restrict attention to those where all players have incentives to perform the choices specified by the strategy if they are called upon to do it. This is done by the concept of *subgame perfect* Nash equilibrium introduced by Selten [1965].

DEFINITION 3 *Let* $\Gamma^e = (N, \mathcal{T}, \mathcal{P}, (C(v))_{v \in V}, u)$ *be an extensive form game with perfect information.*

*For* $v \in V$ *a node of the tree* $\mathcal{T}$, *the subgame at* $v$ *is the extensive form game* $\Gamma^e(v) = (N, \mathcal{T}(v), \mathcal{P}|_{\mathcal{T}(v)}, (C(v))_{v \in \mathcal{T}(v)}, g|_{\mathcal{T}(v)})$ *with game tree* $\mathcal{T}(v)$, *where player sets and payoffs are defined in the obvious way.*

*A strategy array $\sigma = (\sigma_1, \ldots, \sigma_n)$ is subgame perfect if for each node $v$ in the game tree $\mathcal{T}$, the array of restrictions $(\sigma_i|_{\mathcal{T}(v)})_{i \in N}$ to the tree $\mathcal{T}(v)$ is a Nash equilibrium in $\Gamma^e(v)$.*

Here the restriction of a mixed strategy, defined as a probability distribution over pure strategies, to a subtree of $\mathcal{T}$, is defined in the obvious way, restricting all pure strategies to the subtree.

As it could be seen in Example 5.1, the unrealistic threats are connected with nodes which will not be visited when the equilibrium strategies are played. This feature can be extended so as to cover the general case with Nash equilibria in mixed strategies. Recall (cf. Section 4 of this chapter) that $P\{v \mid \sigma\}$ denotes the probability that the node $v$ is visited when the strategies from $\sigma = (\sigma_1, \ldots, \sigma_n)$ are played.

THEOREM 5 *Let $(\sigma_1, \ldots, \sigma_n)$ be a Nash equilibrium in the extensive form game $\Gamma^e = (N, \mathcal{T}, \mathcal{P}, (C(v))_{v \in V}, u)$, and let $v$ be a node in $\mathcal{T}$. If $P\{v | \sigma\} > 0$ then $\sigma|_{\mathcal{T}(v)}$ is a Nash equilibrium in the subgame $\Gamma^e(v)$.*

PROOF: Suppose that $\sigma|_{\mathcal{T}(v)}$ is not a Nash equilibrium in $\Gamma^e(v)$. Then there must be a player $i$ and a mixed strategy $\mu_i$ in $\Gamma^e(v)$ such that
$$U_i^{\Gamma^e(v)}(\mu_i, \sigma_{-i}|_{\mathcal{T}(v)}) > U_i^{\Gamma^e(v)}(\sigma|_{\mathcal{T}(v)}),$$
where
$$U_i^{\Gamma^e(v)}(\mu_i, \sigma_{-i}|_{\mathcal{T}(v)}) = \sum_{v' \in V^i \cap \mathcal{T}(v)} P\{v' \mid (\mu_i, \sigma_{-i}|_{\mathcal{T}(v)})\} u_i(v')$$
is the expected payoff in the game $\Gamma^e(v)$.

In view of Corollary 1 to Theorem 4, we may assume that $\sigma_i$ is a behavior strategy in $\Gamma^e$, assigning to each $v \in P_i$ a probability distribution over the outgoing edges at $v$. Similarly, we may assume that $\mu_i$ is a behavior strategy in $\Gamma^e(v)$. We now define a new behavior strategy $\nu_i$ for player $i$ by
$$\nu_i(v') = \begin{cases} \sigma_i(v') & v' \in P_i \cap [V \setminus \mathcal{T}(v)], \\ \mu_i(v') & v' \in P_i \cap \mathcal{T}(v). \end{cases}$$
If $\mathcal{T}'(v)$ denotes the tree obtained from $\mathcal{T}$ after deleting the subtree $\mathcal{T}(v)$ (and the unique ingoing edge at $v$ if $v$ is not the root of $\mathcal{T}$), we obtain an extensive form game $\Gamma^e_-(v)$ and we have that
$$U_i^{\Gamma^e}(\nu_i, \sigma_{-i}) = U_i^{\Gamma^e(v)}(\mu_i, \sigma_{-i}|_{\mathcal{T}(v)}) + U_i^{\Gamma^e_-(v)}(\sigma_i|_{\mathcal{T}'(v)}, \sigma_{-i}|_{\mathcal{T}'(v)})$$
$$> U_i^{\Gamma^e(v)}(\sigma|_{\mathcal{T}(v)}) + U_i^{\Gamma^e_-(v)}(\sigma_i|_{\mathcal{T}'(v)}, \sigma_{-i}|_{\mathcal{T}'(v)}) = U_i^{\Gamma^e}(\sigma_i),$$
and we have exhibited a strategy which gives a better result to player $i$, contradicting that $\sigma = (\sigma_1, \ldots, \sigma_n)$ is a Nash equilibrium. $\qquad\square$

The result stated in Theorem 5 may be used to introduce a specific class of Nash equilibria which are subgame perfect. A mixed strategy equilibrium in the game $\Gamma$ (in either normal or extensive form) is *completely mixed* if every pure strategy is chosen with a nonzero probability.

COROLLARY 1 *Let $\sigma = (\sigma_1, \ldots, \sigma_n)$ be a Nash equilibrium in the extensive form game $\Gamma^e$. If $\sigma$ is completely mixed, then $\sigma$ is also a subgame perfect Nash equilibrium.*

## 6 Problems

**1.** (The Groves-Ledyard model for allocating a public good, [Groves and Ledyard [1977]]) A community of $n$ individuals must decide how much to produce of a public good, consumed collectively by the individuals. There is an ordinary ("private") good which can be used for consumption *and* as input for producing the public good, and it is assumed that one unit of the private good transforms to $k$ units of the public good. Individuals have concave utility functions $u_i(x_i, y)$ depending on the individual consumption of the private good and collective consumption of the public good. Each individual has an individual endowment $\omega_i$ of the private good.

Show that if the allocation of private and public good is Pareto efficient (in the sense that no other allocation makes everyone as well off and someone better off), then marginal utility of the private good should equal $k$ times marginal utility of the public good for each individual.

We introduce the following game for determining the amount of public good produced: Individual $i$ chooses a proposed contribution $b_i$ to production of public goods, the amount produced is $k \sum_{i=1}^n b_i$, and the final contribution of individual $i$ is $t_i(b_1, \ldots, b_m)$ determined as

$$t_i(b_1, \ldots, b_m) = \frac{1}{n} \sum_{i=1}^n b_i + \left[ \frac{n-1}{n}(b_i - \mu_i)^2 - \sigma_i^2 \right],$$

$$\mu_i = \frac{1}{n-1} \sum_{j \neq i} b_j, \quad \sigma_i = \frac{1}{n-2} \sum_{j \neq i} (b_j - \mu_j)^2.$$

Show that Nash equilibria of the game are Pareto efficient.

**2.** (A simplified version of the "rotten kid theorem" in [Becker [1981]]) A family consists of a mother and a son. The son chooses a level of effort $e$ which determines the income $y_1(e)$ of the mother as well as his own income $y_2(e)$. The mother is altruistic and chooses a transfer $t$ of her income to the son. Both have concave utility functions in income, $U$ and $V$ respectively, so the payoff to the mother given the choices $t$ and $e$ are

$$u_1(t, e) = U(y_1(e) - t) + \alpha V(y_2(e) + t), \quad u_2(t, e) = V(y_2(e) + t).$$

Show that the in the Nash equilibrium, the son chooses $e$ so as to maximize family income $y_1 + y_2$.

Show that the equilibrium value of $t$ is an increasing function of $\alpha$, the degree of altruism, and for sufficiently small $\alpha > 0$, the transfer $t$ may become negative.

**3.** Find a mixed strategy Nash equilibrium in the patent race game (Chapter 1, Problem 3). Does Theorem 1 apply to this game?

**4.** *The centipede game.* Consider a two-person game that may last up to 100 rounds, and where the players alternate in deciding whether to continue or to terminate the game.

The payoffs are as follows: If player 1 terminates the game in round $2n - 1$, for $n = 1, \ldots, 50$), then the payoffs to players 1 and 2 are respectively $2n$ and $2n - 2$. If player 2 terminates the game in round $2n$, for $n = 1, \ldots, 50$, then the payoffs to players 1 and 2 are $2n - 1$ and $2n + 1$. If round 101 is reached (this happens if player 2 decided to terminate at round 100), then the game terminates, with the payoffs 102 and 100 to players 1 and 2.

Sketch the extensive form, and find the Nash equilibria and subgame perfect Nash equilibria.

**5.** (A model of bank regulation, cf. Mailath and Mester [1994]) A bank has two investment possibilities for its deposits of size 1, namely a safe investment yielding an interest rate $r$ with certainty, and a risky investment which yields interest $\rho$ with probability $p$ but where all is lost with probability $1 - p$. It is assumed that the safe investment is better for society than the risky one, in the sense that $p(1 + \rho) < 1 + r$.

The model runs over 2 periods of time, the bank chooses investment at $t = 0$, after which the regulator can choose to close the bank, which means that gains or losses are computed at $t = 1$, or to let it continue. If it continues, then it performs a new round of investments.

The regulator must pay a cost $C$ when closing the bank and when the bank defaults. In addition to this, the regulator must reimburse depositors.

Set up the extensive form game and find the subgame perfect Nash equilibria depending on the values of $C$ and $p$.

# Chapter 6

# Games with Incomplete Information

## 1 Bayesian Nash equilibria

In our discussion of games and their solutions, we have assumed throughout that the preferences of all players are known to all involved parties, so that intuitively, each player is aware of the motives of the others, knows what they are "up to". This may not be the case in many real-world conflicts, indeed the lack of knowledge of what guides other people may be exactly what prevents otherwise reasonable outcomes from materializing. And the uncertainty may pertain not only to the payoffs, but even to the game itself – the players may not know exactly which game they are actually playing, such knowledge may materialize only while playing.

Such situations cannot be treated directly using the methods developed so far. However, the approach to restricted information already introduced is so promising that it might be extended to cover also the case of lacking information about payoffs. As proposed by Harsanyi [1967], the games that may occur, defined by strategy sets and payoff functions of all players, can be embedded in an extensive form game with information sets that exactly describe the incompleteness of information, and our previously considered methods can be applied to this larger game. Intuitively, nature chooses the game to be played, and then the players can select their actions.

**1.1. The Harsanyi construction.** In order to make this precise, we introduce the set of *states* of the game with incomplete information as a set of $\Gamma$ of (usual) games $\Gamma = (N, (A_i)_{i \in N}, (u_i)_{i \in N})$ with the same player set $N$. We have used the notation $A_i$ for $i$'s strategy set in the normal form game and we shall refer to its elements as *actions* of the player in this game. Each player $i$ can observe the set of actions available but may not be aware of the actions open to other players, and consequently may not know even her own payoffs. However, the extensive form game to be constructed must be such that each player has the same set of actions at each information set. To deal with this problem, we introduce for each player $i$ a set of *types* $T_i$ of this player. The set $T = T_1 \times \cdots \times T_n$ of arrays $(t_1, \ldots, t_n)$ of player types is called the *type space*, and we assume that each type $t \in T$ corresponds to a game $G(t) = (N, (A_i(t))_{i \in N}, (u_i(t))_{i \in N})$. The type $t_i$ of player $i$ captures exactly what this player can observe, and the condition on information can be formulated as

$$A_i(t_i, t_{-i}) = A_i(t_i, t'_{-i}), \text{ all } t_{-i}, t'_{-i} \in T_{-i},$$

where $T_{-i} = \times_{j \neq i} T_j$, saying that all type arrays $t$ where player $i$ has the same type $t_i$ give rise to the same action sets for this player.

If nature has to select a state, then there must be given a probability distribution $p$ on the state space $t$. We call this distribution the *common prior*, and for the moment it is taken as given, but we return to it in the next section.

Now we have developed a method for dealing with incomplete information, using only the concepts and notions developed so far. Then game with incomplete information is considered as an extensive form game, where nature has the first move and chooses a state $t$ in accordance with the prior probability distribution $p$. Since the chosen state determines the payoff functions of each player, information sets must be introduced in the extensive form game so as to secure that the player is aware of her own type $t_i$ but not the types of the other players. One may then proceed to the study of solutions – in particular, Nash equilibria – of such games.

DEFINITION 1 *A game with incomplete information (and common prior) is an array* $(N, \Gamma, (T_i)_{i \in N}, \tau, P)$, *where*

(i) *$N$ is a finite set of players,*

(ii) *$\Gamma$ is a set of normal form games $(N, (A_i)_{i \in N}, (u_i)_{i \in N})$ with player set $N$,*

(iii) *for each $i \in N$, $T_i$ is a finite set of types of player $i$, and*

(iv) *$\tau : T \rightarrow \Gamma$ is a surjective map assigning games to type arrays such that if $t, t' \in T$ and $t_i = t'_i$, then $A_i(t) = A_i(t') = A_i(t_i)$,*

(v) *$P$ is a probability distribution over $T = \times_{i \in N} T_i$.*

In our definition of a game with incomplete information, nature chooses the game, but the choice of nature is observable for the players to the extent allowed by the mappings $\tau_i$.

A Nash equilibrium in the game with incomplete information is one where each player chooses a strategy in the extensive form game, or, alternatively, for each of her possible types, an action in the game form giving rise to the game, which is assessed using the conditional probability of the other players' types. For each $i \in N$, let $P(t_{-i}|t_i)$ be this conditional probability (of the type array $t_{-i}$ given the observed type $t_i$), defined (for finite sets of types) as

$$P(t_{-i}|t_i) = \frac{P(t_i, t_{-i})}{\sum_{t'_{-i} \in T_{-i}} P(t_i, t'_{-i})} \tag{1}$$

with $T_{-i} = \times_{j \neq i} T_j$. It may be noticed that the probability $P(t_{-i}|t_i)$ in (1) is found by Bayesian updating from the prior distribution of player $i$ over the types of the others (which is the marginal distribution over $t_{-i}$ obtained from the distribution $P$ over $T$), given that player $i$ has observed her own type $t_i$. Indeed, (1) is Bayes' formula applied to our present case (where own type $t_i$ is observed with certainty).

DEFINITION 2 *Let $(N, \Gamma, (T_i)_{i \in N}, \tau, P)$ be a game with incomplete information. A strategy for player $i \in N$ is a map $s_i : T_i \rightarrow A_i$. A Bayesian Nash equilibrium is an array $(s_1, \ldots, s_n)$ of strategies, such that for each $i$ and each type $t_i \in T_i$,*

$$\sum_{t_{-i} \in T_{-i}} P(t_{-i}|t_i)u_i(s_i(t_i), s_{-i}(t_{-i})) \geq \sum_{t_{-i} \in T_{-i}} P(t_{-i}|t_i)u_i(a_i, s_{-i}(t_{-i})), \qquad (2)$$

*for all actions $a_i \in A_i(t_i)$ of player $i$ (where $s_{-i}(t_{-i}) = (s_j(t_j))_{j \neq i}$ is the array of actions chosen by the other players with types $t_{-i}$).*

As it is seen, the game with incomplete information is treated as an extensive form with an initial random move. However, this initial move is not actually made, rather it is a way of formalizing the beliefs of the players, who are not sure about the game in which they are participating, even if this game is given and unique. This difference of attitudes is reflected in the solution concept, where the action to be chosen by a player who has observed only her own type should be the best possible when taking average over the possible types of all the others (and the actions that they choose given their type).

The Bayesian Nash equilibrium prescribes the choices once (own) types are revealed. We may compare this solution to the Nash equilibria of the extensive form game, which describes the *ex ante* optimal behavior. Fortunately, the equilibria are the same.

THEOREM 1 *Let $(s_1^*, \ldots, s_n^*)$ be a Nash equilibrium of the extensive form game, where nature chooses a game $(N, (A_i)_{i \in N}, (u_i)_{i \in N}) \in \Gamma$ in accordance with the probability distribution $P$ defined on $T$. Then the strategy $(s_i^*(t_i))_{t_i \in T_i, i \in N}$ is a Bayesian Nash equilibrium.*

*Conversely, for each Bayesian Nash equilibrium $(s_1, \ldots, s_n)$ of $(N, \Gamma, (T_i)_{i \in N}, \tau, P)$, the strategy array which prescribes the move $s_i(t_i)$ for player $i$ given that nature has chosen $(t_i, t_{-i}) \in T$, all $i \in N$, is a Nash equilibrium in the extensive form game.*

PROOF: Since $(s_1^*, \ldots, s_n^*)$ is a Nash equilibrium, then for each $i$ the collection of moves $s_i^*(t_i)$ of player $i$, for $t_i \in T_i$, maximizes expected payoff given the choices of the other players, so that

$$\sum_{t_i \in T_i} P(t_i) \sum_{t_{-i} \in T_{-i}} P(t_{-i}|t_i)u_i(s_i^*(t_i), s_{-i}^*(t_{-i})) \geq \sum_{t_i \in T_i} P(t_i) \sum_{t_{-i} \in T_{-i}} P(t_{-i}|t_i)u_i(s_i'(t_i), s_{-i}^*(t_{-i})), \quad (3)$$

where $P(t_i) = \sum_{t_{-i} \in T_{-i}} P(t_i, t_{-i})$. If the collection $(s_i^*(t_i))_{t_i \in T_i, i \in N}$ is not a Bayesian Nash equilibrium, then there must be $i$ and $t_i' \in T_t$ with

$$\sum_{t_{-i} \in T_{-i}} P(t_{-i}|t_i)u_i(s_i(t_i), s_{-i}^*(t_{-i})) > \sum_{t_{-i} \in T_{-i}} P(t_{-i}|t_i)u_i(s_i^*(t_i), s_{-i}^*(t_{-i}))$$

for some action $a_i$, and then the strategy $s_i^0$ with

$$s_i^0(t_i) = \begin{cases} s_i^*(t_i) & t_i \neq t_i^0, \\ a_i & t_i = t_i^0 \end{cases}$$

would violate (3), a contradiction.

To prove the converse, we notice that at the Bayesian Nash equilibrium $(s_1, \ldots, s_n)$, the inequality (2) holds for all $i$ and $t_i \in T_i$, and multiplying the inequality for each $t_i$ by $P(t_i)$ and adding, we obtain that the strategies $(s_i(t_i))_{t_i \in T_i}$, for $i \in N$, form a Nash equilibrium in the extensive form game.                                    $\square$

Having compared Bayesian Nash equilibria to the Nash equilibria in the extensive form games, we might also consider the Nash equilibria in a related extensive form game, namely that where the players are all the types $t_i$ of the original players $i \in N$. Here nature selects an $N$-tuple of players from $T_1 \times \cdots \times T_n$, and these players are active in the remaining play, whereas the types not selected remain passive and obtain a payoff of 0. It may be checked that the Nash equilibria of this $|T_1| \cdot |T_2| \cdots |T_n|$-player game correspond to those of the extensive form game considered above and, by Theorem 1, to the Bayesian Nash equilibria of the game with incomplete information. We leave this to the exercises.

**1.2. Correlated equilibrium and incomplete information.** We may use the concepts developed here to give another interpretation of the correlated equilibria discussed previously (see 3.4). Given a normal form game $\Gamma = (N, (S_i)_{i \in N}, (u_i)_{i \in N})$ and a correlated equilibrium $p$ for $\Gamma$, we may consider the problem of selecting a suitable strategy array $(s_i)_{i \in N}$ as a game with incomplete information $(N, \Gamma, (T_i)_{i \in N}, \tau, p)$, where

- for each $i$, $T_i = S_i$, the set of pure strategies of player $i$ in the game $\Gamma$,
- $\Gamma = \{\Gamma\}$, and $\tau$ is the constant map taking all type arrays to $\Gamma$.

We can easily exhibit a Bayesian Nash equilibrium of the game with incomplete information $(N, \Gamma, (T_i)_{i \in N}, \tau, p)$, namely the array $\mathrm{id}^N = \mathrm{id} \times \cdots \times \mathrm{id}$, where each type chooses the corresponding strategy; indeed the equilibrium condition for the correlated equilibrium, which says that for each $i$ and $s_i \in S_i$,

$$\sum_{s_{-i} \in A_{-i}} p(s_i, s_{-i})[u_i(s_i, s_{-i}) - u_i(s_i', s_{-i})] \geq 0, \tag{4}$$

all $s_i' \in S_i$, is seen to be identical to the equilibrium condition (2) for a Bayesian Nash equilibrium. Conversely, if for the game $\Gamma$ we have that $\mathrm{id}^N$ is a Bayesian Nash equilibrium for a game with incomplete information $(N, \{\Gamma\}, (S_i)_{i \in N}, \tau, P)$, then $P$ is a correlated equilibrium for $\Gamma$.

This gives us an alternative interpretation of the correlated equilibrium based on the subjective beliefs of the players. We shall see below that also the concept of a mixed strategy can be given a new interpretation using the formalism of games with incomplete information.

**1.3. The common prior.** In our treatment of games with incomplete information, we have assumed throughout that the sets of types were finite. This assumption had the advantage of allowing us to define conditional probabilities in a simple way, but it has some disadvantages as well, in particular it fits badly with many prominent applications, for example to the analysis of auctions (see Section 3 below). However, in many cases, including those to be considered below, the extension to infinite type spaces will be straightforward, and therefore we avoid a general treatment of the infinite case with the associated need for additional formalism.

There is, however, another assumption made above which may give rise to some afterthought, namely that of a *common prior* distribution $P$ over types. As mentioned, there is no "real" move by nature, rather the prior probability of types represent the beliefs of the players. But then it may be questioned whether these beliefs can reasonably be assumed to be so well coordinated that they appear as conditional probabilities derived from a common prior distribution. There may well be cases where players have subjective beliefs that do not satisfy this consistency requirement, and we shall have a closer look at this problem in the next section.

## 2   States, beliefs and information

In the previous section, we introduced the Harsanyi approach to games with incomplete information, which formalized the lack of precise knowledge about the payoffs of the other players as an extensive form game with nature having the first move, in which the types of all players are determined. For this approach to work, we needed a probability distribution to guide the choice of nature, and this gave rise to the *prior distribution* of types.

**2.1. Bayesian Nash equilibria without a common prior.** Upon a second thought, we might feel uncomfortable with the approach of the previous section. Formal convenience is important but the theory should also relate to real-world phenomena, and it may indeed be questioned whether the existence of a common prior distribution is a phenomenon which has a useful interpretation in all applications. In particular, one might be tempted to ask where it came from, and if there is an explanation, whether this should not be part of the model.

The definition of a game with imperfect information can in principle be extended rather easily so as to avoid explicit mentioning of a common prior over the space $T$ of admissible utility assignments to players. What is needed to define a Bayesian Nash equilibrium is that for each player $i$ and each type $t_i$ of this player, there are well-defined probability distributions $P_{t_i}$ over $T_i$ such that

$$\sum_{t_{-i} \in T_{-i}} P_{t_i}(t_{-i}) u_i(s_i(t_i), s_{-i}(t_{-i})) \geq \sum_{t_{-i} \in T_{-i}} P_{t_i}(t_{-i}) u_i(a_i, s_{-i}(t_{-i})), \tag{5}$$

holds for the array $(s_1, \ldots, s_n)$ of type-dependent actions.

However, we might be interested in the study of other equilibria than Nash equilibria of the extensive form game with nature as first mover, and here it might be important to include not only beliefs over types, but also over the beliefs $P_{t_i}(t_{-i})$

held by the other types, etc. To capture these features, we return to the basic idea of nature choosing a state, which now is more complex than specifying a particular game, since the description of the game will also contain the specific beliefs of all players. The fact that each player has limited knowledge of the actual state will be formulated using information sets rather than types.

DEFINITION 3  *A game with incomplete information is an array* $(N, \Omega, (\mathcal{I}_i)_{i \in N}, \Gamma, \gamma, P_i)$, *where*

(i) *N is a finite set of players,*

(ii) $\Omega$ *is a finite set of states of nature,*

(iii) $\Gamma$ *is a set of games* $(N, (A_i)_{i \in N}, (u_i)_{i \in N})$ *with player set N,*

(iv) *for each* $i \in N$, $\mathcal{I}_i$ *is a partition of* $\Omega$,

(v) $\gamma : \Omega \to \Gamma$ *assigns a game* $\gamma(\omega) = \left( N, (A_i^\omega)_{i \in N}, (u_i^\omega)_{i \in N} \right)$ *to each state* $\omega$ *so that if* $\omega, \omega' \in I_i$ *for some* $I_i \in \mathcal{I}_i$, *then* $A_i^\omega(t) = A_i^{\omega'}(t'))$, *and*

(vi) *for each i and* $\omega$, $P_i(\omega)$ *is a probability distribution over* $\Omega$ *with support* $I_i(\omega)$, *the element of* $\mathcal{I}_i$ *containing* $\omega$, *and if* $\omega, \omega' \in I_i$ *for some* $I_i \in \mathcal{I}_i$, *then* $P_i(\omega)$ *and* $P_{(\omega')}$ *are identical (as probability distributions).*

The present more general definition of a game with incomplete information differs from Definition 1 mainly in the replacement of a common prior by state-dependent subjective beliefs. The condition (vi) is a consistency requirement on the beliefs: The players must have the same beliefs in all the states which they cannot distinguish, since the players could obtain additional knowledge from observing their own beliefs.

Restricting our treatment to finite sets of states of nature is convenient since this allows us to work with information structures in the form of partitions of $\Omega$, but just as the assumption of a finite set of types it may be too restrictive in practice. In our present case, there is an additional drawback, namely that the very complex nature of the state description, where states among other things define beliefs of players, that is specific probability distributions, makes it unrealistic from the outset that a description in terms of a finite set would be available.

**2.2. Belief spaces.** In order to indicate how the notion of belief can be extended to allow for beliefs on other things than the parameters of the games, such as beliefs over parameters, beliefs over beliefs over parameters etc., we briefly outline the approach of Mertens and Zamir [1985].

We begin by introducing abstract belief spaces as the building block for constructing a state description which contains all the beliefs and derived beliefs.

DEFINITION 4  *A belief space (over S) is an* $(n + 3)$-*tuple* $(C, S, f, (t_i)_{i \in N})$, *where*

(i) *C is a compact space,*

(ii) *S is a set of states, and $f : C \to S$ a continuous map,*

(iii) *for each i, $t_i : C \to \Delta(C)$ maps elements c of C to probability distributions over C, subject to the consistency requirement*

$$c \in C \text{ and } c' \in supp(t_i(c)) \Rightarrow t_i(c') = t_i(c). \tag{6}$$

Here $C$ is a set of full descriptions, including parameters, beliefs, beliefs on beliefs etc., while $S$ contains only the states of nature, the parameters of the game to be selected. Intuitively, one may construct a belief space which captures all the possible belief structures as follows: The trivial belief spaces over $S$ is that where $C_0 = S$ and $f$ is the identity map. Then a more complicated belief space can be constructed letting $C_1$ be the set of probability distributions over $S$, types, and in the next round $C_2$ consists of probability distributions over $C_1$, etc. This intuitive approach has to be suitably refined in order to work. We state without proof the main result obtained by Mertens and Zamir:

THEOREM 2 *Let S be given. Then there exists a belief space $\left(\mathcal{Y}, S, f', (t'_i)_{i \in N}\right)$ over S such that*

$$\mathcal{Y} = S \times (\mathcal{T})^n, \ \mathcal{T}^n = \Delta(S \times (\mathcal{T})^{n-1}),$$

*and this belief space is universal in the sense that for each belief space $(C, S, f, (t_i)_{i \in N})$ over S, the maps f and $t_i$ can be factored through $\mathcal{Y}$, i.e. there is a map $\phi : C \to \mathcal{Y}$ such that*

$$
\begin{array}{ccc}
C & \xrightarrow{\phi} & \mathcal{Y} \\
\downarrow{t_i} & & \downarrow{t'_i} \\
\Delta(C) & \xrightarrow{\hat{\phi}} & \Delta(\mathcal{Y})
\end{array}
$$

*where $\hat{\phi}$ is the induced map on probability distributions.*

PROOF: See Mertens and Zamir [1985], Theorem 2.9.

The universal belief space can be thought of as allowing for the most complex description of beliefs. It retains the structure of the intuitive belief spaces, in the sense that elements of $\mathcal{Y}$ specify the state of nature *and* for each player $i$, the generalized beliefs of this player, which again is a probability distribution on states and (generalized) beliefs of all the other players.

**2.3. Common knowledge and the disagreement theorem.** Having discussed the role of beliefs and the construction of state spaces, we now turn to the informational part of the problem. Information or knowledge as we meet it in the context of games with incomplete information pertains to the extent to which players may observe the state of the nature. Assuming here that the set of possible states $\Omega$ is finite, we can describe the information of a player $i$ as a partition $\mathcal{I}_i$ of $\Omega$, so that $i$ can distinguish between states $\omega$ and $\omega'$ if and only if they belong to distinct elements

$I, I'$ of $\Omega$. An event $E$, that is a subset of $\Omega$, is known to a player $i$ at the (true) state $\omega$ if $I_i(\omega)$, the set in $i$'s information partition which contains $\omega$, is a subset of $E$.

We are interested in moving from the knowledge of any single player to common knowledge, where we for simplicity stay within a 2-player framework: An event is common knowledge if player 1 knows it, player 2 knows it, player 1 knows that 2 knows it, 2 knows that 1 knows it, 1 knows that 2 knows that 1 knows it etc. In order to formalize this notion, we introduce the *meet* $I_1 \wedge I_2$ as the finest partition of $\Omega$ such that all its sets are in both $I_1$ and $I_2$. Then an event $E$ is *common knowledge* at $\omega$ if the element of $I_1 \wedge I_2$ containing $\omega$ is a subset of $E$.

To see that this definition captures the essence of common knowledge as explained intuitively, choose a true state $\omega$, and let $E$ be an event. Then '1 knows $E$' if $I^1 = I_1(\omega)$ is a subset of $E$, and '1 knows that 2 knows $E$' if $E$ contains all the sets $I^2 \in I_2$ which intersect $I^1$. If 1 knows that 2 knows that 1 knows $E$, then $E$ must include all members $I^3$ of $I_1$ that intersect some element $I^2$ of $I_2$ which intersects $I^1$, and so on. Proceeding with this construction, so that in each step we get sets from the partition of the other player intersecting all sets obtained in the previous round (and using that the set of states is finite), we must end with an element of $I_1 \wedge I_2$.

We can use the formal description of common knowledge to prove the *disagreement theorem* of Aumann [1976]. Suppose that $P$ is a given (prior) distribution on $\Omega$; we assume throughout that $P$ assigns nonzero probability to each element of both $I_\infty$ and $I_\epsilon$. and let the *posterior distribution* for player $i$ of an event $B$ given the state $\omega$ be the distribution $Q_i$ defined by

$$Q_i(\omega) = \frac{P(B \cap I_i(\omega))}{P(I_1(\omega))}. \tag{7}$$

THEOREM 3 *Let $\omega \in \Omega$ be arbitrary, and let $t_1, t_2 \in [0, 1]$ be numbers. If it is common knowledge at $\omega$ that $Q_i = t_1$ and $Q_2 = t_2$, i.e. if the set $\{\omega' \mid Q_1(\omega') = t_1, Q_2(\omega') = t_2\}$ contains the set from $I_1 \wedge I_2$ which has $\omega$ as an element, then $t_1 = t_2$.*

PROOF: Let $I \in I_1 \wedge I_2$ be the set containing $\omega$. Then $I$ can be written as a union of disjoint sets from $I_1$, $I = I^1 \cup \cdots \cup I^l$. Since $Q_1$ takes the value $t_1$ on $I$, we have from (7) that

$$P(B \cap I^j) = t_1 P(I^j)$$

for $j = 1, \ldots, l$, and we get that $P(B \cap I) = t_1 P(I)$. Repeating the argument after writing $I = I_1 \cup \cdots \cup I_k$ with $I_1, \ldots, I_k$ disjoint sets from $I_2$, we obtain that $P(B \cap I) = t_2 P(I)$, and we conclude that $t_1 = t_2$. $\square$

The result implies that if it is common knowledge that one player has a particular subjective belief about the states (considered as a posterior distribution given some prior distribution) and another player has different subjective beliefs, then their priors cannot be the same.

## 3 Auctions

**3.1. Types of auctions.** Auctions are used for selling objects to a given set of potential buyers or *bidders*, who are made to signal willingness to buy in some way or another. The intuitive appeal of auctions lies in its ability to distinguish among buyers according to how much they value the object, which ideally will be passed to the bidder which values it highest. Many spectacular auctions have resulted in large revenues for the seller, and auctions are used increasingly also for public procurement contracts, where the job will be given to the bidder who can do it in the cheapest way.

Fundamentally, what auctions try to accomplish is to reveal the values of the bidders. However, values are in general subjective and private, so that they cannot be directly observed, neither can any statement about values be taken as a true report – unless some additional conditions are satisfied. In other words, we are dealing with a game with imperfect information.

Before specifying the game, or rather, the games to be considered, we must introduce *auction designs*, that is the rules under which the auction is performed. These rules specify the messages to be passed from bidders to the seller, the choice of a winner depending on these messages, and the payments from the bidders. Since we fix the auction design, the seller plays no role in the procedure, so the bidders are the players in the game that results from the given auction design.

Here are some of the classical auction designs:

*English or ascending-value auctions:* The price to be paid for the object is initially stated at some minimal level and then increases steadily. The bidders observe the current price and their only signal is that of leaving the auction. When only one remains, the object is passed to this bidder, who pays the current price (at the moment where the last of the other bidders left).

*Dutch or descending-value auctions:* Here the process is reversed, so that the price starts at a high level and decreases, and the first bidder signaling willingness to pay at the current price obtains the object at this price.

The above auctions are *open* in the sense that the messages or signals chosen by any bidder can be observed by the other bidders. One may also consider *closed* auctions, which are in some respects easier to deal with. An example of a closed auction is the following:

*First-price sealed bid auction:* The bidders send a bid (a proposed price to be paid for the object) observable only to the seller. The bidder who has given the highest bid gets the object and pays her bid.

Auctions as those listed here can be treated as games with incomplete information. In a closed auction, the actions of the bidders are the *bids*, and the particular game is specified when the payoffs are given. If we assume that each of the bidders can make a subjective assessment $x_i$ of the *value* of the game, then the payoff of

**Example 6.1. Optimal bidding strategies in second-price auctions.** In the second-price auction, the payment by the winner $t^1(b_1, \ldots, b_n)$ is the second-highest of the bids $b_1, \ldots, b_n$, and the other players pay nothing, $t^2(b_1, \ldots, b_n) = 0$. The particular role played by the second-price auction has to do with the fact that truth-telling is an equilibrium strategy. Indeed, the strategy

$$\beta(x) = x,$$

according to which the bidder states her true value, is a weakly dominating strategy: Suppose that bidder $i$ chooses the strategy $\beta$, while all other players choose some arbitrary strategies. Given that the latter results in bids $b_j$, for $j \neq i$, then there are two cases to consider:

(1) Bidder $i$ wins, i.e. $x > \max_{j \neq i} b_j$. Then $i$ gets the object and pays $\max_{j \neq i} b_j$. Stating another bid $b_i \neq x$ will not change this outcome, unless $b_i < \max_{j \neq i} b_j$, and in this case $i$ loses the auction and gets 0, which is inferior to $x - \max_{j \neq i} b_j > 0$.

(2) Bidder $i$ does not win, i.e. $x < \max_{j \neq i} b_j$, so payoff to $i$ is 0 and can be changed only by stating a bid $b_i > \max_{j \neq i} b_j$, which would then result in a negative payoff (we have neglected the case $x = \max_{j \neq i} b_j$ which under our assumptions occurs with probability 0).

The truth-telling equilibrium is not unique. Consider for example a case with two bidders and with $\omega = 1$. The pair $(\beta_1, \beta_2)$, where $\beta_1(x) = 1$, all $x \in [0, 1]$, and $\beta_2(x) = 0$, all $x \in [0, 1]$, is a Bayesian Nash equilibrium. For all draws of values, bidder 1 wins the object and pays nothing. No alternative strategy will improve the result for bidder 2, given that bidder 1 sticks to $\beta_1$. This equilibrium clearly is not symmetric, and the strategies are not weakly dominating – bidder 1 may have negative payoff if bidder 2 does not stick to $\beta_2$.

The price that the winner with value $x$ must pay in the truth telling equilibrium, the maximum of the other values, depends on the randomly chosen values of the other players, but the average payment of the winner is the conditional mean of the maximal of $n - 1$ random draws, given that this maximum is below $x$.

$$m(x) = \mathsf{E}\left[\max_{j \neq i} X_j \middle| X_j < x_i\right]. \tag{8}$$

In the simple case with 2 bidders having values which are uniformly distributed in $[0, 1]$, we get that the distribution of the second bid, given that it does not exceed $x$, is also uniform with density $\frac{t}{x}$, and its mean is $\frac{x}{2}$. The seller of the object gets the mean value of the smallest of two independent draws. Its distribution is $1 - (1 - t)^2$, its density $2(1 - t)$, and its mean is

$$\int_0^1 2t(1 - t)\, dt = \frac{1}{3}.$$

player $i$ is

$$u_i(b_1, \ldots, b_n) = \begin{cases} x_i - t_i^1(b_1, \ldots, b_n) & \text{if } i \text{ wins the auction,} \\ -t_i^2((b_1, \ldots, b_n) & \text{otherwise,} \end{cases} \tag{9}$$

where $b_1, \ldots, b_n$ are the bids and $t_i^1(b_1, \ldots, b_n)$, $t_i^2(b_1, \ldots, b_n)$ the payments of bidder $i$ in the cases of winning and non-winning, respectively. We may identify the *types* of the players (bidders) with their subjective money value $x_i$ of the object, which then becomes a random variable $X_i$ with values in some interval $[0, \omega]$. For the moment, we shall assume that the values of all bidders are chosen independently and with the same distribution function $F$ for all (in the terminology of auction theory, we assume *private values*). The strategies of a player are the messages or signals or bids that can be chosen. If the player wins the object, then the payoff equals its subjective value minus the payment, and if the player does not win the object, then only the payment (if any) enters the payoff.

Before we proceed to a discussion of equilibria, we notice that the Dutch auction and the first-price auction are strategically equivalent. Indeed, signaling willingness to buy at a certain price in the Dutch auction corresponds to sending this price as a sealed bid, and vice versa. This means that it suffices to study the behavior in one of them, and the sealed bid auction is chosen as the easiest. We can find a sealed-bid counterpart also to the English auction, namely the *second-price auction*. Here the highest bid wins, but the winner pays only the next-highest bid.

A Bayesian Nash equilibrium in an auction specifies for each player and each type of this player a strategy, that is a signal or message or bid, to be chosen. Concentrating on sealed bid auctions, we may write a strategy as a function $\beta : [0, \omega] \to [0, \bar{b}]$, where $[0, \bar{b}]$ is the set of possible bids.

In the model considered here, where all bidders have the same strategic possibilities and are assigned values according to the same probability distribution, it seems natural to consider *symmetric equilibria*, where every bidder uses the same bidding strategy $\beta$. In a symmetric equilibrium, the common bidding strategy must be such that for each bidder $i$ and each possible value $x$ for this bidder, the expected payoff

$$P[\max_{j \neq i} \beta(X_j) < \beta(x)](x - \beta(x))$$

must be maximal. Assuming that $\beta$ is monotonically increasing in $x$, so that higher value results in higher bid, we have that

$$P[\max_{j \neq i} \beta(X_j) < \beta(x)] = G(x),$$

where we have introduced the notation $G$ for the probability distribution function of the maximum of $n - 1$ random values of the other bidders (under our assumptions of independence, we have $G(x) = F^{n-1}(x)$, but the notation is useful also when independence is not assumed).

**Example 6.2. First-price auctions.** Here we have $t_i^1(b_1, \ldots, b_n) = \max_i b_i$, $t_i^2(b_1, \ldots, b_n)$ $= 0$, all $i$. There is no dominating bid strategies in this case, so in order to find a symmetric equilibrium, we must maximize expected payoff

$$G(\beta^{-1}(b))[x - b]$$

in $b$ for each $x$. Assuming differentiability, we may write down first order conditions which reduce to the differential equation

$$G(x)\beta'(x) = xg(x),$$

where $g$ is the density of $G$. This equation has a solution

$$\beta(x) = \frac{1}{G(x)} \int_0^x yg(y)\,dy,$$

giving us the functional form of the symmetric equilibrium. It may be noticed that the optimal bid is the expected value of the next-highest bid. Since the winner pays her bid, she will pay the same amount on average as in the second-price auction (an instance of the *revenue equivalence theorem* discussed below).

In the case of two bidders with values uniformly distributed in $[0, 1]$ we immediately get that the optimal bid given the value $x$ is $\frac{x}{2}$. The seller of the object will get

$$\mathsf{E}\left[\frac{\max\{X_1, X_2\}}{2}\right] = \frac{1}{2}\mathsf{E}[\max\{X_1, X_2\}] = \frac{1}{2} \cdot \frac{2}{3} = \frac{1}{3}$$

as was to be expected.

**3.2. Revenue equivalence.** While the seller has been absent in our formulation of auction designs and the associated games, the reason for considering different auction designs is to choose among them in such a way that seller obtains the best possible result, measured in expected revenue. However, one of the remarkable results of auction theory is the *revenue equivalence theorem*, stating that all standard auctions will yield the same revenue to the seller. To make this statement precise, we define an auction as *standard* if highest bid wins. This is indeed the case for first- and second-price auctions, but an auction where the winner is chosen by a procedure involving randomness, for example with the probabilities of winning proportional to the bids, is not a standard auction.

THEOREM 4 *Consider a standard auction, and let $\beta(\cdot)$ be a symmetric equilibrium with associated expected payment $m(\cdot)$. If $\beta$ is monotonically increasing, differentiable and satisfies $m(0) = 0$, then*

$$m(x) = \int_0^x yg(y)\,dy$$

**Example 6.3. All-pay auctions.** If *all* bidders, not only the winner, pay their bid, we have an *all-pay* auction. Here $t_i^1(b_1, \ldots, b_n) = t_i^2(b_1, \ldots, b_n) = b_i$ for all $i$. One may think of all-pay auctions as a formal representation of corruption: The state must assign some asset to individuals but the government officials can be bribed, and they decide according to the size of the bribe.

Using revenue equivalence we find the symmetric equilibrium bidding strategy immediately as

$$\beta(x) = m(x) = \int_0^x y g(y) \, dy.$$

In the two-bidder case where values are uniformly distributed in $[0, 1]$, we get that

$$\beta(x) = \frac{x^2}{2},$$

and expected revenue is

$$E\left[2\frac{X^2}{2}\right] = \frac{1}{3}.$$

*for all $x \in [0, \omega]$.*

PROOF: Expected payoff when using the equilibrium strategy, given that the value is $x$, can be written as $G(x)x - m(x)$. If the bidder states another value $z$ than the true one, still using the same bidding function, then the bid will be different, namely $\beta(z)$ instead of $\beta(x)$, but also the average payment will differ, being $m(z)$ instead of $m(x)$. Since the optimal bid at $x$ is $\beta(x)$, we have that

$$G(z)x - m(z) \leq G(x)x - m(x), \text{ all } x \in [0, \omega],$$

and first order conditions for a maximum yields

$$m'(x) = g(x)x, \text{ all } x.$$

Solving the differential equation with initial value $m(0) = 0$ gives the desired result.

□

The revenue equivalence theorem is important, since it shows that expected revenue is independent of auction design, at least under the stated assumptions. Upon a closer look, these assumptions are however quite restrictive. First of all, we consider only symmetric equilibria, and there may be other, non-symmetric equilibria. Next, the assumed symmetry of the bidders matters, since bidders with different value distributions will typically have different optimal bidding strategies. But the most crucial of the assumptions is that of private values. Once this assumption is relaxed, auction theory loses much of the simplicity which characterizes the study of private value auctions.

**3.3. Auctions with interdependent values.** When the value that buyers attach to the object are not drawn independently, we cannot identify the type of a bidder with her valuation; indeed, the way in which a potential buyer values the object will depend on how nature has selected all the other buyers. The usual approach to this problem is to introduce values for each bidder as utilities depending on random *signals* received by all bidders. The signals $X_1, \ldots, X_n$ are drawn randomly from some signal set, say $[0, 1]^N$, in accordance with a distribution function $F : [0, 1]^N \to [0, 1]$. Given the realized signals $x_1, \ldots, x_n$, the bidders will attach a value $v_i(x_1, \ldots, x_n)$; however, when bidding for the object, only some of the signals have been observed by the bidder, namely her own and what may be revealed by the choices of the other bidders, given that they adhere to their bidding strategies.

---

**Example 6.4. English auctions.** In the English auction, the bidders leave the auction when the quoted price is considered as too high. In the case of private values, it can be decided once and for all when to leave the auction, and the optimal decision is to leave when the price is equal to the value (not surprising in view of the equivalence of English and second-price auctions when values are private).

In the general case, the fact that some player left the auction gives some additional information about the expected payoff, so the strategies must take this into account. We consider the bidding strategy in a symmetric equilibrium: Let $\beta^N(x_i)$ be the optimal move given the signal $x_i$ when all the individuals are present. Since only this signal is available, a qualified guess about the signals of the others would be that they are also $x_i$, and therefore the bidder should leave the auction when $p = v(x_i, \ldots, x_i)$.

Now only $N - 1$ bidders are left, and each of these bidders have obtained an additional signal, namely that of the bidder, say individual $N$ who left the auction, since the observed $p^N$ at which this happened corresponds to a particular signal $x_N$. Repeating the argument from before, the bidders should leave the auction when $p = v(x_i, x_i, \ldots, x_i, x_N)$, and the new exit of a bidder, say $N-1$, will give the additional observation $x_{N-1}$. Proceeding in this way, we end with a 2-bidder auction, where bidder $i$, for $i = 1, 2$, chooses to leave when $v(x_i, x_i, x_3, \ldots, x_{N-1})$ equals the price. At this moment the auction terminates, and the remaining bidder gets the object at this price.

It is easily shown (see exercises) that the strategies outlined here are indeed optimal, so that they constitute an equilibrium for the English auction.

---

We consider here only a special case, namely that of *common values*, where the valuation functions of all bidders are the same,

$$v_i(x_1, \ldots, x_n) = v(x_1, \ldots, x_n), \text{ all } i \text{ and } (x_1, \ldots, x_n),$$

so that the value assigned to the object will be the same for all bidders once all signals are revealed. At the time of bidding, each bidder knows only her own signal. The best estimate of the value for a player, say player $i$, observing the signal $x$ is therefore $E[v(X)|X_i = x]$. Suppose that the auction design is first-price sealed bid. Then after an announcement that player 1 wins the auction (and all have used the same bidding strategy increasing in signals), there is some additional piece of information, namely that all other bidders had signals which were below $x$, and therefore

$$E[v(X) \mid X_i = x, X_j \leq x, j \neq i] < E[v(X) \mid X_i = x]$$

(given suitable weak assumptions on the distribution $F$). This is the case of *bad news*, the winner realizes that the value is lower than what was assessed originally, and unless this is taken into account when bidding, the winner is hit by *winners curse*, paying more for the object than what it is worth.

## 4   Mechanisms and the revelation principle

In our treatment of auctions, we established the celebrated *revenue equivalence* result which says that the seller will obtain the same expected revenue using any standard auction. Thus, from the seller's point of view, all these auctions are equally good. This will not necessarily be the case if the auction is not standard, so that highest bid will not necessarily be remunerated by getting the object (also, revenue equivalence will not hold if we relax the assumption of private values, but we do not consider this extension here). An analysis of what will happen in this case takes us to the field of *mechanisms*, game forms of a particular type, where the revenue to be obtained may depend on the design.

A mechanism is a generalized auction, in the sense that the individuals do not necessarily submit bids, but send a message which carries some information about preferences and willingness to pay, and where the object may be assigned to a player using a lottery. As with auctions, the rules also determine a payment by one or all of the individuals. We keep the formulation of payoffs being given by a subjective value $x_i$ of the object, drawn independently for each player according to a probability distribution $F$ on an interval $T_i = [0, \omega]$, so that payoff receiving the object with probability $q_i$ and paying $t_i$ is

$$q_i x_i - t_i.$$

For $M_1, \ldots, M_n$ given spaces (of messages to be chosen by the players), we obtain a set of normal form games $\left(N, (M_i)_{i \in N}, (u_i(x_i))_{i=1}^n\right)$, where for each $i \in N$

$$u_i(x_i)(m_1, \ldots, m_n) = q_i(m_1, \ldots, m_n)x_i - t_i(m_1, \ldots, m_n).$$

Here $q_1, \ldots, q_n$ and $t_1, \ldots, t_n$ are the outcome functions of the mechanism, assigning to each $N$-tuple of messages a probability distribution over individuals and a vector of payments. We write the mechanism as $((M_i)_{i \in N}, q, t)$.

As in our treatment of auctions, we may identify the type of individual $i$ with the subjective value $x_i$ of the object. A Bayesian Nash equilibrium for the mechanism considered is therefore a type-dependent choice of message $m_i(x_i)$, for each individual $i$, which yields the best expected payoff given the strategies of the other players,

$$E\left[q_i(m_i(x_i), m_{-i}(X_{-i})x_i - t_i(m_i(x_i), m_{-i}(X_{-i}) \,\middle|\, X_i = x_i\right]$$

$$\geq E\left[q_i(m_i', m_{-i}(X_{-i})x_i - t_i(m_i', m_{-i}(X_{-i}) \,\middle|\, X_i = x_i\right], \quad (10)$$

for all $m_i' \in M_i$, where the expectation is taken over all type arrays $(X_1, \ldots, X_n)$ with $X_i = x_i$.

The reason for studying mechanisms is to select a suitable design for achieving certain results such as maximizing the revenue of a seller or distributing a service in an optimal way given the individual preferences. Searching among all conceivable mechanisms may be a demanding task, but fortunately this search may be restricted to a particular class of mechanisms, called *direct*, where individuals report a type (which may be the true type or any other admissible type). Moreover, given any Bayesian Nash equilibrium of the original mechanism, the direct mechanism can be selected in such a way that reporting the truth is a Bayesian Nash equilibrium and gives the same outcome (and therefore the same payoff). This result is known as the *revelation principle*.

THEOREM 5 (Revelation principle) *Let* $\mathcal{M} = ((M_i)_{i \in N}, q, t)$ *be a mechanism and* $(m_i(\cdot))_{i \in N}$ *a Bayesian Nash equilibrium in* $\mathcal{M}$. *Then there is a direct mechanism* $((T_i)_{i \in N}, q', t')$ *such that true reporting of type for each individual is a Bayesian equilibrium, and*

$$q_i(m_1(x_1), \ldots, m_n(x_n)) = q_i'(x_1, \ldots, x_n), \quad t_i(m_1(x_1), \ldots, m_n(x_n)) = t_i'(x_1, \ldots, x_n) \quad (11)$$

*for each individual* $i$ *and type* $x_i$.

PROOF: Define $q'$ and $t'$ by (11). Then truthtelling in the mechanism $((T_i)_{i \in N}, q', t')$ yields the same outcome as applying the Bayesian equilibrium strategies, and it follows from inserting into (10) that truthtelling is indeed a Bayesian equilibrium in $((T_i)_{i \in N}, q', t')$.  $\square$

Since mechanisms have allocation procedures which may go far beyond the principle that highest bid wins, we cannot expect the revenue equivalence result for auctions to carry over to mechanisms. However, a somewhat weaker result, which in its turn implies revenue equivalence for standard auctions, can be shown to hold.

We use some more notation: For the mechanism $\mathcal{M} = ((T_i)_{i \in N}, q, t)$, let

$$\bar{q}_i(x_i) = E[q_i(x_i, X_{-i})], \, m_i(x_i) = E[t_i(x_i, X_{-i})]$$

be the expected probability of getting the object and the expected payment, given that own value is $x_i$, and

$$U_i(x_i) = \bar{q}_i(x_i)x_i - m_i(x_i)$$

**Example 6.5. Vickrey-Clarke-Groves mechanisms.** Consider the following direct mechanism: The object is assigned to the individual having reported the largest $x_i$, that is

$$q_i(x) = \begin{cases} 1 & x_i = \max_{j \in N} x_j, \\ 0 & \text{otherwise}, \end{cases}$$

and the payments $t_i$ are determined from the reported types $x_1, \ldots, x_n$ by the rule

$$t_i(x_1, \ldots, x_n) = \sum_{j \neq i} q_j(0, x_{-i}) x_j - \sum_{j \neq i} q_j(x_i, x_{-i}) x_j,$$

so that individual $i$ pays the difference between the (reported) value to the others of the allocation in the case that $i$ does not participate and the actual value. Thus, the second-price auction is a special case of the Vickrey-Clarke-Groves mechanism, since for a winner, the value to the others in the case of her non-participation is the next-highest of the reported values, and the value to the others is 0 when the winner is participating.

It is easily seen that truthful reporting of values is a weakly dominating strategy: If the reports of the others (true or false) are $x_j$, $j \neq i$, and the true value for $i$ is $x_i$, then

$$q_i(x_i, x_{-i}) x_i - t_i(x_i, x_{-i})$$

$$= q_i(x_i, x_{-i}) x_i - \sum_{j \neq i} q_j(0, x_{-i}) x_j + \sum_{j \neq i} q_j(x_i, x_{-i}) x_j$$

$$= \sum_{j \in N} q_j(x_i, x_{-i}) x_j - \sum_{j \neq i} q_j(0, x_{-i}) x_j$$

$$\geq \sum_{j \in N} q_j(x_i', x_{-i}) x_j - \sum_{j \neq i} q_j(0, x_{-i}) x_j$$

$$= q_i(x_i', x_{-i}) x_i - \sum_{j \neq i} q_j(0, x_{-i}) x_j + \sum_{j \neq i} q_j(x_i', x_{-i}) x_j = q_i(x_i', x_{-i}) x_i - t_i(x_i', x_{-i}),$$

for all $x_i'$, where we have used the definitions of $q$ and $t$.

It is seen that payment demanded from the individual $i$ does not depend on the report $x_i$, which matters only for the allocation of the object.

---

THEOREM 6 *Let* $((T_i)_{i \in N}, q, t)$, $((T_i)_{i \in N}, q', t')$ *be two direct mechanisms, where* $T_i = [0, \omega_i]$ *for* $i \in N$. *Assume that truthful reporting is an equilibrium strategy in both mechanisms. If* $q = q'$, *so that the allocation parts are the same in the two mechanisms, and if* $m_i(0) = m_i'(0)$ *for each* $i$, *then*

$$\mathsf{E}\left[t_i(x_i, X_{-i})\right] = \mathsf{E}\left[t_i'(x_i, X_{-i})\right],$$

*for each* $i \in N$, *i.e. expected payment of an individual, given her type, is the same in the two mechanisms.*

PROOF: Let $i \in N$ and let $x = (x_i, x_{-i})$, $x' = (x_i', x_{-i})$ be two value vectors differing only in their $i$th component. Since truthful reporting is an equilibrium strategy, we have that

$$\bar{q}_i(x_i')x_i' - m_i(x_i') \geq \bar{q}_i(x_i)x_i' - m_i(x_i). \tag{12}$$

Rewrite the right-hand side of the last equation as

$$\bar{q}_i(x_i)x_i' - m_i(x_i) = [\bar{q}_i(x_i)x_i - m_i(x_i)] + \bar{q}_i(x_i)(x_i' - x_i),$$

and insert into (12) to get the inequality

$$U_i(x_i') \geq U_i(x_i) + \bar{q}_i(x_i)(x_i' - x_i). \tag{13}$$

Using again that truthful reporting is optimal, we can write

$$U_i(x_i) = \max_{x'}\{q_i(x_i')x_i - m_i(x_i')\},$$

showing that $U_i$ is a pointwise maximum of affine functions and therefore convex. Then (13) is the sub gradient inequality for the convex function $U_i$, and $U_i(x_i)$ may be recovered from $\bar{q}_i$ as

$$U_i(x_i) = U_i(0) + \int_0^{x_i} \bar{q}_i(z)\, dz.$$

Using the definition of $U_i$, this can be rewritten as

$$\bar{q}_i(x_i)x_i - m_i(x_i) = -m_i(0) + \int_0^{x_i} \bar{q}_i(z)\, dz$$

or

$$m_i(x_i) = m_i(0) + \bar{q}_i(x_i)x_i - \int_0^{x_i} \bar{q}_i(z)\, dz. \tag{14}$$

It is seen that the expected payment given the type depends only on its value at 0 and on the allocation part of the mechanism. Consequently, if another mechanism has the same allocation part and expected payment at 0, then it has the same expected payment for each individual. $\qquad\square$

Since there are mechanisms which give rise to different expected payments, and consequently to different expected revenue to seller, the logical next problem would be to search for the mechanisms which maximize revenue. This problem was (along with those outlined above) dealt with in Myerson [1981], and we consider some of its aspects in the exercises.

## 5 Sequential equilibria

We return here to the problem of selecting Nash equilibria which are credible in the sense of not being based on unrealistic threats. One of the solutions to this problem was to consider subgame perfect equilibria, which however were defined only for games with perfect information. Indeed, the notion of a subgame is not well-defined when considering a node which together with other nodes form a nontrivial information set for a player. Thus, for games with imperfect information, and indeed for games with incomplete information where the initial choice by nature remains unobserved, there is a need for an extension of the idea of a subgame, so that it may be thought of as starting with an information set instead of a node.

Our treatment of games with incomplete information suggests an approach, namely to attach a *belief* in the form of a probability distribution to each information set where the player should move. Given this subjective assessment of, how likely the individual nodes in the information set are, the player can evaluate behavior strategies over the game starting from this information set against those chosen by the other players.

Formally, let $\Gamma^e = (N, \mathcal{T}, \mathcal{P}, (C(v), h_v)_{v \in V}, (\mathcal{I}_i)_{i \in N}, (p_v)_{v \in P_0}, u)$ be an extensive form game. A *belief system* in $\Gamma^e$ is a map $\mu : \cup_{i \in N} \mathcal{I}_i \to \Delta(\mathcal{I}_i)$ which to each information set $I \in \cup_{i \in N} \mathcal{I}_i$ assigns a probability distribution over the nodes of $I$. If $\sigma = (\sigma_1, \ldots, \sigma_n)$ is an array of behavior strategies, then $\sigma$ is *rational* at the information set $I_i \in \mathcal{I}_i$ for the belief system $\mu$ if

$$\sum_{v \in I_i} \mu(I_i, v) U_i^{\Gamma^e(v)}(\sigma) \geq \sum_{v \in I_i} \mu(I_i, v) U_i^{\Gamma^e(v)}(\sigma_i', \sigma_{-i}) \tag{15}$$

for all mixed strategies $\sigma_i'$ of player $i$. Thus, a strategy array is rational at $I_i \subset P_i$ for the belief system $\mu$ if the strategy $\sigma_i$ maximizes expected payoff given the belief $\mu(I_i, \cdot)$ and the choices $\sigma_{-i}$ of the other players. It is *sequentially rational* for the belief system if it is rational at every information set $I \in \cup_{i \in N} \mathcal{I}_i$.

In the particular case of perfect information, information sets are singletons, and there is only one belief system, namely the probability distribution assigning full weight to the single element in the information set. It is easily seen that a subgame perfect Nash equilibrium is sequentially rational for this trivial belief system.

Returning to the general case, we need some additional conditions on the belief system in order to obtain equilibria which can be considered as a reasonable generalization of the notion of subgame perfectness. One such condition almost suggests itself: The belief at an information set should correspond to the probabilities of reaching the nodes of this information set in a play of the game according to the behavior strategies $\sigma_1, \ldots, \sigma_n$. This will work at least for those information sets that are reached with nonzero probabilities in the course of the play, so that the belief system $\mu$ should satisfy the condition

$$\mu(I, v) = \frac{P\{v \mid \sigma\}}{\sum_{v \in I} P\{v \mid \sigma\}} \tag{16}$$

for each information set $I$ and node $v \in I$, and in this case we say that $\mu$ is *consistent* with $\sigma$. It may be noticed that the set of belief systems is defined as $\prod_{I \in \cup_{i \in N} \mathcal{I}_i} \Delta(I)$, which is a compact set, so that the limit $\mu^0$ of a sequence $(\mu_t)_{t=1}^{\infty}$ of belief systems is again a belief system, but it may not satisfy the consistency condition (16).

If the belief system $\mu$ satisfies (16) for the strategy array $\sigma$, and the latter are sequentially rational for $\mu$, then the pair $(\mu, \sigma)$ can be used as a generalized subgame perfect equilibrium, at least under the condition that $P\{U \mid \sigma\} = \sum_{v \in I} P\{v \mid \sigma\}$ for all information sets $I \in \cup_{i \in N} \mathcal{I}_i$. The information sets which are never visited during the play are however not irrelevant, in particular if we want to keep the intuitive reasoning behind the selection of equilibria which would not be upset if the players make small mistakes. The possibility of such mistakes means that no information sets can be totally ruled out.

To capture this situation, we shall consider pairs $(\mu, \sigma)$ which are limits of a sequence of belief systems and strategy arrays such that the beliefs are consistent and the strategies sequentially rational, and all information sets are visited. This leads to the notion of a *sequential equilibrium*:

DEFINITION 5 *Let* $\Gamma^e = (N, \mathcal{T}, \mathcal{P}, (C(v), h_v)_{v \in V}, (\mathcal{I}_i)_{i \in N}, (p_v)_{v \in P_0}, u)$ *be an extensive form game. A sequential equilibrium in* $\Gamma^e$ *is a pair* $(\mu, \sigma)$, *where* $\mu$ *is a belief system and* $\sigma$ *a strategy array, so that* $(\mu, \sigma)$ *is the limit of a sequence* $(\mu^t, \sigma^t)_{t=1}^{\infty}$ *of pairs satisfying for each t the conditions:*

 (i) $\sigma^t$ *is an array of completely mixed behavior strategies which is sequentially rational for* $\mu^t$,

 (ii) $\mu^t$ *is consistent with* $\sigma^t$.

What is obtained here is indeed an extension of subgame perfectness for games with perfect information, also with regard to the role played by nodes not visited in the course of the play. This is shown by the following result, where we need the property of perfect recall (cf. Section 5.4).

THEOREM 7 *Let* $\Gamma^e = (N, \mathcal{T}, \mathcal{P}, (C(v), h_v)_{v \in V}, (\mathcal{I}_i)_{i \in N}, (p_v)_{v \in P_0}, u)$ *be an extensive form game. If* $\sigma$ *is a Nash equilibrium in completely mixed behavior strategies, and* $\mu$ *is consistent with* $\sigma$, *then* $(\mu, \sigma)$ *is a sequential equilibrium.*

PROOF: If $\sigma$ is completely mixed, then $P\{I \mid \sigma\} > 0$ for every information set $I$, and (16) induces a well-defined probability distribution on each information set $I$. Reasoning as in the proof of Theorem 5.5, we have that since $\sigma$ is a Nash equilibrium, $\sigma_i$ must be the best choice at each $v \in I_i$ given $\mu$, so that $\sigma$ is sequentially rational for the belief system $\mu$.                                                                                      □

It is intuitive that sequential equilibria should be Nash equilibria, but it is not entirely obvious, so a proof must be given. Here we shall need that the game has perfect recall.

THEOREM 8 *Let* $\Gamma^e = (N, \mathcal{T}, \mathcal{P}, (C(v), h_v)_{v \in V}, (\mathcal{I}_i)_{i \in N}, (p_v)_{v \in P_0}, u)$ *be an extensive form game with perfect recall. If* $(\mu, \sigma)$ *is a sequential equilibrium, then* $\sigma$ *is a Nash equilbrium.*

PROOF: Let $(\mu^t, \sigma^t)$ be an element in a sequence converging to $(\mu, \sigma)$ such that $\sigma^t$ is completely mixed and sequentially rational for $\mu^t$ and $\mu^t$ is consistent with $\sigma^t$. Choose a player $i$ arbitrarily. We need to show that the strategy $\sigma_i$ cannot be improved by any other choice given the beliefs at the moment where $i$ moves for the first time.

Consider all information sets $I_i$ which are *initial* for player $i$ in the sense that if a path in $\mathcal{T}$ passes through a path $v \in I_i$ then segment $[v_0, v]$ does not intersect $P_i$ (so that $v$ is a first move of $i$). Since $\Gamma^e$ has perfect recall, every path which intersects $P_i$ must also intersect an initial information set for $i$. If we denote by $V^0$ the subset of terminal nodes which are connected to the root $v_0$ by paths that do not intersect $P_i$, then we can write the expected payoff for player $i$ at the strategy array $\sigma$ as

$$U_i(\sigma^t) = \sum_{I_i \ initial} \sum_{v \in I_i} \mu^t(I_i, v) U_i^{\Gamma^e(v)}(\sigma^t) + \sum_{v \in V^0} P\{v \mid \sigma^t\} u_i(v). \qquad (17)$$

Since $\sigma^t$ is sequentially rational for $\mu^t$, the inequality (15) must hold for every information set $I_i$, and since the probability of selecting a play which does not intersect $P_i$ is independent of the strategy choices of player $i$, we get that the expression in (17) is no smaller than

$$U_i(\sigma'_i, \sigma^t_{-i}) = \sum_{I_i \ initial} \sum_{v \in I_i} \mu^t(I_i, v) U_i^{\Gamma^e(v)}(\sigma'_i, \sigma^t_{-i}) + \sum_{v \in V^0} P\{v \mid (\sigma'_i, \sigma^t_{-i})\} u_i(v). \qquad (18)$$

for all behavior strategies $\sigma'_i$ of player $i$. Since $\sigma^t$ converges to $\sigma$, we obtain from (17) and (18) that

$$U_i(\sigma_i) \geq U_i(\sigma'_i, \sigma_{-i})$$

for every behavior strategy $\sigma'_i$ of player $i$, so that $\sigma$ is indeed a Nash equilibrium.

$\square$

## 6 Global games

While the general model of incomplete information, with a hierarchy of beliefs for each player, is sufficiently complex to be almost intractable in realistic situations, certain simple cases may be both tractable and interesting. One such case is that of *global games* as introduced by Carlsson and van Damme [1993] and further developed in Morris and Shin [2003], which is followed in the exposition below.

We consider a situation with a continuum of players, each having to choose an action $a \in \{0, 1\}$. The payoff depends on an unknown parameter for which each agent received a signal $x \in \mathbb{R}$. All players have the same payoff function, namely $u : \{0, 1\} \times [0, 1] \times \mathbb{R} \to \mathbb{R}$, where $u(a, \lambda, x)$ is the payoff if action is $a$, the signal received is $x$, and the fraction of players choosing $a = 1$ is $\lambda$. Since there are only two possible actions, we may analyze net payoff

$$\Pi(\lambda, x) = u(1, \lambda, x) - u(0, \lambda, x),$$

so that formally the action does not enter.

We assume that the unknown parameter or state $\theta$ is uniformly distributed over $\mathbb{R}$ (representing initial ignorance of players; this is not ordinary probability distribution, but it is only used to find conditional probabilities). The signal of any player $i$ is

$$x_i = \theta + \sigma \varepsilon_i,$$

where $\sigma > 0$ and $\varepsilon_i$ is random with continuous density $f$ on $\mathbb{R}$. If player $i$ observes the signal $x_i$, then by Bayes' formula the posterior probability is a constant multiplied by

$$f\left(\frac{x_i - \theta}{\sigma}\right)$$

and the constant can be found as the integral of this quantity, which is $\dfrac{1}{\sigma}$.

We add some assumptions to make sure that the model is well-behaved: First of all, we shall assume that $\Pi(\lambda, x)$ is non-decreasing in $\lambda$ ("action monotonicity") and in $x$ ("state monotonicity"). Next, we assume that there is a unique $\theta^*$ such that

$$\int_0^1 \Pi(\lambda, \theta^*) d\lambda = 0; \tag{19}$$

we may interpret $\theta^*$ as the state where the player is indifferent between the two actions if $\lambda$ has the uniform density in $[0, 1]$. Moreover, we assume that there are $\underline{\theta}$ and $\overline{\theta}$ such that $\Pi(\lambda, x) < 0$ when $x \le \underline{\theta}$ and $\Pi(\lambda, x) > 0$ when $x \ge \overline{\theta}$. Finally, we shall need an assumption of more technical character, namely that

$$\int_0^1 \Pi(\lambda, x) g(\lambda) \, d\lambda$$

is continuous as a function of densities $g$ on $[0, 1]$ and signals $x$ (here the space of density functions has the weak topology for probability measures on $[0, 1]$).

We show that there is a unique symmetric Bayesian Nash equilibrium in this game, actually the result is somewhat stronger.

THEOREM 9 *In the global game model defined above, there is an essentially unique strategy $s$ which emerges by iterated elimination of dominated strategies, namely that which assigns the action $s(x) = 0$ to all signals $x < \theta^*$ and $s(x) = 1$ to $x > \theta^*$ (and either 0 or 1 to the signal $\theta^*$), where $\theta^*$ is the state defined in (19).*

PROOF: The expected payoff at signal $x$, given that all players with signals $\le y$ will choose action 0, is

$$\overline{\Pi}(x, y) = \int_{-\infty}^{\infty} \Pi\left(1 - F\left(\frac{y - \theta}{\sigma}\right), x\right) \frac{1}{\sigma} f\left(\frac{x - \theta}{\sigma}\right) d\theta.$$

Define for each nonnegative integer $n$ the pair of numbers $(\underline{\xi}_n, \overline{\xi}_n)$ inductively by $\underline{\xi}_0 = -\infty, \overline{\xi}_0 = \infty$, and

$$\underline{\xi}_{n+1} = \min \left\{x \,\middle|\, \overline{\Pi}(x, \underline{\xi}_n) = 0\right\}, \quad \overline{\xi}_{n+1} = \max \left\{x \,\middle|\, \overline{\Pi}(x, \overline{\xi}_n) = 0\right\}.$$

We claim that a strategy $s$ survives $n$ rounds of deletion of dominated strategies if and only if

$$s(x) = \begin{cases} 0 & x < \underline{\xi}_n, \\ 1 & x > \overline{\xi}_n. \end{cases}$$

To prove the claim, we use induction: For $n = 0$ the claim is trivially true; suppose that it holds for $n \geq 0$. Then, if action 0 is a best reply to some strategy remaining after the previous rounds of deletions, it must be a best reply to a strategy which chooses 0 below $\overline{\xi}_n$ and 1 above $\overline{\xi}_n$. But $\overline{\xi}_{n+1}$ is exactly the highest signal for which this occurs. The argument for $\underline{\xi}_{n+1}$ is similar.

The sequences $(\underline{\xi}_n)_{n=0}^{\infty}$ and $(\overline{\xi}_n)_{n=0}^{\infty}$ are easily seen to be respectively increasing and decreasing, so that they converge to some $\underline{\xi}$ and $\overline{\xi}$. By our assumptions, we have that $\overline{\Pi}(\underline{\xi}, \underline{\xi}) = 0$ and $\overline{\Pi}(\overline{\xi}, \overline{\xi}) = 0$.

---

**Example 6.6.** The following is a simple version of the Diamond-Dybvig model of financial intermediation. There is a continuum of depositors having a contract normalized to the amount of 1 with the bank, giving them the right to receive either $c_1 > 1$ at $t = 1$ (if "impatient") or $c_2$ at $t = 2$ (if "patient"). The bank uses the deposits for investing in a project with a return $R(\theta)$ at $t = 2$ depending on a parameter $\theta$. If the proportion of impatient depositors is $\pi$, then the remaining depositors face the decision problem of whether to withdraw at $t = 1$ or to wait until $t = 2$, running the risk that the bank may go bankrupt at $t = 1$ if the withdrawals exceed the reserves.

We consider only the situation of the $1 - \pi$ patient depositors. If $\theta$ is known and if $R(\theta) > 1$, then the optimal choice of $c_1$ maximizes

$$\pi U(c_1) + (1 - \pi)U\left(\frac{1 - \pi c_1}{1 - \pi}R(\theta)\right),$$

where $U$ is the common utility of the depositors. If, however, $\theta$ is not known, we have a global game with the two actions $a = 0$ (withdrawal at $t = 1$) and $a = 1$ (withdrawal at $t = 2$), and with the payoffs as in the table, where $\lambda$ is the proportion of depositors withdrawing at $t = 1$.

| | $\lambda \leq \dfrac{c_1 - 1}{c_1(1 - \pi)}$ | $\lambda \geq \dfrac{c_1 - 1}{c_1(1 - \pi)}$ |
|---|---|---|
| $a = 0$ | $\dfrac{1 - \pi c_1}{(1 - \pi)(1 - \lambda)c_1}$ | $c_1$ |
| $a = 1$ | $0$ | $\left(c_1 - \dfrac{c_1 - 1}{\lambda(1 - \pi)}\right)R(\theta)$ |

**Example 6.6, continued.**

Assuming that $R(\theta)$ is increasing in $\theta$, we get that for small enough $\theta$ it is a dominating strategy to withdraw at $t = 1$. Suppose that we also know that for large enough $\theta$, withdrawing at $t = 2$ is dominating. We then have that

| | $\lambda \leq \dfrac{c_1 - 1}{c_1(1 - \pi)}$ | $\lambda \geq \dfrac{c_1 - 1}{c_1(1 - \pi)}$ |
|---|---|---|
| $u(1, \lambda, \theta)$ | $U(0)$ | $U\left(\left(c_1 - \dfrac{c_1 - 1}{\lambda(1 - \pi)}\right)R(\theta)\right)$ |
| $u(0, \lambda, \theta)$ | $U\left(\dfrac{1}{1 - \lambda(1 - \pi)}\right)$ | $U(c_1)$ |

so that

$$
\Pi(\lambda, \theta) = \begin{cases} U(0) - U\left(\dfrac{1}{1 - \lambda(1 - \pi)}\right) & \lambda \leq \dfrac{c_1 - 1}{c_1(1 - \pi)} \\ U\left(\left(c_1 - \dfrac{c_1 - 1}{\lambda(1 - \pi)}\right)R(\theta)\right) - U(c_1) & \lambda \geq \dfrac{c_1 - 1}{c_1(1 - \pi)}. \end{cases}
$$

The threshold value $\theta^*$ of the state parameter can now be found from (19), solving the equation

$$
\int_0^{\lambda^*} [U(0) - U(s_1)]\, d\lambda + \int_{\lambda^*}^1 [U(s_2 R(\theta)) - U(c_1)]\, d\lambda = 0, \tag{20}
$$

with $\lambda^* = \lambda \leq \dfrac{c_1 - 1}{c_1(1 - \pi)}$, $s_1 = \dfrac{1}{1 - \lambda(1 - \pi)}$ and $s_2 = c_1 - \dfrac{c_1 - 1}{\lambda(1 - \pi)}$. Implicit differentiation in (20) gives that $\theta^*$ is increasing in $c_1$.

We now show that $\theta^*$ as defined above is the only solution to $\overline{\Pi}(x, x) = 0$. For this, we introduce the probability that $\leq \lambda$ of the other players observe a signal $\geq y$ given the signal $x$ observed by the player,

$$
P(\lambda, x, y) = \int_{-\infty}^{y - \sigma F^{-1}(1 - \lambda)} \frac{1}{\sigma} f\left(\frac{x - \theta}{\sigma}\right) d\theta.
$$

Writing $z = \dfrac{x - \theta}{\sigma}$, this integral can be rewritten as

$$
\int_{\frac{x - y}{\sigma} + F^{-1}(1 - \lambda)}^{\infty} f(z)\, dz = 1 - F\left(\frac{x - y}{\sigma} + F^{-1}(1 - \lambda)\right).
$$

For $x = y$, we get that $P(\lambda, x, x) = \lambda$, so that it is the distribution function of the uniform distribution. Therefore,

$$
\overline{\Pi}(x, x) = \int_0^1 \Pi(\lambda, x)\, d\lambda,
$$

and from uniqueness of $\theta^*$, we conclude that $\overline{\Pi}(x, x) = 0$ implies that $x = \theta^*$.    □

Summing up, we see that there is a rather convincing solution to this game, even though the players know very little about each other. This lack of precise knowledge seems even to be helpful in predicting outcomes, since it allows for a unique solution to the problem, whereas multiple equilibria will arise in models where the players are assumed to know more about each other.

## 7 Problems

**1.** Show that the bidding strategy $\beta(x)$ defined by

$$\beta(x) = \frac{1}{G(x)} \int_0^y g(y)dy,$$

where $G$ is the probability distribution function of the highest of $N-1$ independent draws with distribution $F$, and $g$ its density function, defines a symmetric equilibrium of the first-price auction under private values.

Use this to show that first and second price auctions have the same expected payment $m(x)$ in a symmetric equilibrium.

**2.** (Adverse selection in insurance) Suppose that a non-profit insurance company must provide a community with insurance against a loss of size $L$. The individuals are assumed to have different probability $p$ of loss, distributed uniformly between $p_{\min}$ and $p_{\max}$, and all individuals are risk averse. The individual loss probabilities cannot be observed by the company (or alternatively, the company is not allowed to use this information), so it must fix a premium which is the same for all.

Draw the curve representing the reimbursement per individual if those taking insurance are exactly the individuals with probability $\geq p$. Show that if this curve intersects the one showing the willingness to pay of insurance depending on $p$, then there will be individuals opting out of the insurance scheme.

**3.** (The Principal-agent model) An employer and a worker must negotiate a contract. The worker chooses an effort level $e$ from a finite set, which results in a probability distribution $p(e)$ over a finite set of money outcomes $\{y_1, \ldots, y_r\}$. Effort cannot be observed by the employer, so the contract takes the form of transfers $t_1, \ldots, t_r$ depending on the outcome. The employer wants to maximize expected profits and the utility of the worker has the quasi-linear form $v(t) - w(e)$. The participation constraint for the agent demands that expected utility should be nonzero.

Show that if the contract $(t_1^*, \ldots, t_r^*)$ with associated effort level $e^*$ maximizes the employer profits and is incentive compatible, then

$$\frac{1}{v'(t_h^*)} = -\lambda - \sum_{e' \in K(e^*)} \mu(e') \frac{p_h(e^*) - p_h(e')}{p_h(e^*)}, h = 1, \ldots, r,$$

where $K(e^*)$ is the set of effort levels which are as good as $e^*$ for the agent, and $\lambda$ and $\mu(e'), e' \in K(e^*)$ are positive constants.

**4.** Efficiency in auctions. Consider a two person auction with interdependent values, where each agent observes signals and both individuals have the same value function $v(x, y)$, where $x$ is the own signal and $y$ is the other agent's signal.

Show that in a second price auction, bidding $v(x, x)$ is a Bayesian Nash equilibrium strategy.

Show that for suitable specification of $u$, the second price auction may not be efficient (there are pairs of signals such that the winner has lowest value of the object).

Show that if $v(x, y)$ satisfies a *single-crossing* condition,

$$\frac{\partial v}{\partial x}(x, y) \geq \frac{\partial v}{\partial x}(y, x)$$

for all $(x, y)$ (my value depends more on my signal than does the other agent's value), then the auction is efficient.

**5.** (The "lemon" problem, cf. Akerlof [1970]) An individual wants to buy an object from a seller. The quality $q$ of the commodity is known to the seller but unknown to the buyer who knows only that it is uniformly distributed in the interval $[0, Q]$. The seller has utility $\tau_S q$ if retaining the object and $p$ if selling it at price $p$, whereas the buyer has utility $\tau_B q - p$ if buying the object at the price $p$ and 0 otherwise, and the constants $\tau_S, \tau_B$ are known to both.

Initially the rules for trading in the market are such that the buyer proposes a price, and the seller either accepts or rejects. Find a Bayesian Nash equilibrium.

Now the rules are changed so that the seller proposes a price and the buyer can accept or reject. Describe the equilibria of this new game.

# Chapter 7

# Choosing Among Nash Equilibria

## 1  Introduction

Having dealt at length with Nash equilibria in the previous chapters, it is time for a closer look at the concept. We have already seen that there are cases where some Nash equilibria may be discarded, since they are based on threats which would not be carried out if it came to the matter. This situation gave rise to the concept of a subgame perfect Nash equilibrium; in other words, we could *select* some Nash equilibria as more serious candidates for a solution to the game than the remaining ones. However, the method works only in extensive form games, and we would like to extend the selection procedure so as to cover general normal form games, or at least normal form games with finite sets of pure strategies.

Selecting the intuitively reasonable Nash equilibria in a given normal form game may however not be a simple matter, and attempts at finding a simple – or even less simple – definition of such a selection fail. In this situation it seems reasonable to retrace our steps and consider in more detail what should be demanded from a "reasonable" equilibrium. This procedure leads us to the consideration of stable sets of equilibria which satisfy many of the desiderata; as was to be expected, the simplicity and to some extent the intuitive content is lost in the process.

In the final section of this chapter, we consider a somewhat different approach towards selection of equilibria. Instead of setting up criteria by which the analyst may pick the right equilibria, we are interested in cases where the players themselves might find ways to select a particular equilibrium strategy array, without violating the non-cooperative character of the game. This leads to the study of *preplay communication;* we shall consider the so-called cheap-talk equilibria, where the given game is extended by adding one or several phases of communication, sending messages which formally have no influence on the final payoff, once the original game is played. It is seen that the approach towards selection of equilibria has shifted; the original game has been inserted in another game and this game may have many other Nash equilibria, but some of these may be of special interest since they are related to desirable equilibria in the original game.

The study of preplay communication can be further refined so as to consider a pure signaling game with a sender and a receiver, and this is the topic of the final part of the section, where we consider the model of strategic information transmission by Crawford and Sobel [1982].

## 2   Some Nash equilibria are better than others

Let $\Gamma = (N, (S_i)_{i \in N}, (u_i)_{i \in N})$ be a game, where the sets $S_i$ of strategies are finite, with $|\Sigma_i| = m_i$, $i \in N$. We shall be interested in solution concepts which pick some, but not necessarily all, Nash equilibria in $\Gamma$. The need for selecting among the Nash equilibria can be seen from simple examples such as Example 7.1.

---

**Example 7.1.** Consider the two-player game

|       |   L    |   R    |
|-------|--------|--------|
| **T** | $(1,1)$ | $(0,0)$ |
| **B** | $(0,0)$ | $(0,0)$ |

There are two Nash equilibria in pure strategies, namely $(T, L)$ and $(B, R)$, but only $(T, L)$ seems reasonable, while $(B, R)$ appears as unstable – if a player expects the other player to choose her other pure strategy with some probability, however small, it would be better to choose the other pure strategy.

---

In this section, we take a look at some early attempts to single out what could be considered as reasonable equilibria. In order to exclude those Nash equilibria which are outright unreasonable, we make use of the notion of a best response introduced earlier. We need also some additional terminology.

A pure strategy $s_i$ of player $i$ is a best reply to the array $\sigma \in \times_{i \in N} \Delta(S_i)$ of mixed strategies if

$$u_i(s_i, s_{-i}) \geq u_i(s'_i, s_{-i}), \text{ all } s'_i \in \Sigma_i.$$

The set of best responses of player $i$ to $\sigma$ is denoted $B_i(\sigma)$, and $B(\sigma) = B_1(\sigma) \times \cdots \times B_n(\sigma)$. Finally, for $\sigma_i$ a mixed strategy of player $i$, the carrier for player $i$ of $\sigma_i$ is the set

$$C_i(\sigma_i) = \{s_i \in \Sigma_i \mid \sigma_i(s_i) > 0\}$$

of all pure strategies which have positive weight in $\sigma_i$, and the carrier of the strategy array $\sigma$ is $C(\sigma) = C_1(\sigma_1) \times \cdots \times C_n(\sigma_n)$.

Following Harsanyi [1973], we define a *quasi-strong equilibrium* in $\Gamma$ as a strategy array $\sigma$ which satisfies the condition

$$C(\sigma) = B(\sigma). \tag{1}$$

Thus, $\sigma$ is a quasi-strong equilibrium if it assigns positive weight to exactly those pure strategies which are best responses. It is easily checked that quasi-strong equilibria are Nash; indeed, if $C(\sigma) \subseteq B(\sigma)$, then each of the pure strategies in the carrier of $\sigma$ is a best reply, which is exactly the Nash equilibrium condition. Thus, (1) is a stability condition which resembles that of Nash equilibria, but it goes further; in a Nash equilibrium, we have that $C(\sigma) \subseteq B(\sigma)$ but we do not necessarily have equality.

The equilibrium $(B, R)$ in Example 7.1 is not a quasi-strong equilibrium, since the set of best responses of player 1 to $R$ is $\{T, B\}$, and the carrier of $(B, R)$ is just $\{B\}$ for player 1 and $\{R\}$ for player 2. Thus, the concept does quite well in this particular example, but not so in other, equally simple examples, cf. Example 7.2.

---

**Example 7.2.** Consider the game

|   | L | R |
|---|---|---|
| T | $(1, 1)$ | $(1, 1)$ |
| B | $(1, 1)$ | $(0, 0)$ |

Here $(T, L)$ is not quasi-strong, since $C(T, L) = (T, L)$, where as $B(T, L) = \{T, B\} \times \{L, R\}$. Moreover, if we take any mixed strategy (say of player 2) which puts nonzero weight on both strategies, then $L$ is the only pure strategy which is a best response. Arguing similarly for player 1 we see that there are no quasi-strong equilibria in this example, a somewhat unfortunate situation.

---

Since the quasi-strong equilibria are not in general well-behaved, they are not serious candidates for the 'reasonable' Nash equilibrium solution. We have to add some new aspects to the discussion.

Before leaving this topic, a brief comment on the terminology: The equilibrium was called quasi-strong since Harsanyi had introduced another concept of a strong equilibrium, namely a mixed strategy $\sigma$ such that $C(\sigma) = \{\sigma\} = B(\sigma)$. This condition is very restrictive since there should be pure strategy Nash equilibria, which is very often not the case. This notion of a strong equilibrium should not be confused with another notion, dealing with cooperative games and to be encountered later (cf. Chapter 10 below).

## 3  Perfect and proper equilibria

**3.1. Perfect equilibria.** It is natural to begin the discussion of selections in the context of normal form games with what comes closest to the notion of a subgame perfect equilibrium in an extensive form game. A suitable formalization of this may be obtained when we interpret the notion of subgame perfectness as a consequence of the assumption that players *make errors,* known in the literature as the *trembling hand hypothesis.* Even though players have chosen specific strategies, they may end up doing something different from what these strategies prescribe, not due to an intentional deviation but just as an error.

In our context of games with a finite number of pure strategies, we introduce a *tremble* as a vector $\eta \in \mathbb{R}_{++}^{m_1+\cdots+m_n}$, assigning a number $\eta_i(s_i) > 0$ to each pure strategy $s_i \in S_i$ of player $i$, for $i = 1,\ldots,n$. We may think of $\eta_i(s_i)$ as the probability that $s_i$ is chosen by player $i$ even when some other choice was intended.

For $\eta \in \mathbb{R}_{++}^{m_1+\cdots+m_n}$ a tremble, define the game $\Gamma^\eta = (N, (\Sigma_i^\eta)_{i\in N}, (u_i)_{i\in N})$ with strategy sets

$$\Sigma_i^\eta = \{\sigma_i \in \triangle(S_i) \mid \sigma_i(s_i) \geq \eta_i(s_i), \text{ all } s_i \in S_i\}$$

(that is all the mixed strategies for player $i$ in $\Gamma$ which are such that the pure strategy $s_i$ is chosen with probability at least $\eta_i(s_i), s_i \in S_i$), and where $u_i$ is the same as in $\Gamma$ (or, to be precise, its mixed extension). We notice in passing that the game $\Gamma^\eta$ satisfies the standard conditions for existence of a Nash equilibrium, at least for sufficiently small vectors $\eta$, so that all $\Sigma_i^\eta$ are nonempty, convex and compact. Thus, each $\Gamma^\eta$ has at least one Nash equilibrium $\sigma^\eta$, which by the definition of $\Gamma^\eta$ is *completely mixed* (in the sense that $\sigma_i^\eta(s_i) > 0$ for all $s_i \in S_i$).

We can now define perfectness as something which retains the equilibrium properties even if players make small errors.

DEFINITION 1 *A Nash equilibrium $\sigma$ in $\Gamma$ is a perfect equilibrium if there is a sequence $(\eta_t)_{t=1}^{\infty}$ of trembles converging to the zero vector, and Nash equilibria $\sigma^{\eta_t}$ in $\Gamma^{\eta_t}$ for each $t$, such that the sequence $(\sigma^{\eta_t})_{t=1}^{\infty}$ converges to $\sigma$.*

Since Nash equilibria exist for each tremble $\eta$, the question of whether there exists perfect equilibria (for a game of the type we consider here) can be answered in the affirmative: Choose a sequence of trembles converging to 0, and for each $\eta$ in the sequence, choose a Nash equilibrium in $\Gamma^\eta$. The sequence of Nash equilibria obtained in this way is not necessarily convergent, but since it consists of mixed strategies and thus belongs to a compact set, it has a convergent subsequence. This subsequence may then be used instead of the original sequence, and its limit is indeed a perfect equilibrium.

In the above definition, we have used arbitrary trembles. It may be useful to restrict attention to trembles which are more easy to work with, namely such where all $\eta_i(s_i)$ have the same size $\varepsilon$. We define an $\varepsilon$-*perfect equilibrium* in $\Gamma$ as a mixed strategy combination $\sigma$ which is completely mixed and satisfies

$$u_i(s_i, \sigma_{-i}) < u_i(s_i', \sigma_{-i}) \text{ implies } \sigma_i(s_i) \leq \varepsilon$$

for all pure strategies $s_i$ and $s'_i$ of player $i$, $i \in N$. Thus, if the pure strategy $s_i$ is inferior to the pure strategy $s'_i$, then $i$ will put a very low weight on $s_i$ (since using it reduces average payoff to $i$), at least below $\varepsilon$.

We have the following characterization of perfect equilibria:

THEOREM 1 *Let $\sigma$ be a strategy array in $\Gamma$. Then the following are equivalent:*

(i) *$\sigma$ is a perfect equilibrium,*

(ii) *$\sigma$ is a limit of $\varepsilon$-perfect equilibria, for $\varepsilon$ going to 0,*

(iii) *$\sigma$ is the limit of a sequence $(\sigma^\varepsilon)_{\varepsilon \to 0}$ of completely mixed strategies, such that $\sigma$ is best response to $\sigma^\varepsilon$ (that is $\sigma \in B(\sigma^\varepsilon)$) for $\epsilon$ small enough.*

PROOF: (i)⟹(ii). If $\sigma$ is a perfect equilibrium, there is a sequence of Nash equilibria $\sigma^\eta$ in $\Gamma^\eta$ converging to $\sigma$ for $\eta$ going to 0. For each tremble $\eta$, let

$$\varepsilon^\eta = \max_{s_i \in S_i, i \in N} \eta_i(s_i)$$

be the largest of all the numbers $\eta_i(s_i)$. Clearly $\sigma^\eta$ is completely mixed, and since it is a Nash equilibrium in $\Gamma^\eta$, we have that a pure strategy $s_i$ for $i$ which is inferior to another pure strategy $s'_i$ must have as small probability as possible in $\sigma_i^\eta$, meaning that $\sigma_i^\eta(s_i) \le \varepsilon^\eta$. We conclude that $\sigma^\eta$ is a $\varepsilon^\eta$-equilibrium, and now (ii) follows.

(ii)⟹(iii). Let $s_i$ be a pure strategy for $i$ such that $\sigma_i(s_i) > 0$. Since $\sigma$ is a limit of the mixed strategies $\sigma^\varepsilon$, we have that $\sigma_i^\varepsilon(s_i) > \varepsilon$ for $\varepsilon$ small enough. But then $s_i$ is a best response to $\sigma^\varepsilon$, since otherwise $\sigma_i^\varepsilon(s_i) \le \varepsilon$. But this shows that $\sigma$ is a best reply to $\sigma^\varepsilon$, as it has positive weight only on pure strategies which are best responses.

(iii)⟹(i). We show that $\sigma^\varepsilon$ are equilibria in $\Gamma^\eta$ for suitable trembles $\eta$. For each $i$ and $s_i \in S_i$, if $\sigma_i(s_i) = 0$, choose $\eta_i^k = \sigma_i^\varepsilon(s_i)$, and if $\sigma_i(s_i) > 0$, let $\eta_i(s_i)$ be an arbitrary very small positive number, e.g. $\eta_i(s_i) = \min_{s'_i \in S_i} \sigma_i^\varepsilon(s'_i)$. If $s_i$ is not a best response of $i$ to $\sigma^\varepsilon$, then $\sigma_i(s_i) = 0$, so by our definition of $\eta$, we have that $\sigma_i^\varepsilon(s_i) = \eta_i(s_i)$. But this means that $\sigma^\varepsilon$ is an equilibrium in $\Gamma^\eta$, and we are done. □

It is seen that for an equilibrium to be perfect, it should be the limit of a sequence of equilibria obtained for *some* tremble. Another, seemingly related concept is that of strict perfectness.

DEFINITION 2 *An equilibrium $\sigma$ in $\Gamma$ is strictly perfect if for all sequences of trembles $\eta$ converging to 0 there is a sequence of equilibria $\sigma^\eta$ in $\Gamma^\eta$ converging to $\sigma$.*

The concept of strict perfectness is quite restrictive, actually so as to exclude existence in some otherwise rather standard games. Consider the game $\Gamma$ with matrix

|   | L | M | R |
|---|---|---|---|
| T | (1,1) | (1,0) | (0,0) |
| B | (1,1) | (0,0) | (1,0) |

Here both $(T, L)$ and $(B, L)$ are pure strategy equilibria. Also notice that both $M$ and $R$ are dominated, so all equilibria must have player 2 choosing $L$. Assume that $s$ is a strictly perfect equilibrium in $\Gamma$. Now choose a sequence of trembles such that $\eta_2(M) > \eta_2(R)$; in each equilibrium $\sigma^\eta$ of $\Gamma^\eta$ we then have that $B$ is not a best response to $\sigma^\eta$, meaning that $\sigma_1^\eta(B) = 0$, and in the limit we get that $\sigma_1(B) = 0$. Choosing another sequence $(\hat\eta)$ with $\hat\eta_2(M) < \hat\eta_2(R)$ we similarly obtain that $\sigma_1(T) = 0$. But we cannot have a mixed strategy for player 1 with zero weight on all the pure strategies, and we conclude that there is no strictly perfect equilibrium.

That the notion of perfectness does not solve all problems which initiated the search for a selection, can be seen from the Example 7.3.

---

**Example 7.3.** Consider the game

|   | L | C | R |
|---|---|---|---|
| T | $(1, 1)$ | $(0, 0)$ | $(-1, -2)$ |
| M | $(0, 0)$ | $(0, 0)$ | $(0, -2)$ |
| B | $(-2, -1)$ | $(-2, 0)$ | $(-2, -2)$ |

This is the game of the Example 7.1, where we have added a dominated strategy for each player. Now the strategy array $(M, C)$, which would not have been a perfect equilibrium without the new strategies, has become perfect. Indeed, for a sequence of trembles with higher value on $R$ than on $L$, the strategy $T$ is no longer a best reply and gets minimal weight, meaning that in the limit its weight is 0, and similarly for the column player, if the tremble assigns higher value to $B$ than to $T$, then $\sigma$ will have zero weight on $L$. Since dominated strategies are never best replies, there must be full weight on $M$ and $C$.

It is seen that the equilibrium $(T, L)$, which is a perfect equilibrium in the game without the two added strategies, remains so in the new game.

---

**3.2. Proper equilibria.** In the interpretation of the perfect equilibria, it is presumed that players make errors with some small probability. These errors may involve arbitrary pure strategies, including some which may be considered as quite harmful to the player. Thus, there is no distinction between strategies when it comes to selection by error, and this feature may be criticized, since we rather expect players to be more careful when an erroneous choice will inflict a big loss than when the difference in payoff is minor.

Such considerations lead to the notion of a *proper equilibrium* introduced by Myerson [1978]. Inspired by the previous discussion, we begin with the notion of $\varepsilon$-properness: The mixed strategy array $\sigma^\varepsilon$ is an $\varepsilon$-proper equilibrium, for $\varepsilon > 0$, if it is completely mixed and

$$u_i(s_i, \sigma^\varepsilon_{-i}) < u_i(s'_i, \sigma^\varepsilon_{-i}) \Rightarrow \sigma^\varepsilon_i(s_i) < \varepsilon\sigma^\varepsilon_i(s'_i).$$

A strategy array is a proper equilibrium if it is the limit of a sequence $(\sigma^\varepsilon)_{\varepsilon\to 0}$ for $\varepsilon$ going to 0.

The existence problem is taken care of rather easily.

THEOREM 2 *Every normal form game with finite sets of pure strategies has a proper equilibrium.*

PROOF: We show that there exist $\varepsilon$-proper equilibria for suitably small $\varepsilon$.

Let $0 < \varepsilon < 1$, and for each $i \in N$ let $\eta_i(s_i)$, for $s_i \in S_i$, be given by

$$\eta_i(s_i) = \frac{\varepsilon^{m_i}}{m_i},$$

where $m_i = |S_i|$ is the cardinality of $S_i$, the set of pure strategies of player $i$. We let $\eta = (\eta_1, \ldots, \eta_n)$, and let $\Sigma^\eta_i$ denote the set of mixed strategies $\sigma_i$ of player $i$ such that $\sigma_i(s_i) \geq \eta_i(s_i)$ for all $s_i \in S_i$. Finally $\Sigma^\eta = \Sigma^{\eta_1}_1 \times \cdots \times \Sigma^{\eta_n}_n$.

We now define the correspondences $\Phi_i : \Sigma^\eta \twoheadrightarrow \Sigma^{\eta_i}_i$, for $i \in N$, by

$$\Phi_i(\sigma) = \{\hat{\sigma}_i \in \Sigma^{\eta_i}_i \mid u_i(s_i, \sigma_{-i}) < u_i(s'_i, \sigma_{-i}) \Rightarrow \hat{\sigma}_i(s_i) < \varepsilon\sigma_i(s'_i), \text{ all } s_i, s'_i \in S_i\}.$$

We check that $\Phi_i(\sigma) \neq \emptyset$ for all $\sigma \in \Sigma^\eta$. Indeed, let $v_i(\sigma, s_i)$ be the number of pure strategies in $S_i$ such that $u_i(s_i, \sigma_{-i}) < u_i(s'_i, \sigma_{-i})$, and let

$$\hat{\sigma}_i(s_i) = \frac{\varepsilon^{v_i(\sigma,s_i)}}{\sum_{s'_i \in S_i} \varepsilon^{v_i(\sigma,s'_i)}}.$$

Then $\hat{\sigma}_i(s_i) \geq \eta_i(s_i)$ for small enough $\varepsilon$, so that $\Phi_i(\sigma) \neq \emptyset$. It is easily seen that the graph of $\Phi_i$, that is the set of all $(\sigma, \hat{\sigma}_i)$ such that $\hat{\sigma}_i \in \Phi(\sigma)$, is closed, and that $\Phi_i(\sigma)$ is a convex set for each $\sigma$.

Now we may apply Kakutani's fixed point theorem to the correspondence $\Phi : \Sigma^\eta \twoheadrightarrow \Sigma^\eta$ defined by $\Phi(\sigma) = \Phi_1(\sigma) \times \cdots \times \Phi_n(\sigma)$ for all $\sigma \in \Sigma^\eta$ to get a strategy array $\sigma$ such that $\sigma_i \in \Phi_i(\sigma)$ for each $i \in N$. It is easily checked that $\sigma$ is indeed an $\varepsilon$-proper equilibrium. $\qquad\square$

One of the drawbacks of the perfect equilibria is that adding dominated strategies may enlarge the set of perfect equilibria. Unfortunately, this may happen also with proper equilibria, as it can be seen from the game

|   | L | R |
|---|---|---|
| T | $(1,1,1)$ | $(0,0,1)$ |
| B | $(0,0,1)$ | $(0,0,1)$ |

I

|   | L | R |
|---|---|---|
| T | $(0,0,0)$ | $(0,0,0)$ |
| B | $(0,0,0)$ | $(1,1,0)$ |

II

where player 1 chooses row, player 2 column, and player 3 matrix. Here the second strategy of player 3 is dominated, and it might therefore be considered as superfluous. In that case, only $(T, L, I)$ is a reasonable equilibrium, being the unique proper equilibrium in the game where player 3 has only the first matrix. But actually $(B, R, I)$ is also a proper equilibrium.

**3.3. Essential equilibria.** While the refinements discussed above used the concept of trembles, thereby restricting the choices of the players to certain subsets of mixed strategies, there is another way of getting rid of undesired Nash equilibria, or rather another type of undesirability to be considered. We might allow all strategies, mixed or pure, but instead apply the strategies to games which are almost the same but have small perturbations of the payoffs. If some Nash equilibrium disappears if the payoff structure is changed ever so little, we may have some reservations towards this equilibrium.

These considerations lead to the concept of an *essential equilibrium*. To define it, we need to introduce the notion of *distance* between two games (with the same player set and the same set of pure strategies). If $\Gamma = (N, (S_i)_{i \in N}, (u_i)_{i \in N})$ and $\Gamma' = (N, (S_i)_{i \in N}, (u'_i)_{i \in N})$ are two games with the same sets of players and strategies, but possibly with different payoff functions, then we define the distance between $\Gamma$ and $\Gamma'$ as

$$d(\Gamma, \Gamma') = \max_{i \in N, s \in S} \|u_i(s) - u'_i(s)\|.$$

One can easily check that this notion of distance satisfies (i) $d(\Gamma, \Gamma') = d(\Gamma', \Gamma)$, (2) $d(\Gamma, \Gamma') \geq 0$ with $d(\Gamma, \Gamma') = 0$ if and only if $\Gamma = \Gamma'$, and also (iii) $d(\Gamma, \Gamma'') \leq d(\Gamma, \Gamma') + d(\Gamma', \Gamma'')$ for all $\Gamma, \Gamma', \Gamma''$ (where $\Gamma'' = (N, (S_i)_{i \in N}, (u''_i)_{i \in N})$ ) showing that $d(\cdot, \cdot)$ is a metric on the set of games with the fixed player and strategy sets. Notice also that the set of mixed strategy arrays can be considered as a subset of a suitable Euclidean space, so that there is also a well defined notion of distance between such strategy arrays.

DEFINITION 3 *An equilibrium $\sigma$ in $\Gamma$ is an essential equilibrium if for every $\varepsilon > 0$ there exists some $\delta > 0$ such that for every game $\Gamma'$ with $d(\Gamma, \Gamma') < \delta$ there is a Nash equilibrium $\sigma'$ in $\Gamma'$ such that $\|\sigma - \sigma'\| < \varepsilon$.*

The essential equilibria are points of (lower hemi-)continuity of the Nash equilibrium correspondence – each neighbourhood of the equilibrium will contain equilibria of the games which are sufficiently close to the game considered.

Although the concept of essential equilibria has some intuitive content – there is an equilibrium near by even if we change the payoff slightly – it is not easy to work with, and even existence is not trivial. It also turns out that essential equilibria are not necessarily better than non-essential ones, as is shown by the example below.

|   | L | C | R |
|---|---|---|---|
| T | $(1,1)$ | $(0,0)$ | $(0,0)$ |
| M | $(0,0)$ | $(2,2)$ | $(2,2)$ |
| B | $(0,0)$ | $(2,2)$ | $(2,2)$ |

Here $(T, L)$ is an essential equilibrium (it will remain so even if all payoffs are changed as long as these changes are small). But there are many other Nash equilibria, including some which are better, such as $(M, C)$. This equilibrium is not essential, however, since arbitrarily small positive changes in the payoff of player 1 at the strategy $B$ will upset the equilibrium. The fact that there is a non-essential equilibrium which is better is not in itself a strong argument against the concept, but in our case there is an additional reason to discard $(T, L)$ – in an actual play there would be some informal agreement among players of never using $T$, respectively $L$, since avoiding them would improve payoff to both.

Due to this and other shortcomings, the essential equilibria have only theoretical interest, and there has not been much discussion of them in the literature.

## 4   Strategic stability

**4.1. A general approach to refinements: Background.** From the discussion of perfect or proper equilibria, we have seen that some Nash equilibria may be quite unreasonable, and we can avoid many of them by application of a trembling-hand principle in some concrete form. However, doing this poses some new problems, and none of the explicit proposals for a refinement considered so far was entirely convincing.

The particular cases of refinements considered so far, perfect, proper or essential equilibria, all had some commen structure: They were obtained as limits of mixed strategies having an almost-equilibrium character, and therefore could be interpreted as equilibria which were robust against some kind of trembling-hand perturbation, according to which players make small errors, possibly with some structure, rather than just choosing the exact equilibrium strategy. Unfortunately, the interpretation also had some flaws; some small errors from a given equilibrium strategy may well result in best replies which eventually lead to a different equilibrium.

Such shortcomings of the proposed solution concepts eventually led to a more general approach, initiated in Kohlberg and Mertens [1986] and leading to the introduction of *stable* solutions. For this, we reorient our search for a "reasonable" Nash equilibrium in at least two directions:

- the concept of a solution must be redefined to mean *a set* of strategy combinations rather than, as we have had it previously, a single strategy combination. This takes care of the problem that sequences of almost-equilibria starting close to each other may have quite different limits,

• the choice of solution will be guided by a list of criteria, properties to be satisfied by a solution, replacing the previous vague notions of "reasonableness".

The Kohlberg-Mertens approach puts the intuitive discussion of trembling hands on more precise formal foundation, showing that the many different notions of perturbed games can be given a common expression. This comes however at a cost: While the simple notions of trembling hands can be given an interpretation in terms of uncertainty of players, no such interpretation is available for the more general concept, or such an interpretation would be extremely far-fetched. We should consider stability as a technical concept which gives solutions with reasonable properties.

**4.2. Axioms for reasonability of equilibria.** As in previous sections, we consider a game $\Gamma = (N, (S_i)_{i \in N}, (u_i)_{i \in N})$ , where the set $S_i$ of pure strategies of player $i$ is finite, all $i \in N$. We write the set of mixed strategies of player $i$ as $\Sigma_i$, and we let $\Sigma = \times_{i \in N} \Sigma_i$ be the set of mixed strategy combinations. A *solution* to $\Gamma$ is a nonempty subset of the set of (mixed strategy) Nash equilibria of $\Gamma$.

Since the approach to reasonable Nash equilibria conducted so far was largely unsuccessful, it might be better to specify exactly what is meant by "reasonableness", setting up a list of properties (axioms) to be satisfied by the solution that we are looking for. Below we list some such conditions.

(i) (Existence) Each game $G$ has a solution.

(ii) (Connectedness) A solution of any game is a connected subset of the set of mixed strategy arrays.

(iii) (Admissibility) In any element of a solution no player uses a weakly dominated strategy.

(iv) (Invariance) A solution of a game is a solution of any game having the same reduced normal form (cf. Box 7.4).

(v) (Sequentiality) A solution of an extensive form game contains at least one sequential equilibrium.

(vi) (Iterated dominance) A solution of a game contains a solution of any game obtained by deleting a weakly dominated strategy.

Here (i) seems uncontroversial, since our aim is to select from the Nash equilibria, and we can hardly imagine situations where *all* Nash equilibria are unreasonable. The connectedness property would seem to be only of technical importance, but if the solution set contained several subsets which were separated from each other, we might prefer to consider each of these subsets as a "reasonable" solution. Condition (iii) says that weakly dominated strategy arrays should not occur in a solution, which appears as a natural property.

**Box 7.4. Payoff equivalence and reduced normal form.** In many cases, normal form games may have strategies which could be deleted without changing the strategic aspects of the game. This may occur when the game has been derived from an extensive form: In the following extensive form game below, Player 1 has two moves and Player 2 only one.

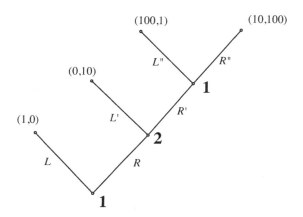

The normal form of this game looks as follows:

|      | $L'$      | $R'$       |
|------|-----------|------------|
| $LL''$ | $(1,0)$ | $(1,0)$    |
| $LR''$ | $(1,0)$ | $(1,0)$    |
| $RL''$ | $(0,10)$ | $(100,1)$ |
| $RR''$ | $(0,10)$ | $(10,100)$ |

Here Player 1 has the two pure strategies $LL''$ and $LR''$ which give exactly the same payoff, no matter what Player 2 would do, not only to Player 1 but also to the other player.

Actually, a pure strategy may be deleted also when the player has a *mixed strategy* which gives exactly the same payoff to all players. In general, if $\Gamma = (N, (S_i)_{i \in N}, (u_i)_{i \in N})$ is a game with finite sets $S_i$ of pure strategies, then two strategies $\sigma_i$ and $\sigma_i'$ of player $i$ are *payoff equivalent* if

$$u_j(\sigma_i, \tau_{-i}) = u_j(\sigma_i', \tau_{-i})$$

for all $j \in N$ and all $\tau_{-i} \in \Sigma_{-i} = \bigtimes_{k \neq i} \Sigma_k$ (where as usual, $\Sigma_i$ denotes the set of mixed strategies of player $i$). When a pure strategy of a player is payoff equivalent to some other strategy, it may be deleted, and a a game is *reduced* if there are no pure strategies that can be deleted in this way.

**Example 7.4, continued.**

It is intuitive, but not quite obvious that normal form games can be reduced in an essentially unique way to a reduced normal form game. This can be done in a series of *reductions:* A game $\Gamma' = (N, (S_i')_{i \in N}, (u_i')_{i \in N})$ is a reduction of $\Gamma = (N, (S_i)_{i \in N}, (u_i)_{i \in N})$ if there is an affine and surjective map $f = (f_i)_{i \in N}$ from $\Sigma$ to $\Sigma'$ such that the diagram

$$
\begin{array}{ccc}
\Sigma_1 \times \cdots \times \Sigma_n & \xrightarrow{u} & \mathbb{R} \\
f_1 \downarrow \qquad f_n \downarrow & & \parallel \\
\Sigma_1' \times \cdots \times \Sigma_n' & \xrightarrow{u'} & \mathbb{R}
\end{array}
$$

commutes. For further detail, see e.g. Vermeulen and Jansen [1998].

The invariance property (iv) is worth mentioning separately – it says that a solution is related to a normal form game, so that different extensive form games resulting in the same normal form game will have the same solutions. In other words, we consider the normal form game as the basic entity in game theory. This does not mean that extensive form games are unimportant, but the choice of solution is determined from the normal form only. The properties (v) and (vi) state that the solution should capture some of the features of the intuitively appealing solution concepts previously encountered (subgame perfect Nash equilibria, strategy arrays obtained by iterative elimination of weakly dominated strategies).

The above list is by no means exhaustive. In particular, one might want to add something to (v) and (vi) – equilibria obtained through backwards induction also seem quite intuitve and should therefore belong to a solution. However, making the list of desiderata too long means that it will be difficult if not impossible to satisfy all its properties.

In addition, the specification of a solution alone by its properties (some or all of the items in the list) does not give much information of how to find it. In order to get a better feeling of this, we follow Hillas [1990] who extends the previously employed idea of sequences of equilibria of games to a more general approach.

**4.3. Best response correspondences.** As usual, the best response correspondence associated with the game $\Gamma = (N, (S_i)_{i \in N}, (u_i)_{i \in N})$ takes each (mixed) strategy array $\sigma$ to

$$
B^\Gamma(\sigma) = \{\sigma' \mid \forall i, u_i(\sigma_i', \sigma_{-i}) \geq u_i(\sigma_i'', \sigma_{-i}), \text{ all } \sigma_i'' \in \Sigma_i\}.
$$

An array of mixed strategies is a Nash equilibrium if it is a fixed point of $B^\Gamma$.

The best response correspondence is closed (in the sense that the set $\{(\sigma, \sigma') \mid \sigma' \in B^\Gamma(\sigma)\}$ is closed). Indeed, let $(\sigma^t)_{t=1}^\infty$ and $(\hat{\sigma}^t)_{t=1}^\infty$ be sequences of mixed strategy arrays in $\Gamma$ converging to some $\sigma^0$ and $\hat{\sigma}^0$, respectively, and such that $\hat{\sigma}^t \in B(\sigma^t)$

for all $t$. Then $\hat{\sigma}^0$ must be in $B^\Gamma(\sigma^0)$, since otherwise there would be some $i$ and $\sigma'_i$ such that $u_i(\sigma'_i, \sigma^0_{-i}) > u_i(\hat{\sigma}_i, \sigma^0_{-i})$, and by continuity of $u_i$, we would also have that $u_i(\sigma'_i, (\sigma^t)_{-i}) > u_i(\hat{\sigma}_i, \sigma^t_{-i})$ for sufficiently large $t$, a contradiction.

Since the graph of $B^\Gamma$ is closed, we have that $B^\Gamma(\sigma)$ is a closed and hence compact subset of $\Sigma$ (the set of mixed strategy arrays in $\Gamma$) for each $\sigma$. It is clear that mixing two best responses also gives a best response, so $B^\Gamma(\sigma)$ is convex for each $\sigma$. We shall use these nice properties of the best response correspondence in the following.

In our discussion of trembling hand equilibria, we have worked with best-response correspondences which are slightly changed since the best response to any strategy array is constrained to suitable sets of completely mixed strategies. Thus we may think of the equilibrium concepts considered previously as those which are fixed points also for slightly perturbed best response correspondences. We therefore introduce a notion of distance between two correspondences $\Phi_1, \Phi_2$ : $\Sigma \rightrightarrows \Sigma$ from the set of mixed strategy arrays to itself with closed graph and convex values, which may or may not be best reply correspondences themselves), using the Hausdorff distance between graphs of $\Phi_1$ and $\Phi_2$, that is the smallest number $\delta$ such that

$$\text{Graph}\,\Phi_1 \subseteq (\text{Graph}\,\Phi_2)_\delta, \text{Graph}\,\Phi_2 \subseteq (\text{Graph}\,\Phi_1)_\delta,$$

where $(\text{Graph}\,\Phi_j)_\delta$ for $h = 1, 2$ consists of all points $(\tau, \tau') \in \Delta^N \times \Delta^N$ which have distance $< \delta$ to some point $(\sigma, \sigma') \in \text{Graph}\,\Phi_j$.

We may define a *stable set of equilibria* of $\Gamma$ as a minimal (with respect to inclusion) set $S$ with the following property:

(∗) $S$ is a closed set of Nash equilibria of $\Gamma$ such that for any game $\Gamma'$ with the same reduced normal form as $\Gamma$ and for every $\varepsilon > 0$ there is $\delta > 0$ such that any correspondence which is within distance $\delta$ of $B^{\Gamma'}$ has a fixed point within distance $\varepsilon$ of $S$.

Our first concern about stable sets of equilibria are – as always – the existence problem, which is taken care of here:

THEOREM 3 *Let* $\Gamma = (N, (S_i)_{i \in N}, (u_i)_{i \in N})$ *be a game with finite sets of pure strategies. Then there are stable sets of equilibria of* $\Gamma$.

PROOF: We need only show that there is a set of strategy arrays in $\Gamma$ which satisfies property (∗), since choosing a minimal one among such sets will give us a stable set. Our candidate is the set $S$ of all Nash equilibria of $\Gamma$ (which is also the set of Nash equilibria of any game having the same reduced form as $\Gamma$).

Let $\varepsilon > 0$ be given, and suppose now that $(\Phi^t)_{t \in \mathbb{N}}$ is a sequence of correspondences with closed graph and nonempty convex values such that $\delta(\Phi^t, B^\Gamma) < \dfrac{1}{t}$. For each $t$, there is at least one fixed point $\sigma^t$ for $\Phi^t$, i.e. $(\sigma^t, \sigma^t)$ belongs to the graph of $\Phi^t$. Without loss of generality, we may assume that the sequence $(\sigma^t)_{t \in \mathbb{N}}$ converges to some $\sigma^0 \in \Sigma$. By our construction, $(\sigma^0, \sigma^0)$ belongs to $B^\Gamma$, which means that $\sigma^0$ is a Nash equilibrium of $\Gamma$, so that it belongs to $S$. This means that $S$ satisfies (*), and consequently, there are stable sets of equilibria in $\Gamma$. $\qquad\square$

Once the existence problem is taken care of, we may proceed towards verifying that stable sets of equilibria satisfy the remaining conditions (ii)-(vi) stated above, as indeed they do. We refer to Hillas [1990] for details, see also Problems 4 and 5 below.

## 5   Cheap talk and signaling games

In games with several equilibria, there is usually no convincing reason for choosing one of these as more likely than the other equilibria as final outcome. However, from an intuitive point of view, one might expect that preplay communication phase might facilitate this choice. This intuition is enforced by the consideration of *cheap talk*, adding an initial phase of communication, even when payoff of the extended game do not depend on strategies chosen in the initial phase.

**5.1.  Cheap talk.** The way in which cheap talk may influence the equilibria is illustrated by the example used by Farrell (1987) when introducing the concept. We consider a two-person game with payoff matrix

|     | In         | Out      |
|-----|------------|----------|
| In  | $(-L,-L)$  | $(M,B)$  |
| Out | $(B,M)$    | $(0,0)$  |

which can be interpreted as payoffs to participants in conflict where to firms must decide whether or not to enter into a given market. This game has three Nash equilibria, namely the two equilibria where one firm chooses "In" and the other firm "Out", and the symmetric mixed strategy equilibrium where each player chooses "In" with probability $p$ defined such that the two pure strategies have the same expected payoff, that is

$$p(-L) + (1-p)M = pB,$$

from which we get that $p = \dfrac{M}{B+L+M}$.

Now the game is enlarged by one initial round of communication, where each player may choose "In" or "Out", after which the original game is played, with the same payoff matrix as before. An equilibrium strategy in the enlarged game prescribes a choice of communication strategy, followed by a choice of strategy in the original game, depending on the choices observed in the initial round. Following Farrell (1987), we are particularly interested in (a) symmetric equilibria, and (b) equilibrium strategies which reflect some aspects of real-life communication, namely the following:

- if one player says "In" and the other "Out" in the communication phase, then the player saying "In" will choose "In" when it comes to playing the original game, and
- if both players say "In" or both say "Out", then they will play the symmetric equilibrium strategies in the original game.

It is easily checked that the enlarged game possesses such equilibria. Assume that each player chooses "In" with probability $q_1$ in the communication phase. If the strategy has property (b), then

$$q_1 u_1 + (1 - q_1)M = q_1 B + (1 - q_1)u_1,$$

where $u_1$ is the payoff of the symmetric equilibrium of the original game. Solving for $q_1$ and using that $u_1 = pB$ we get that

$$q_1 = \frac{M - u_1}{M - 2pB + B} \quad \text{and} \quad 1 - q_1 = \frac{B}{M - 2pB + B}(1 - p).$$

If $M > 2pB$, then $q_1 > p$, and for large enough $M$ we get that each player will claim to enter but then randomize subsequently. However, the overall probability of a coordination failure (both enter or both stay out) has been reduced, since it was $p^2 + (1-p)^2$ in the original game but has now changed to $(q_1^2 + (1-q_1)^2)(p^2 + (1-p)^2)$.

Having added one initial round of communication, we may proceed adding further similar rounds. Doing this, we obtain a $(T+1)$-stage game, where the initial $T$ stages are communication rounds, where each player must say either "In" or "Out", and the final stage is the original game. Following the above reasoning, we look for symmetric strategies where players repeat their choices if one says "In" and the other "Out", and move to the equilibrium strategy of the $(T-S)$-stage game if in $S$th stage both players have said 'In" or both "Out". If $u_{T+1}$ is expected payoff of the equilibrium strategy where players choose "In" with probability $q_T$ in the first communication round, then

$$u_{T+1} = q_T u_T + (1 - q_T)M = q_T B + (1 - q_T)u_T = u_T + q_T(B - u_T), \tag{2}$$

so that $u_{T+1}$ is a weighted average of $u_T$ and $B$, from which we get that $u_T$ is increasing and bounded from below by $B$. It may be seen that $u_T$ converges to $B$, since by (2),

$$u_{T+1} - u_T = q_T(B - u_T), \tag{3}$$

and if $u_T$ converged to some $u^* < B$, then the righthand side in (3) would be $>$ than some positive number, a contradicting that the sequence $(q_T)_{T=1}^{\infty}$ converges.

As it can be seen, adding communication rounds we get new symmetric equilibria which are better for both players than the symmetric equilibrium of the original game. However, there are also many other equilibria. For example, if players choose statements "In" and "Out" with equal probability in each round together with the mixed strategy $p$ in the final phase independent of previous messages, then we are back where we were – the communications are clearly non-informative in the sense that the choice of the players are not influenced by them.

So, while preplay communication opens up for better coordination, it comes with a selection problem, and the considerations of the previous sections will in general not be very helpful here. To see what is at stake, we consider in the next section a particular case of communication which is simplified as far as possible while retaining the main idea of this section, namely the use of moves which have no impact on the payoff of the players.

**5.2. A pure signaling game.** This consideration of pre-play communication, where some moves are added to the original game, without changing the ultimate payoff, will take us to the problems of signaling in order to see whether the players can communicate there intents to each other in a way that makes the selection of a designated equilibrium more likely.

Signaling in itself is not an altogether simple matter, however. This is shown in a by now classical work by Crawford and Sobel [1982], who consider what may be seen as the simplest possible setup for strategic signaling. There are two players, one being a *sender S*, who makes an observation and sends a signal, the other one a *receiver R*. The latter player chooses a decision after having received the signal.

We assume that the observation (which is available only to $S$) matters for both sender and receiver. Let this observation $m$ be random, distributed in the interval $[0, 1]$ with distribution function $F$ and density $f$. Signals $n$ (which are the pure strategies of $S$) are chosen from $[0, 1]$, and the decision (which is $R$'s strategy) is a real number $y$. The payoff functions (in terms of utility) are

$$U^S(y, m), \ U^R(y, m)$$

for sender and receiver, respectively. We assume that the utility functions are strictly concave in $y$, allowing for a unique maximizing choice.

An equilibrium in this game is a pair $(q(n|m), y(n))$, where $q(n|m)$ is a mixed strategy over $N$ for each $m$, and $y(n)$ is a function of $n$, such that

(i) for each signal value $n$ with $q(n|m) > 0$, (so that the mixed strategy may choose $n$), $y(n)$ solves the problem

$$y(n) = \max_{n'} U^S(y(n'), m),$$

(incentive compatibility, given the strategy of $R$ the sender cannot gain from sending another signal than $n$),

(ii) for each $n$, $y(n)$ solves

$$\max_y \int U^R(y, m) p(m|n) \, dm,$$

where $p(m|n)$ is the posterior density of $m$ given the signal $n$, found using Bayes' formula

$$p(m|n) = \frac{q(n|m)f(m)}{\int_0^1 q(n|t)f(t) \, dt}.$$

The equilibrium conditions are standard: We allow for mixed strategies, and an equilibrium strategy of $S$ must be such that each pure strategy with nonzero weight gives the best possible payoff (given the choice of the other player). For the player $R$, receiving a signal $n$ gives some hints as to the true value of $m$, since $R$ may update the general knowledge about likelihood of $m$, described by $f(m)$, using Bayes' formula. Having done that, the decision must be one which maximizes expected utility given the updated probabilities.

If the two players have identical utility functions, their common interest would suggest that truthful signaling is optimal, and indeed one may easily check that there is a Nash equilibrium where $S$ behaves in this way. The interesting case is however where there is a conflict of interest, and in the following, we assume that $R$ and $S$ have different utility functions. We define

$$y^j(m) = \arg\max_y U^j(y, m), j = S, R$$

as the utility maximizing decisions for each of the players, given $m$ (only $R$ is actually choosing a decision, but it is useful to consider the favorite choices of both players).

We formulate the non-coincidence of utilities with this new notation, namely as

$$y^S(m) \neq y^R(m), \text{ all } m. \tag{4}$$

We shall make use of this assumption to show that the set of decisions which can be made in an equilibrium is finite. Indeed, suppose that $u$ and $u'$ are two such decisions (results of the equilibrium strategies $y(n)$ for a suitable signal $n$) with $u < u'$. Since $u$ is an equilibrium outcome, the observation $m$ of $S$ must be such that $U^S(u, m) \geq U^S(u', m)$, and similarly the observation $m'$ at which $u'$ is equilibrium choice must satisfy

$$U^S(u', m') \geq U^S(u, m').$$

Now by continuity of $U^S$ there is some observation $\hat{m}$ such that

$$U^S(u, \hat{m}) = U^S(u', \hat{m}), \tag{5}$$

and with our assumptions on well-behavedness of utility functions, we have that $y^S(\hat{m})$ must satisfy

$$u < y^S(\hat{m}) < u' \tag{6}$$

and that observations above $\hat{m}$ will not result in equilibrium decision $u$, observations below $\hat{m}$ not in $u'$.

Turning now to $R$, we have that $y^R(\hat{m})$ must satisfy $u \leq y^R(\hat{m}) \leq u'$, but as $y^R(m) \neq y^S(m)$ for all $m$, there is $\varepsilon > 0$ such that

$$|y^R(m) - y^S(m)| \geq \varepsilon, \text{ all } m \in [0, 1].$$

In view of the above, we may now conclude that $u' - u \geq \varepsilon$, so there is a minimal distance between decisions induced in equilibrium. As a consequence, we get that there are only finitely many possible decisions in equilibrium.

This discreteness of equilibrium choices can be elaborated upon so as to give us a candidate for the equilibrium. The simplest case is that where $S$ always sends the same signal which contains no information, that is the mixed strategy $q_1(n, m)$ is the uniform distribution on $[0, 1]$. Since the signal now carries no information about $S$'s information, the best reply of $R$ is

$$y_1 = \arg\max_y \int_0^1 U^R(y, m) f(m) \, dm$$

(Bayesian updating with $q(\cdot|m)$ uniform in $[0, 1]$ for all $m$ does not change the prior distribution). This is an equilibrium – since $R$ always chooses the same decision, there will be nothing gained for $S$ by using any other signaling strategy, and $R$ cannot do better since no information is obtained from $R$ – but it does not quite satisfy our intuitive demands to an equilibrium with transmission of information.

The idea behind our construction of this trivial equilibrium may however be used to give more interesting equilibria. Let us try an equilibrium with 2 possible decisions. This can be achieved if $[0, 1]$ is partioned into two intervals $[a_0, a_1]$ and $[a_1, a_2]$, with $a_0 = 0, a_2 = 1$, so that

$$q_2(\cdot|m) = \frac{f(m)}{a_{i+1} - a_i}$$

for $i = 0, 1$ is the uniform distribution on $[a_i, a_{i+1}]$ if $m$ belongs to this interval, and

$$y_2^1(a_1) = \arg\max_y \int_0^{a_1} U^R(y, m) f(m) \, dm, \quad y_2^2(a_1) = \mathrm{argmax}_y \int_{a_1}^1 U^R(y, m) f(m) \, dm$$

are the best responses of $R$ to the strategy of $S$. This will again be an equilibrium if

$$U^S(y_2^1(a_1), a_1) = U^S(y_2^2(a_1), a_1), \tag{7}$$

so that $S$ will not want to send messages in another interval than that containing the signal $m$. However, for this to be the case, $a_1$ must be chosen in a particular way.

To see that such a choice of $a_1$ is possible, note that there is $\underline{a}_1$ close to 0, such that

$$U^S(y_2^1(\underline{a}_1), \underline{a}_1) < U^S(y_2^2(\underline{a}_1), \underline{a}_1),$$

and $\bar{a}_1$ close to 1 such that

$$U^S(y_2^1(\bar{a}_1), \bar{a}_1) > U^S(y_2^2(\bar{a}_1), \bar{a}_1).$$

When $a_1$ moves from $\underline{a}_1$ to $\bar{a}_1$, by continuity of $y_2^1(\cdot)$, $y_2^2(\cdot)$ and $U^S$ it will attain a value such that (7) holds.

The reasoning used in the above two cases can be generalized, even if the technique must be slightly refined. For later use, we define

$$y(a, b) = \mathrm{argmax}_y \int_a^b U^R(y, m) f(m) \, dm$$

for $a \leq b$ as the optimal choice of $R$ in the interval $[a, b]$ given a uniform posterior in this interval.

THEOREM 4 *Assume that R and S are different in the sense that (4) is satisfied. Then there is $\overline{N} \geq 1$ and for each $N \leq \overline{N}$ an equilibrium $(y_N, q_N)$ and numbers $0 = a_0 < a_1 < \cdots < a_{N-1} < a_N = 1$ such that:*

(i) $y_N^i(n) = \arg\max_y \int_{a_i}^{a_{i+1}} U^R(y, m) f(m)\, dm,\ n \in (a_i, a_{i+1}),\ for\ i = 0,\ldots,, N-1,$

(ii) $U^S(y_N^i, a_i) = U^S(y_N^{i-1}, a_{i-1}),\ i = 1,\ldots, N-1,$

(iii) *if* $m \in (a_i, a_{i+1})$, *then*

$$q_N(n, m) = \begin{cases} \dfrac{1}{a_{i+1} - a_i} & n \in [a_i, a_{i+1}], \\ 0 & n \notin [a_i, a_{i+1}]. \end{cases}$$

PROOF: (sketch) We show that there are numbers $0 = a_0 < a_1 < \cdots < a_N$ such that (ii) is satisfied, when $y_N^i$ for $i = 0,\ldots, N-1$ and $q(m|n)$ are as defined in (i) and (iii).

We now construct a game with $N - 1$ players, with player $i$ choosing a point $a_i$ on the interval $[0, 1]$. The payoff $\pi_i(a_1, \ldots, a_{N-1})$ is defined as follows: Let the coordinates of $a_{-i}$ be ranked according to size,

$$a^{(1)} \leq \cdots \leq a^{(N-2)},$$

and define the payoff to player $i$ at $(a_i, a_{-i})$ as

$$\pi_i(a_i, a_{-i}) = -\left| U^S(y_N(a^{(i-1)}, a_i), a_i) - U^S(y_N(a_i, a^{(i)}), a_i) \right|. \tag{8}$$

We notice that $\pi_i(\cdot, a_{-i})$ can be increased by moving $a_i$ towards $a^{(i-1)}$ when $U^S(y_N(a^{(i-1)}, a_i), a_i) > U^S(y_N(a_i, a^{(i)}), a_i)$, and by moving in the opposite direction when $U^S(y_N(a^{(i-1)}, a_i), a_i) < U^S(y_N(a_i, a^{(i)}), a_i)$, so that

$$U^S(y_N(a^{(i-1)}, a_i), a_i) = U^S(y_N(a_i, a^{(i)}), a_i)$$

when $\pi_i$ is maximal for $a_i$ given $a_{-i}$.

Since the strategy sets are compact, the payoff functions $\pi_i$ are continuous, and the sets

$$\{a_i' \mid \pi_i(a_i', a_{-i}) > \pi_i(a_i, a_{-i})\}$$

are convex for each $(a_i, a_{-i})$, we conclude by Theorem 3.1 that the game has a Nash equilibrium $(a_1^0, \ldots, a_{N-1}^0)$. Since $\pi_i$ is maximal at $a_i$ given $a_{-i}$, we have that

$$U^S(y^N(a_{i-1}, a_i), a_i) = U^S(y^N(a_i, a_{i+1}), a_i),\ i = 1,\ldots, N-1,$$

which shows that $a_0, a_1, \ldots, a_N$ define an equilibrium. $\square$

The theorem shows that there are partition equilibria with different fineness of partitions. It can be shown [Crawford and Sobel, 1982] that all equilibria are of this type. Thus, communication induces some degree of inaccuracy, due to the differences in preferences of sender and receiver, and this cannot be avoided.

Since there are multiple equilibria, the final outcome of the information transmission game may not be obvious. However, it seems reasonable that the equilibrium with the finest partition will be selected, since expected payoff is better for each player in this equilibrium than in any other.

## 6   Problems

**1.** Does a subgame perfect Nash equilibrium in an extensive form game correspond to a perfect Nash equilibrium in the associated normal form game?

**2.** Show that perfect Nash equilibrium strategies may be weakly dominated by other strategies which give rise to a non-perfect Nash equilibrium.

**3.** Consider the game

|   | L | C | R |
|---|---|---|---|
| T | (2, 2) | (1, 1) | (0, 0) |
| M | (1, 1) | (1, 1) | (1, 1) |
| B | (0, 0) | (1, 1) | (1, 1) |

Show that the $(T, L)$, $(M, C)$ and $(B, R)$ are Nash equilibria, but that only $(T, L)$ and $(M, C)$ are perfect and proper. Compare the two equilibria; are they equally reasonable?

**4.** (A unique Nash equilibrium which is unreasonable, cf. Aumann and Maschler [1972]) Consider the game

|   | L | R |
|---|---|---|
| T | (2, 2) | (4, 1) |
| B | (4, 1) | (3, 3) |

Find the unique Nash equilibrium (in mixed strategies) of this game. Show that the average payoff to player 1 is 10/3 in this equilibrium.

Find the *maximin* (mixed strategy) of player 1, and show that this strategy guarantees a payoff of 10/3, no matter what player 2 chooses. Compare the two strategy choices. What is best for player 1?

**5.** [Kohlberg and Mertens, 1986] Show that the game in extensive form (where player 2 has a single information set)

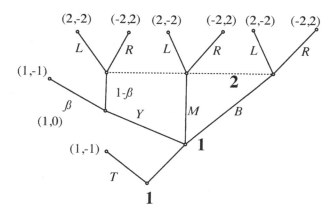

has the normal form

|   | L | R |
|---|---|---|
| T | $(1,-1)$ | $(1,-1)$ |
| M | $(2,-2)$ | $(-2,2)$ |
| B | $(-2,2)$ | $(2,-2)$ |

(the move labelled $Y$ corresponds to choosing a mixture of $T$ and $M$). A solution satisfying (iv) of Section 4 of this chapter should not depend on $\beta$. Show that the extensive form game has a unique sequential equilibrium where player 2 chooses $L$ with a probability which depends on $\beta$. What are the implications of this?

**6.** [Kohlberg and Mertens, 1986] Let $\Gamma$ be the game

|   | L | R |
|---|---|---|
| T | $(3,2)$ | $(2,3)$ |
| M | $(1,1)$ | $(0,0)$ |
| B | $(0,0)$ | $(1,1)$ |

Use this game to show that there can be no single-valued solution concept which satisfies (i), (iii) and (vi) of Section 4 of this chapter.

**7.** (Gilligan and Krehbiel [1987], Tadelis [2013]) Two players 1 and 2 constitute a committee. Player 1 knows the state of the world $\theta$ which is either $-w$ or $w$, and player 2 knows only that the states are equally likely, but determines the action $a \in \mathbb{R}$. The payoffs of the two players are

$$v_1(\theta, a) = -(\theta + b - a)^2, b > 0,$$
$$v_2(\theta, a) = -(\theta - a)^2,$$

so that under full information, the optimal action for player 1 is $a = \theta + b$, while for player 2 it is $a = \theta$.

Show that if player 2 has no information, her expected payoff is maximized at $a = 0$, which is called the *status-quo* decision. The rules of the game are such that player 1 observes $\theta$ and sends a message to player 2, who then chooses $a$. But this choice may be further constrained: We consider two decision rules, namely an *open rule*, where player 2 has no constraints, and a *closed rule*, where player 1 makes a recommendation for the choice of $a$, and player 2 must then either accept this recommendation or keep the status quo.

Show that under the open rule, there is an equilibrium with truthful reporting giving rise to the action $a = \theta$ if and only if $b \leq w$, and in a closed rule, a truthful reporting equilibrium with $a = \theta + b$ exists if and only if $b \leq w$. Show that the closed rule can yield truthful reporting also in cases where the open rule cannot.

# Chapter 8

# Repeated Games

## 1 Introduction: Repetition enhances cooperation

The intuition behind the theory of repeated games – which to some extent is confirmed by the results of the theory - says that if a given game is repeated sufficiently many times, then the consideration of what will happen in future cases may influence the behavior of the players, making them choose their actions in a way which is mutually beneficial to a much larger extent than if the game was played only once.

That this intuitive reasoning needs some further precision follows from the fact that one could just as well argue in the opposite direction. Consider for example the Prisoners' dilemma game with the following payoff function.

|   | L | R |
|---|---|---|
| T | (2, 2) | (−1, 3) |
| B | (3, −1) | (0, 0) |

Here the equilibrium in the game played only once (in the following to be called the one-shot game) is $(B, R)$, where player 1 plays second row, player 2 second column. If we repeat this game 17 times – and both players know that it is to be repeated exactly these 17 times – then this is also what will happen last time, since nothing will be gained in the future by foregoing some gain here and now. But since both players know what will happen in the last instance of the game, then effectively last instance is not number 17, but number 16, to which the same argument applies! It follows that $(B, R)$ will be played repeatedly 17 times, so that repeating the game adds little new in terms of solutions.

This example is however a rather special one; in which way it is special will become clearer as we proceed. At this point it suffices to notice that what will result from repeting a gain is not trivial. This justifies our subsequent formalization of the situation, and we shall derive some general results about repeated games.

Results of the type in which we are interested – namely that rather many of the payoff configurations that are available in the one-shot game can be obtained as equilibrium payoffs in the repeated game – are known as *folk theorems*, since they had the status of being known to researchers in the field in some more intuitive version before they were ever written down. This situation changed after the publication of several papers on repeated games, starting with Rubinstein [1979], and there is by now a large amount of contributions and several surveys, such as Sorin [1992], which is followed in our exposition below.

We start our discussion with a normal form game $\Gamma = (N, (S_i)_{i \in N}, (u_i)_{i \in N})$. As previously, we denote by $\Sigma_i$ the set of mixed strategies of player $i$, and we use the same notation $u_i$ for the affine extension of $u_i$ to $\Sigma = \times_{j \in N} \Sigma_j$. We introduce some further concepts which will be used as we proceed. Let $D$ be the convex hull of the set of payoff vectors that can be obtained as convex combinations of pure strategies,

$$D = \text{conv}\left(\left\{u(s_1, \ldots, s_n) \in \mathbb{R}^N \mid s_i \in S_i, i = 1, \ldots, n\right\}\right).$$

We may think of $D$ as containing all payoff vectors obtainable by a *correlated* strategy choice of the players (assigning a probability to each strategy combination).

Define the *minimax payoff* for player $i$ as the worst damage that the other players can inflict on $i$, given that $i$ counteracts their choice as far as possible,

$$v_i = \min_{\sigma_{-i} \in \Sigma_{-i}} \max_{\sigma_i \in \Sigma_i} u_i(\sigma_i, \sigma_{-i}).$$

We write the strategy combination for the complement of $i$ which realizes this minimax (or an arbitrarily selected one of them, if there are several), as $\hat{\sigma}_{-i}$. It will play a role as a punishment strategy selected by the others if $i$ does not perform according to certain rules.

The set of payoff vectors in $D$, which are as good for any player as the minimax, is

$$E = \{d \in D \mid d_i \geq v_i, \text{ all } i\}.$$

Payoff vectors in $E$ are *individually rational*: if player $i$ gets less than $v_i$, then she may change her strategy without fear of punishment by the others.

*Types of repetitions of* $\Gamma$. In order to give a precise description of the game which results from repeating $\Gamma$ a number of times, we need to define strategy sets and payoffs. With regard to the strategy sets, the problems are mainly notational. For the payoff function, however, there are several alternative definitions.

Looking first at the strategies in the repeated game, the obvious new aspect of the repetitions as compared to the one-shot game is the possibility of making the choice of strategy in the $t$th repetition dependent on what happened in the past. We introduce the notion of a *t-history*, for $t \geq 1$ to describe a finite sequence $h^t = (s^1, \ldots, s^t)$ of strategy choices of the players, so that every $s^\tau$ can be written as $s^\tau = (s_i^\tau)_{i \in N}$ with $s_i^\tau \in S_i$, $i \in N$. Notice that we are here interested only in *pure* strategies, since the history of the play at $t$ is what is observed by the players and used for determining the choice of $\Gamma$-strategy in the $(t + 1)$th repetition, and mixed strategies cannot be observed as such, what can be seen is only the pure strategies which actually result from playing mixed strategies. If $T$ is the total number of repetitions (allowing for the case $T = \infty$), then a $T$-history describes what happened while the players carried out the repeated game, and it is consequently called a *play* of the repeated game. We denote the set of $t$-histories by $H^t$ and let $H = \cup_{t=1}^T H^t$ be the set of all histories. We shall also need a set of 0-histories, which is just a singleton $\{*\}$.

As we have already hinted, strategies in the repeated game $\Gamma^T$ are instructions for the choices of $\Gamma$-strategies depending on the observed history. Thus, a strategy for player $i$ is a sequence $\mathfrak{s}_i = (\mathfrak{s}_i^t)_{t=1}^T$, where $\mathfrak{s}_i^1 \in \Sigma_i$, so that first choice is an ordinary mixed strategy in $\Gamma$ (or equivalently a map from $\{*\}$ to $\Sigma_i$), whereas for $t \geq 2$,

$$\mathfrak{s}_i^t : H^{t-1} \to \Sigma_i$$

is a map which to each $(t-1)$-history $h^{t-1}$, describing what has happened in the repeated game up to repetition $t - 1$, assigns a mixed strategy in $\Gamma$ as the choice of player $i$.

When each player has chosen a strategy $\mathfrak{s}_i$, the result is a play $\xi(\mathfrak{s})$, where $\mathfrak{s} = (\mathfrak{s}_i)_{i \in N}$, determined inductively by

$$\xi_1(\mathfrak{s}) = (\mathfrak{s}_1^1, \ldots, \mathfrak{s}_n^1),$$
$$\xi_t(\mathfrak{s}) = (\mathfrak{s}_1^t(\xi_1(\mathfrak{s}), \ldots, \xi_{t-1}(\mathfrak{s})), \ldots, \mathfrak{s}_n^t(\xi_1(\mathfrak{s}), \ldots, \xi_{t-1}(\mathfrak{s}))).$$

We have thus determined the strategy sets of the repeated game $G^T$ (for $T$ finite or infinite). It remains to define the payoff. For this we need a method for reducing the flows of payoffs in each period to a single number. There is no unique way of doing this, so the further analysis will depend on the method chosen.

*The finite (but long) repetition $\Gamma^K$.* The conceptually simplest way of weighing together several payoffs is obtained in the case where $T$ is finite, so that the game is repeated only a finite number of times. We shall then be concerned with payoffs which can be obtained for long enough sequences of repetitions.

The obvious way of defining average payoff for player $i$ over $t$ repetitions is

$$\mathfrak{u}_i^t(\mathfrak{s}) = \frac{1}{t} \sum_{\tau=1}^t u_i(\xi_\tau(\mathfrak{s}))$$
$$= \frac{1}{t} \sum_{\tau=1}^t u_i(\mathfrak{s}_1^\tau(\xi_1(\mathfrak{s}), \ldots, \xi_{\tau-1}(\mathfrak{s})), \ldots, \mathfrak{s}_n^\tau(\xi_1(\mathfrak{s}), \ldots, \xi_{\tau-1}(\mathfrak{s}))),$$

that is as the average over time of the payoff obtained in each of the periods. The expression looks simpler than it is, since the choice is among mixed strategies, so that for each $t$ one has to find the expected value of the given mixed strategies, and after than average is taken over the $t$ rounds of the game. The payoff in $\Gamma^K$ is defined as $\mathfrak{u}^K$.

We shall have much more to say about finite repetitions later, our main interest in such games is what happens when $n$ becomes large enough.

*The discounted repeated game* $\Gamma^\lambda$. Here $G$ is repeated infinitely often, and the payoff is discounted using a fixed rate of interest, after which one looks at the annuity which has the same discounted present value. Writing the *discount factor* as $1 - \lambda$ with $\lambda$ a parameter (this approach has the advantage that periods are treated more and more equally when $\lambda \to 0$), we have that the associated rate of interest $r$ satisfies $1 - \lambda = 1/1 + r$, from which we get that

$$r = \frac{\lambda}{1 - \lambda},$$

and payoff is now defined as

$$u^\lambda(\mathfrak{s}) = \frac{\lambda}{1 - \lambda} \sum_{t=1}^{\infty} (1 - \lambda)^t u_i^t(\mathfrak{s}),$$

where the quantity outside the summation is the interest rate found above. The results to be found pertains to the situation where the parameter $\lambda$ is small, so that the future has almost as much weight as the present time.

*The non-discounted infinite repetition* $\Gamma^\infty$. This last variation over the theme appears as complicated, but it is in many situations the one which is most easy to work with. Strictly speaking it is not a game in the usual sense, since we do not define a unique payoff for every strategy combination of the supergame. We actually restrict ourselves to defining *equilibria,* which are strategy arrays $\mathfrak{s} = (\mathfrak{s}_1, \ldots, \mathfrak{s}_n)$ with the property that the sequence $(u^t(\sigma))_{t=1}^{\infty}$ converges to a vector $u^\infty(\sigma)$, and such that for every player $i$ and every strategy $t_i$ of player $i$,

$$\liminf_t u_i^t(t_i, \mathfrak{s}_{-i}) \leq u_i^\infty(\mathfrak{s}),$$

so that a player who deviates from the strategy $\mathfrak{s}$ will have a smaller payoff from a certain point of time (here we use the *overtaking* criterion known from the economic theory of growth).

The three types of repetitions will be considered in more detail in the following, where we shall be interested in the connection between equilibria in the one-shot game and in the repetitions. Before doing so, it will be useful to check which payoff vectors can be obtained in the repetitions, as equilibria or not.

LEMMA 1 *Let $d \in D$. Then the following holds:*

    *(1) for every $\varepsilon > 0$ there is $t$ and a strategy array $\mathfrak{s}$ such that $\|u^t(\mathfrak{s}) - d\| \leq \varepsilon$,*
    *(2) there are strategy combinations $\mathfrak{s}^\lambda$ such that $u^\lambda(\mathfrak{s})$ converges to $d$ for $\lambda \to 0$,*
    *(3) there is a strategy combination $\mathfrak{s}$ such that $\lim_{t \to \infty}(\mathfrak{s}) = d$.*

*The supergame strategies in (1)-(3) can be chosen so as to use only pure strategies from $\Gamma$.*

PROOF: (1) Write $d \in D$ as

$$d = \sum_{j=1}^{k} \mu_j d_j,$$

where $d_j = u(s_1^j, \ldots, s_n^j)$ is a payoff vector in $\mathbb{R}^N$ which can be obtained by a suitable choice of pure strategies by the players, so that the coefficients $\mu_j$ are nonnegative and $\sum_{j=1}^k \mu_j = 1$. Choose positive rational numbers $q_j$ with $\sum_{j=1}^k q_j = 1$ such that

$$\left\| \sum_{j=1}^k q_j d_j - d \right\| \le \varepsilon;$$

we may assume that all the numbers $q_j$ are written as

$$q_j = \frac{r_j}{M}, \ j = 1, \ldots, k,$$

with $M > 0$ the common denominator. If the game $G$ is repeated $M$ times, so that $(s_1^j, \ldots, s_n^j)$ is chosen $r_j$ times, for $j = 1, \ldots, k$, then the payoff of this finite repetition is

$$\frac{1}{M} \sum_{j=1}^k r_j u(s_1^j, \ldots, s_n^j),$$

which is the desired result.

(2) To show that we can get $d$ as a limit of $u^\lambda$ for suitable sequences of strategy combinations, we use the same technique as in (1) above, having rounds of $r_j$ of the strategy combinations giving $d_j$ as payoff. When $\lambda$ is sufficiently close to 0, so that there is almost no discounting, so in average one will be close to $d$ after every round. Since the set of all the $d$ which can be obtained as a limit of payoffs from $D^\lambda$ for $\lambda \to 0$ is closed, we get that $d$ must belong to this set.

(3) This follows immediately from (2). $\qquad\qquad\square$

## 2 Nash equilibria in repeated games

We can now start the discussion of the basic problem, namely the question of which payoff vectors that can emerge as equilibrium payoffs in repetitions of the game $\Gamma$.

**2.1. A folk theorem for $\Gamma^\infty$.** We begin with the game $\Gamma^\infty$ which perhaps is less intuitive than the other repetitions, but which is the easiest to deal with when it comes to deriving results.

THEOREM 1 *Let $d \in E$ be an individually rational payoff vector. Then $d$ can be obtained as Nash equilibrium payoff in $\Gamma^\infty$.*

PROOF: By Lemma 1.(1), $d$ can be obtained by a strategy array the play of which results in a sequence $(s^t)_{t=1}^\infty$ of pure strategy arrays $s^t = (s_1^t, \ldots, s_n^t)$. We define equilibrium strategies $\mathsf{s}_i$ for player $i = 1, \ldots, n$ in the following way: Put $\mathsf{s}_i^1 = s_i^1$ for all $i$, and suppose that $\mathsf{s}^\tau$ has been defined for all $\tau < t$, where $t > 1$. If $h^t = (s^1, \ldots, s^{t-1})$, then $\mathsf{s}^t(h) = s^t$; if $\hat{h}^t$ is a history which differs from $(s^1, \ldots s^{t-1})$, let $\tau^0$ be the smallest $\tau \in \{1, \ldots, t-1\}$ for which $\hat{h}^\tau \ne (s^1, \ldots, s^\tau)$, choose $i^0$ as the smallest $i$ for which $(s_i^1, \ldots, s_i^\tau)$ and let

$$\mathsf{s}_i^t(h) = \hat{\sigma}_{-i^0}, \ i \ne i^0,$$

so that the first player deviating from the sequence $(s^t)_{t=1}^\infty$ is punished by its complement selecting a strategy which keeps the payoff of $i^0$ to the minimax.

It remains to show that $\mathsf{s}$ as defined here (known as a "trigger strategy') is actually a Nash equilibrium, and for this we use the definition of equilibrium given in the description of $\Gamma^\infty$. It follows from Lemma 1.(1) that $\mathsf{u}_i^t(\mathsf{s})$ tends to $d$ for $t \to \infty$; if $i$ is any player and $t_i$ a strategy for player $i$, we let $t^0$ denote the first time where the play of $(t_i, \mathsf{s}_{-i})$ differs from $(s^t)_{t=1}^\infty$; if there is no such date, then the results for $i$ using $t^i$ are the same as if no deviation took place. By the description of $\mathsf{s}_j$ for $j \neq i$, these strategies will prescribe the choice of $\hat\sigma_{-i}$ at every $t > t^0$, so the contribution to the overall payoff is $\leq v_i$ from $t = t^0 + 1$ and onwards. But this means that the sequence $\mathsf{u}_i^t(t_i, \mathsf{s}_{-i})_{t=1}^\infty$ is smaller than every number greater than $v_i$ for sufficiently large $t$, and since $t_i$ was chosen arbitrarily, we have that $\mathsf{s}$ is a Nash equilibrium.                                                                           $\square$

The technique of the proof is easily explained: We can support the payoff $d$ using strategies which consist of choosing exactly such moves that the payoff becomes $d$, and punishing every deviating player by sending her down to the minimax in all future rounds. This may hurt also the other players, but we are not considering such aspects for the moment, where we are concerned only about Nash equilibria. For such, the threat of being sent down to the minimax is sufficiently deterring.

**2.2. Folk theorems for the discounted repeated game.** Here it is not quite as easy as in the above case to exhibit a Nash equilibrium, since we have to make sure that a given deviation does not pay. What can go wrong can be seen from the following example;

|   | $L$ | $R$ |
|---|---|---|
| $T$ | $(1,0,0)$ | $(0,1,0)$ |
| $B$ | $(0,1,0)$ | $(1,0,1)$ |

Here there are three players, but the third player has only one strategy and is therefore a dummy player in the game. The first two players have a minimax of $1/2$ (what the other player can squeeze them down to by a suitable choice of strategy). There is clearly a strategy in the $\lambda$-repetition which gives the payoff $(1/2, 1/2, 1/4)$, but if one tries to get it by a suitable choice of strategies, one has to involve cases where either player 1 or player 2 can improve, and in this case they cannot really be punished since anyway the outcome of the original strategy was the lowest possible.

This shows that the result for discounted repetitions is less general than our first result. More precisely, it looks as follows:

THEOREM 2 *Assume that there is $d^0 \in D$ with $d_i^0 > v_i$ for $i = 1, \ldots, n$. Then for each $d \in E$ with $d \geq d^0$ there are Nash equilibria in $\Gamma^\lambda$ wth payoff $d^\lambda$ such that $d^\lambda \to d^0$.*

PROOF: Let $\lambda$ be such that there is a payoff vector $d^\lambda$ in $\Gamma^\lambda$ with $d_i^\lambda > v_i$ for all $i$. Choose a sequence of pure strategies $(s^t)_{t=1}^\infty$ which leads to $d^\lambda$ and construct strategies $\mathsf{s}_i = (\mathsf{s}_i^t)_{t=1}^\infty$ as in the proof of Theorem 1.

We show that $\mathsf{s}$ is a Nash equilibrium in $\Gamma^\lambda$, at least if $\lambda$ is chosen sufficiently close to 0: If a player $i$ attempts a deviation, then there will be $t^0$ such that the period $t$ payoff for $i$ becomes $v_i < d_i^\lambda$ for all $t > t^0$. This means that the gain in period $t^0$ (which is upper bounded, since $\Gamma$ has only finitely many pure strategies) must counterweigh the discounted sum of the differences $d_i^\lambda - v_i$, which it cannot, at least when $\lambda$ is sufficiently small. □

**2.3. The finite repetition $\Gamma^K$:** To get a folk theorem in this situation, we must accept further assumptions. Actually, we need that for each player $i$, there must be Nash equilibria in the finite game such that $i$ gets something better than her minimax. The idea is then to construct a sequence of strategy choices such that the payoff gets close to the desired payoff; this sequence should then be followed by a number of rounds for each of the equilibria which gives a player $i$ something better than minimax. In this way we obtain that each player has something to lose if deviating, namely the difference between the later equilibrium payoff and the minimax.

We shall see how this works in details. To get rid of cumbersome details, the assumptions have been strengthened in comparison with previous formulations, so that we now demand existence of equilibria in *pure* strategies with the above properties.

THEOREM 3 *Assume that the game $G$ is such that for each player $i$ there is a Nash equilibrium $\sigma^{[i]} = (\sigma_1^{[i]}, \ldots, \sigma_n^{[i]})$ such that $u_i(\sigma^{[i]}) > v_i$. Let $d \in D$ be an arbitrary payoff vector. Then there is a sequence $(d^\nu)_{\nu=1}^\infty$ such that $d^\nu$ is a Nash equilibrium payoff in $\Gamma^\nu$ for $\nu$ sufficiently large, and $d^\nu \to d$.*

PROOF: From the assumptions we have, that $E$ contains payoff vectors which are strictly greater than $v$. Choose a sequence $(s^t)_{t=1}^\infty$ of pure strategies in $\Gamma$ such that $u(s^t) > v$ and converges to $d$. For arbitrary $\varepsilon > 0$ we choose a number $m$ of repetitions such that $\|u(s) - d\| < \varepsilon$. We now consider $\Gamma^{m+Rn}$, for $R \in \mathbb{N}$, and in this game we define a strategy $\mathsf{s}$, which gives the play $(s^t)_{t=1}^m$ followed by $R$ rounds consisting of $n$ repetitions, where the Nash equilibrium strategies $\sigma_1^{[i]}, \ldots, \sigma_n^{[i]}$ are played.

The strategy $\mathsf{s}$ is constructed as previously, ie. by $\mathsf{s}_i^1 = s_i^1$ for all $i$, and

$$\mathsf{s}_i^t(h^t) = \begin{cases} s_i^t & t \le m, \\ \sigma_i^{[j]} & t = m + kj, \ j = 1, \ldots, n, \ k = 1, \ldots, R, \end{cases}$$

for $h^t = (s^\tau)_{\tau=1}^{t-1}$, while $\sigma_i^t(h') = \hat{\sigma}_{-i^0}$ for every history $h'$ different from $h$, where $i^0$ is the first to deviate.

We check that $\mathsf{s}$ is a Nash equilibrium. Assume that player $i$ changes to some $t_i$. The first $t$ at which $t_i$ prescribes a different choice than $\mathsf{s}$ must be among the initial $m$ repetitions, since after this only Nash equilibria are chosen, and no player can improve in this situation. But if the other players switch to the punishment strategies immediately, a gain can be obtained at most in one round by deviating. This gain should then be weighted against the loss: By playing $\hat{\sigma}_i$ the player $i$ is sure to obtain at least $v_i$ and subsequently there will be $R$ rounds with a loss in each round, namely the difference between the equilibrium payoff and $v_i$. Since there is an upper bound to the one-round gain, the numbers $R$ can be determined independently of $m$ so as to make sure that the deviation results in a loss. It follows that $\mathsf{s}$ is a Nash equilibrium in $\Gamma^{m+Rn}$.

It remains to notice that when $m$ is large enough, then the contribution from the last $Rn$ rounds becomes small, so the sequence of Nash equilibrium payoffs from $\Gamma^{m+Rn}$ tends to $d$ for $m \to \infty$.                                    $\square$

The results of this section all tell the same story, namely that everything (strictly speaking, everything that gives each player at least as much as minimax) can be obtained as non-cooperative (Nash) equilibrium, or, as it is often formulated, repetition enhances cooperation.

One may also see the results from another, more destructive point of view: When everything is obtainable as a result of equilibrium behavior, then the theory becomes next to meaningless, since it cannot be used for projections into the future of what will happen given the observations today. In other words, the non-cooperative game theory loses its explanatory power when the game is repeated.

Before one accepts these rather drastic consequences of the theory, it may be worthwhile considering whether the results hinge on our specific choice of solution concept (Nash equilibrium), against which we may have other reservations. Therefore, one has to repeat the story above using a more realistic solution concept.

## 3   Payoffs in subgame perfect Nash equilibria

As we have seen, the Nash equilibrium is not an ideal choice of solution concept, in particular when we deal with dynamic games, since they allow for equilibrium strategies which involve unrealistic threats. This objection against the solution concept can be directed against all the previous results, where the equilibrium strategies of the repeated games contain threats of punishment that may be very harmful not only the deviating player but also to the players carrying out the punishment.

In order to avoid that such unrealistic threats are used in the equilibrium strategies, it seems natural to switch attention to subgame perfect Nash equilibria. The subgames of the repeated games arise in a very natural way, namely as the games starting at any date $t$. In this section we shall see which modifications must be made of the previous results when we replace Nash with subgame perfect Nash equilibrium.

**3.1. The non-discounted repetition $\Gamma^\infty$.** The simplest situation is once again $\Gamma^\infty$.

The following result is the first of the "second-generation" folk theorems.

THEOREM 4 *Let $d \in D$ be such that $d_i \geq v_i$, all $i$. Then there is a subgame perfect Nash equilibrium $s$ in $\Gamma^\infty$ with $u^\infty(\sigma) = d$.*

PROOF: As previously we choose a play $s = (s^t)_{t=1}^\infty$ such that $u^t(s^1, \ldots, s^t)$ converges to $d$. We then need to construct strategies $s = (s^t)_{t=1}^\infty$ which give rise to the play $s$. This is done as follows: We let $s^1 = s^1$, and for $t > 1$ we put $s(s^1, \ldots, s^{t-1}) = s_i^t$, $i = 1, \ldots, n$.

Now, let $h^t$ be a $t$-history which is equal to $(s^1, \ldots, s^t)$ except in $s_i^t$ for some $i$. We let $s_j^\tau(h) = \hat{\sigma}_j$ for every history $h$ which continues $h^t$, for $\tau$ in $[t + 1, R(t)]$, where $R(t)$ is determined in such a way that player $i$'s gain over all the periods from $t$ to $t + R(t)$ is smaller than $d_i + \dfrac{1}{t}$ (so that convergence to towards $d_i$ is assured). For histories, where the minimax has been chosen from some $\tau$ and onwards, $s^t$ is chosen again for $t > \tau + R(\tau) + 1$.

This choice of strategies constitutes a subgame perfect Nash equilibrium: A deviating player $i$ is punished with her minimax for a period which is long enough to secure that the average of payoffs gets below $d_i$. To make sure that the other players will accept participating in this punishment it should not cost them anything, and indeed it does not: They may lose something in the each period where they are minimaxing player $i$, but the punishment phase terminates after $R(t)$ rounds, and a loss over a fixed number of rounds will go to zero when averaged over sufficiently many periods. If they should want to take an extra gain during the period where they were supposed to participate in punishing a deviator, this again will not show up in the average after sufficiently many periods. In total, we have that $s$ is a subgame perfect Nash equilibrium. □

This result is on one hand rather strong, since it tells us that restricting the solution concept considerably, which is what we have done by moving from Nash to subgame perfect Nash equilibria, does not change the fundamental picture, which is that everything can emerge as result of equilibrium behavior. On the other hand the proof suggests that this result is obtained mainly as a by-product of our way of constructing the repeated game, since what happens in any period of finite duration becomes unimportant. It may therefore be expected that the result will look different if we consider other ways of constructing the repeated game.

**3.2. The discounted repetition $\Gamma^\lambda$.** To obtain a payoff vector in $E$ as the outcome of a subgame perfect Nash equilibrium (for $\lambda \to 0$), we need – in contrast to the previous case – to pay attention to the loss suffered by the players participating in a punishment of deviators, which should be balanced by a gain in the phase following the termination of punishment (something which happened automatically in the non-discounted case). We will show the idea behind the general result by considering a special case, which however differs from the general one mainly by omission of some tedious detail.

THEOREM 5 *Assume that* $\Gamma$ *is such that all* $d \in D$ *can be obtained using pure strategies, and that* $E$ *has nonempty interior. Then every* $d \in D$ *can be obtained in the limit as a result of a subgame perfect Nash equilibrium in* $\Gamma^\lambda$, *for* $\lambda \to 0$.

PROOF: Choose a play $(s^t)_{t=1}^\infty$ such that $u^\lambda(s) = d$ (the existence of such a play is straightforward when enough pure strategies are available, since one may choose $\Gamma$-strategies giving $d$ in each round). We now have to define a strategy which gives us this play when nobody deviates, and which leads to punishment of deviators. We restrict ourselves to describing the choices of $\Gamma$-strategies necessary after deviations, and the construction of the strategy from these choices follows the same pattern as above.

Assume that player $i$ deviates at period $t$ (and has smallest index among the deviators at time $t$). Then $\hat{\sigma}_{-i}$ is played in $R$ rounds, where $R$ is so large that the gain at the time of deviation is smaller than the discounted loss of $d_i^\lambda - v_i$ over the $R$ periods.

Having thus designed punishments, we are left with the problem that it may hurt the punishing players to punish player $i$. We therefore introduce a bonus-phase following the $R$ punishment rounds; this phase may cover all the future rounds, and its result is, seen in isolation some $b^i \in E$. The different bonus-phases (depending on the deviator $i$ giving rise to it) must satisfy the condition

$$b_j^i > b_i^i$$

(and to be able to choose $m$ points $b^i \in E$ with this property we need that $E$ has nonempty interior). This condition assures that the final result for a player of fulfilling her role punishing a deviator is better than what is obtained by being a deviator (through not fulfilling this role).

It remains only to make sure that the gain obtained in the second phase by following the rules is big enough to cover the loses due to participation in punishment. Here we use that the tails gets higher and higher weight as $\lambda$ goes to 0. Sooner or later the gain, how small it may be, in each round of the tail, will be big enough to counterbalance the finitely many losses in the punishment phase.  $\square$

### 3.3. The finite repetition $\Gamma^K$.
For the finite repetitions there is a result which connects to the general comments made in the introduction about backwards induction:

LEMMA 2 *If* $\Gamma$ *has exactly one Nash equilibrium payoff* $d$, *then only* $d$ *is attainable as outcome of a subgame perfect Nash equilibrium of* $\Gamma^n$, *for arbitrary* $n$.

The proof follows the ideas of the intuitive reasoning about Prisoners' dilemma. In the final round only $d$ is possible, meaning that the next-to-final round effectively is the last one, but then also here one can get only $d$, etc.

It is therefore clear that a folk theorem in the same style as for the other repetitions demands the existence of several Nash equilibria in $\Gamma$. The result below

**Table 8.1. Folk theorems for repeated games of different types**

| Type of repetition | Nash equilibria | Subgame perfect Nash |
|---|---|---|
| Final repetition ($\Gamma^h$) | For each $i$ equilibria better than minimax | (1) for each $i$ two equilibria (2) $E$ has nonempty interior |
| Discounted repetition ($\Gamma^\lambda$) | There is a payoff better for all than minimax | $E$ has nonempty interior |
| Non-discounted repetition ($\Gamma^\infty$) | | |

uses the same method as Theorem 3, and therefore the assumptions are of same type, only stronger. We will again omit the complications arising from the need for mixed strategies; the result holds also when they are included.

THEOREM 6 *Assume that all $d \in E$ can be obtained choosing pure strategies in $\Gamma$. If for each player $i$, there are Nash equilibria $\sigma^{[i,1]}$ and $\sigma^{[i,2]}$ in $\Gamma$ with $u_i\left(\sigma^{[i,1]}\right) > u_i\left(\sigma^{[i,2]}\right)$, and $E$ has nonempty interior, then every point in $E$ can be obtained as a limit of subgame perfect Nash equilibrium payoffs in $\Gamma^K$.*

PROOF (sketch): As earlier the game is split in phases, of which the last consists of sufficiently many plays of each of the strategy arrays $\sigma^{[1,1]}, \ldots, \sigma^{[n,1]}$. Since they are all Nash equilibria in $\Gamma$, there cannot be any deviations in this phase. We also use the Nash equilibria, namely in the case where there were deviations by player $i$ during the next-to-last phase. Again, these are Nash equilibrium strategies, so that there will be no incentive for deviations.

All this concerns only the final part of the game; in the essential part we follow the ideas of the previous theorem. A play is chosen which brings $u^m$ close to $d$. If player $i$ deviates, she is punished the necessary number of rounds with $\hat{\sigma}_{-i}$. The punishment is followed by a bonus-phase which achieves that player $j$ prefers to participate in punishing rater than being punished herself. And all this is concluded by the phase of playing of equilibria discussed above. The lengths of the different phases are determined beginning with the first one (the gain by deviation should be offset by the losses of the punishments), followed by the second one (the bonus-phase should be so long that the gain by adhering to the rules is greater than the losses incurred while punishing), and concluding by the last phase (equilibria should be played in so long a period that the gain from deviating during the bonus-phase is offset by the difference between the good and the bad equilibrium. The details can be filled in following the procedure of the previous theorems.  □

The different types of folk theorems are summarized in Table 8.1, which also indicates the special assumptions which must be made in some of the cases.

It can be seen that the non-discounted case is the one which gives the most striking results: Everything better than minimax can be obtained not only as Nash, but also as subgame perfect Nash equilibrium payoff in the repeated game. For the other types of repetition further assumptions are necessary, but at least for the discounted case these assumptions are not very restrictive.

## 4   Repeated games with limited rationality

In the previous sections, we have shown that the basic statement of the folk theorem, namely that every payoff of the one-shot game can be obtained as an equilibrium payoff when the game is repeated infinitely often, can indeed be shown to hold, at least under some minor additional assumptions. However, the equilibrium strategies might be quite complicated, and if players have less than perfect analytical and computational capacities, then the equilibrium payoffs of the repeated games might be a much smaller set.

Before we enter into the discussion of games where each of the players has limited computational capacity, it may be useful to consider the specific tools used in the analysis, since they are not quite commonplace in economics or game theory. The concepts used are taken from the theory of finite automata, which has a long record of applications in many sciences, and it is therefore a well-established field of theory from which the ideas are chosen.

**4.1. Automata.** An automaton can be introduced as a general system (time independent and with a finite state description), but it may be easier to start just with the formal definition and develop whatever we need from there.

A *finite automaton* is given as a quintuple $\mathfrak{A} = (A, B, Q, \delta, \lambda)$, where $A$ and $B$ are finite sets, referred to as input and output alphabets, $Q$ a finite set of states, and $\delta, \lambda : Q \times A \to Q$ are maps, described as the next-state and the output functions, respectively.

To get a more precise understanding of what automatas can be used for, we take a closer look at the alphabets $A$ and $B$ and the finite sequences of elements of either $A$ og $B$. If $I$ is an alphabet (such as $A$ or $B$), a *word* over $I$ is defined as a finite sequence of elements (naturally referred to as letters) from $I$, whereby each letter may occur several times. If $I = \{x_1, x_2\}$, then

$$x_1, \ x_2 x_1 x_2, \ x_1 x_1 x_2 x_1$$

are all words over $I$. The *length* of a word is the number of letters in the word. For practical reasons we allow also the *empty word* $v$, which has no letters; it has length 0.

The set of all words (that is finite sequences) over $I$ is denoted $I^*$. Two elements from $I^*$ can be multiplied in the obvious way, namely by writing one word after the other, so that the product of $w_1 \in I^*$ and $w_2 \in I^*$ is written as $w_1 w_2$. If for example $w_1 = x_1 x_2 x_1$ and $w_2 = x_1 x_1 x_2 x_1$, then

$$w_1 w_2 = x_1 x_2 x_1 x_1 x_1 x_2 x_1.$$

We have $vw = wv = w$ for all $w$ (so that $v$ is neutral element for the multiplication). It is seen from the example above that multiplication is not commutative in general.

A subset of $I^*$ (which is the set of all words) is called a *language*. Returning to the automaton and the two maps $\delta$ and $\lambda$, then they can be extended to $Q \times A$ in a straightforward way: We put $\delta(q, v) = q$ for all $q \in Q$, and then we define $\delta$ inductively by the length of the words: For pairs $(q, w)$, where the length of $w$ is 1, $\delta(q, w)$ is already defined, and if *delta* is defined on all $(q, w)$ with length of $w \geq k$ for some $k \geq 1$, and the length of $w'$ is $k + 1$, then $w'$ can be written uniquely as $w' = wx$ for some word $w$ of length $k$ and some letter $x$, and we let

$$\delta(q, w') = \delta(q, wx) = \delta(\delta(q, w), x).$$

The map $\lambda$ is extended in similar way.

---

**Example 8.1.** The following is an example of an automaton: We let $A = \{a_1, a_2\}$, $B = \{b_1, b_2, b_3\}$, and $Q = \{q_1, q_2, q_3\}$. The maps $\delta$ and $\lambda$ are given by the table below (separated by a bar '|'):

|       | $q_1$       | $q_2$       | $q_3$       | $q_4$       |
|-------|-------------|-------------|-------------|-------------|
| $a_1$ | $q_2|b_2$   | $q_3|b_2$   | $q_4|b_1$   | $q_4|b_2$   |
| $a_2$ | $q_1|b_2$   | $q_1|b_2$   | $q_1|b_2$   | $q_1|b_3$   |

The table is read as follows: If the automaton in the state $q_1$ gets the input letter $a_1$, then it shifts its state to $q_2$ and writes $b_2$ as output.

---

Traditionally, automata are illustrated as graphs where nodes correspond to states and edges to input-output combinations. The reader may try out this type of illustration on the example above.

In the literature, distinction is made between Mealy and Moore automata. The first kind of automata is what has been introduced above. We have a Moore automaton when output depends only on state, not on current input. Such automata are simpler to deal with, and they can do the same jobs as Mealy automata. In the following we give a more precise content to this statement.

For an arbitrary (Mealy or Moore) automaton of the form $\mathfrak{A} = (A, B, Q, \delta, \lambda)$ and for any state $q \in Q$, we can define the response function $r_q^{\mathfrak{A}} : A^* \to B^*$ by

$$r_q^{\mathfrak{A}}(v) = v,$$

$$t_q^{\mathfrak{A}}(aw) = \lambda(q, a)r_{\delta(q,a)}(w), \quad w \in A^*.$$

Now, let $\mathfrak{A}' = (A, B, Q', \delta', \lambda')$ be another automaton (with the same input and output alphabets). Two states $q \in Q$ and $q' \in Q'$ are *equivalent* if their response functions $r_q^{\mathfrak{A}}$ and $r_{q'}^{\mathfrak{A}'}$ are identical. Furthermore, $\mathfrak{A}$ and $\mathfrak{A}'$ are *equivalent automata* if for each $q \in Q$, there is an equivalent state $q' \in Q'$, and vice versa.

We have the following:

THEOREM 7 *For every Mealy automaton* $\mathfrak{A} = (A, B, Q, \delta, \lambda)$ *there is an equivalent Moore automaton* $\mathfrak{A}' = (A, B, Q', \delta', \lambda')$ *with* $|Q'| \leq |Q||B|$.

PROOF: Let $Q' = Q \times B$, then the last statement of the theorem is true. Define $\delta'$ by

$$\delta'((q, b), a) = (\delta(q, a), \lambda(q, a))$$

(so that $\delta'$ does not depend on the $b$-component in $q'$), and let $\lambda'$ be given by

$$\lambda'((q, b), a) = b.$$

It is seen that $\lambda'$ depends only on the state, so that $\mathfrak{A}'$ is a Moore automaton.

We must show that this automaton is equivalent with $\mathfrak{A}$. Choose $q_0 \in Q$ and $w = a_1 a_2 \cdots a_k \in A^*$ arbitrarily, and let $q_i = \delta(q_{i-1}, a_i)$, $b_i = \lambda(q_{i-1}, a_i)$. For arbitrary $b_0 \in B$ we have that $q_0$ and $(q_0, b_0)$ are equivalent states: We have that

$$\delta'((q_{i-1}, b_i), a_i) = (\delta(q_{i-1}, a_i), \lambda(q_{i-1}, a_i)) = (q_i, b_i)$$

and $\lambda'((q_{i-1}, b_i), a_i) = b_i$ for $i = 1, \ldots, k$, so that

$$r_{q_0}^{\mathfrak{A}}(a_1 a_2 \cdots a_k) = f_{(q_0, b_0)}^{\mathfrak{A}'}(a_1 a_2 \cdots a_k) = b_1 b_2 \cdots b_k,$$

from which we conclude that $q_0$ and $(q_0, b_0)$ are equivalent.

It is now easy to see that for each state in $\mathfrak{A}$ there is an equivalent state in $\mathfrak{A}'$, and vice versa.                                                                           □

**4.2. Decomposition of automata and Krohn-Rhodes complexity.** This subsection represents a digression from our main theme, but having introduced automata with the purpose of using them in the context of game theory we take a brief glance of some of the more remarkable results about their structure.

Let $\mathfrak{A} = (A, B, Q, \delta, \lambda)$ and $\mathfrak{A}' = (A', B', Q', \delta', \lambda')$ be automata. A *morphism* from $\mathfrak{A}$ to $\mathfrak{A}'$ is a triple $(f, h, g)$ of maps,

$$f : A \to A', h : Q \to Q', g : B' \to B$$

(notice that $g$ goes from $B'$ to $B$), such that

$$\delta(q, a) = \delta'(h(q), f(a)),$$
$$\lambda(q, a) = g(\lambda'(h(q), f(a))).$$

This means that the automaton $\mathfrak{A}$ is emulated by $\mathfrak{A}'$ – if we have $\mathfrak{A}'$, then we can make it behave exactly as $\mathfrak{A}$ by translating its input according to $f$, letting states in $\mathfrak{A}$ correspond to states in $\mathfrak{A}'$ as indicated by $h$, and translating output back again using $g$.

The point of all this is that it may be possible to emulate a complicated automaton by another one with a simpler structure. There are two standard ways in which one may achieve such an emulation:

(1) Let $\mathfrak{A}'$ be an automaton with the same input and output alphabets as $\mathfrak{A}$, but with a smaller number of states, that is $|Q'| < |Q|$. Clearly, there is a limit to

the simplification that can be obtained in this way, indeed, for each automaton $\mathfrak{A}$ there is a minimal (with respect to number of states) automaton which emulates $\mathfrak{A}$, and there exists an algorithm which determines this minimal automaton in a finite number of steps.

(2) The other possibility is more interesting, as it involves a new – but straight-forward – idea, namely that of building an automaton from specific automata using a few simple methods of combination. This is known as *decomposition* of the given automaton.

Before giving the main result about decomposition of automata, we take a closer look at the ways in which two automata $\mathfrak{A}$ and $\mathfrak{A}'$ may be composed. Here again, two ways of connecting automata to obtain a new automaton almost suggest themselves: In the *serial connection*, written $\mathfrak{A} \oplus \mathfrak{A}'$, the output of the first automaton is the input of the next one, that is $A' = B$. The states of the combined automaton are pairs $(q, q')$ of states in the original automata, and the next-state and output maps are defined as

$$\delta_{\mathfrak{A} \oplus \mathfrak{A}'}((q, q'), a) = (\delta(q, a), \delta'(q', \lambda(q, a)))$$
$$\lambda_{\mathfrak{A} \oplus \mathfrak{A}'}((q, q'), a) = \lambda'(q', \lambda(q, a)).$$

The *parallel connection* of $\mathfrak{A}$ and $\mathfrak{A}'$, written $\mathfrak{A} \otimes \mathfrak{A}'$ pairs of input are sent to pairs of output, and its states are pairs of states in the original automata. The mappings are now

$$\delta_{\mathfrak{A} \otimes \mathfrak{A}'}((q, q'), (a, a')) = (\delta(q, a), \delta'(q', a')$$
$$\lambda_{\mathfrak{A} \otimes \mathfrak{A}'}((q, q'), (a, a')) = (\lambda(q, a), \lambda'(q', a')).$$

Having now some suggestions as to how to combine simple elements in order to get more complicated automata, we may introduce the building blocks. In order to do this, we need to consider some abstract algebra.

In the previous section, we introduced the response function $r_q^{\mathfrak{A}}$ which to each automaton $\mathfrak{A} = (A, B, Q, \delta, \lambda)$ and each state $q \in Q$ gave us a function from $A^*$ to $B^*$. A related construction, this time departing from $\delta$, is the following: For each word $w \in A^*$ we consider the map $(w) : Q \to Q$ defined inductively by

$$(v)(q) = q,$$
$$(w'a)(q) = \delta((w')(q), a).$$

It may now be checked that $(w_1 w_2)(q) = (w_2)((w_1)(q))$, so that the product of two words in $A^*$ obtained by concatenation (writing one word after the other) is trans-lated to composition of the associated maps. This makes the set $A^*$ a *semigroup* of mappings from $Q$ to itself $Q$ (recall that a semigroup is a set $S$ with a composition $\circ$ which is associative, so that $s_1 \circ (s_2 \circ s_3) = (s_1 \circ s_2) \circ s_3$ for all $s_1, s_2, s_3 \in S$). This semigroup contains an identity, namely $(v)$.

The advantage of considering words over $A$ as mappings from $Q$ to itself is that the relation to the specific input alphabet is loosened, the automaton is character-ized by the semigroup of maps from $Q$ to $Q$. Also, it is easy to point out what constitutes a very well-behaved automaton, namely that the associated semigroup should be well-behaved in some sense, and the most obvious choice here is that the semigroup should be a *group*, meaning that each $s \in S$ should have an inverse $s^{-1}$ with $s \circ s^{-1} = s^{-1} \circ s = e$, where $e$ is the identity of the group; in our case to each word $w$ there should be a word $w'$ such that $(w) \circ (w') = (w') \circ (w) = (v)$. Returning to the interpretation as automata, it means that if in a state $q$ we input the word $w$, thereby getting to a state $q'$, then there should be a word $w'$ which inputted at $q'$ will take the automaton back to $q$. Automata for which the associated semigroup is a group are called *permutation automata*.

We need another building block, which is quite different from the previous one. It is known as a *flip-flop* (or *two-state reset automaton*), and its next-state function is as follows:

|       | $q_1$ | $q_2$ |
|-------|-------|-------|
| $a_1$ | $q_1$ | $q_2$ |
| $a_2$ | $q_1$ | $q_2$ |
| $a_3$ | $q_2$ | $q_2$ |

It is seen that there are three input letters and two states, and the three input letters correspond to either the identity map or the two constant maps from $Q$ to $Q$.

We now have the requisites for a formulation of the structure theorem for finite automata, due to Krohn and Rhodes [1965], given here without proof.

THEOREM 8 *Any automaton can be emulated by an automaton constructed from serial and parallel connections of automata which are either permutation automata or flip-flops.*

The insights obtained by this result are far-reaching, since it shows that all automata have rather simple structure, something which was by no means obvious from the outset. Also, it makes it possible to introduce a complexity measure of automata, namely by counting the number of building blocks in the above construction. The resulting measure is however quite difficult to use, and there are only few examples of applications, e.g. Futia [1977].

**4.3. Prisoners' dilemma with automata.** We now return to the theory of repeated games. Our point of departure is the folk theorem in its different versions, which says that all individually rational payoff vectors in the original game can be obtained in equilibria of the infinitely repeated game, even if one demands subgame perfectness. Intuitively this is a somewhat unsatisfactory result: It may be acceptable that repetition gives rise to new equilibrium payoffs, but it is more difficult to accept that all payoffs can occur in equilibrium. This points to the need for introduction of new aspects in the analysis. One such aspect is that of bounded rationality, according to which the players can plan only strategies that follow a certain, not too complicated pattern.

An obvious way of allowing for this is to assume that players must leave the implementation of the strategies to (Moore) automata. The role of the players is then reduced to creating or choosing the automaton which takes care of the moves, and to monitor the play; if the play is unsatisfactory, the automaton can be replaced by another one.

We consider here a two-person game $\Gamma = (S_1, S_2, u)$, and we assume that $S_1$ and $S_2$ are finite. From this game we form the supergame by infinite repetition, but we leave it to finite automata to take care of the moves in the repeated games. Let $\mathfrak{A}_i$ be the automaton of player $i$, $i = 1, 2$. The input alphabet for $\mathfrak{A}_i$ is $S_j$, $j \neq i$, that is the set of (pure) strategies of the other player in the stage game, and the output alphabet is $S_i$, the pure strategies of player $i$. As usual, we have a set $Q_i$ of states of the automaton, the "next-state" map $\delta_i : Q_i \times S_j \to Q_i$ updates the state, and the output map $\lambda_i : Q_i \to S_i$ takes care of the move. Finally, each automaton has an initial state $q_i^0$, $i = 1, 2$.

When the game is played using finite automata, there are only finitely many possible combinations of states and outputs. This means that the same states must be repeated sooner or later. The play is therefore cyclical, in the sense that after some initial number of rounds one enters a sequence of moves stretching from $t_1$ to $t_2$ which thereafter is repeated again and again. Consequently, average payoff in the long run may be computed by restricting attention to any of these cycles, so that

$$u(s) = \frac{1}{t_2 - t_1 + 1} \sum_{t=t_1}^{t_2} u(s^t),$$

where $s = (s^t)_{t=1}^{\infty}$ is a play giving rise to cycles from $t_1$ of length $T = t_2 - t_1 + 1$. We also write this average payoff as $u(\mathfrak{A}_1, \mathfrak{A}_2)$ to emphasize that it originates from a play carried out by the two automata.

Clearly the choice of the automaton, which plays the game as representative of the player, must be guided by the payoff which this automaton can give rise to. But there may be other considerations as well. Here we follow the idea of Rubinstein [1986] and assume that players want to use as small automata as possible, when size of automata is measured by the number of states in the automaton. Intuitively this may be compared to the memory size of a computer; however, considerations of memory size should be considered as a second order concern, which comes into play only when the payoff is not reduced as a result of replacing the automaton. From a formal point of view, players have lexicographic preferences $>_L$ over pairs of payoff and machine capacity.

To formulate the equilibrium concept, we must also allow for the possibility that a player $i$ at any stage $t$ can replace her automaton by another one, which then proceeds the play against the automaton $\mathfrak{A}_j(q_j^t)$ of the other player, where $q_j^t$, the initial state of the automaton, is the state reached at the time of replacement.

Formally a pair $(\mathfrak{A}_1^0, \mathfrak{A}_2^0)$ of automata is an equilibrium if there is no $i$, $t$ and $\mathfrak{A}_i'$

such that

$$\left(u(\mathfrak{A}'_i, \mathfrak{A}^0_j(q^t_j)), -|\mathfrak{A}'_i|\right) >_L \left(u(\mathfrak{A}^0_i(q^t_i), \mathfrak{A}^0_j(q^t_j)), -|\mathfrak{A}^0_i|\right)$$

(the minus sign in second coordinate indicates that the number of states should be as small as possible). The concern for machine capacity seems reasonable, in particular since it comes into play only when payoff considerations are irrelevant. However, it turns out to play an important role in deriving results, as we shall see below.

We shall restrict attention to a particular two-person game, namely Prisoners' dilemma, here in the form

|   | E | T |
|---|---|---|
| E | (2, 2) | (−1, 3) |
| T | (3, −1) | (0, 0) |

We are interested in finding the set of payoff vectors that can be achieved in a (semi-perfect) equilibrium. The key to the answer is in the following two small lemmas.

LEMMA 3 Let $(\mathfrak{A}^0_1, \mathfrak{A}^0_2)$ be an equilibrium. If $\{t_1, \dots, t_2\}$ represents a cycle, then each of the automata $\mathfrak{A}^0_1$ and $\mathfrak{A}^0_2$ runs through $t_2 - t_1 + 1$ different states.

PROOF: Assume that one of the automata, say $\mathfrak{A}^0_1$, repeats a state at two different stages $k_1 < k_2$ during the cycle; we may assume that $t_1 \le k_1$ and $k_2 \le t_2$, since a cycle can be taken to start at any stage. Since $\mathfrak{A}^0_1$ is back at the same state in $k_2$, we have that player 2 if she wishes can change her automaton so that it performs at $k_2$ as in $k_1$, in $k_2 + 1$ as in $k_1 + 1$ etc. Since we have an equilibrium, such a change of automaton cannot improve player 2's payoff, so

$$\frac{1}{k_2 - k_1} \sum_{t=k_1}^{k_2-1} u(s^t_1, s^t_2) \le u^*_2,$$

where $u^*_2$ is the equilibrium payoff of player 2. If strict inequality < holds, then player 2 could consider the play which starts at $k_2$, passes the end of the cycle and then goes to $k_1$. Also this sequence can be made into a cycle as indicated above, and if the above average was $< u^*_2$, then the new one must be $> u^*_2$, contradicting equilibrium.

We conclude that average payoff over the play from $k_1$ to $k_2$ is the same as before, namely $\pi^*_2$. But it is part of the equilibrium condition that no player can replace the automaton with another one using fewer states, and the replacement machines considered above can be made so as to have no more states than the minimum of $k_2 - k_1$ and $(t_2 - t_1) - (k_2 - k_1)$. This gives a contradiction, which proves that there can be no repetitions of states in the equilibrium.                                    □

LEMMA 4 *Let $(\mathfrak{A}_1^0, \mathfrak{A}_2^0)$ be an equilibrium. Then there cannot be a stage $j > t_1$ such that $\mathfrak{A}_i^0$ prescribes different strategies from $\{T, E\}$ in $j$ and $j + 1$ whereas $\mathfrak{A}_j^0$ prescribes the same strategy from $\{T, E\}$.*

PROOF: If player 2's automaton has to choose the same strategy in $j$ and $j + 1$, then from Lemma 1 we know that it uses different states in $j$ and $j + 1$, whereby the state at $j + 1$ is only used as a counter. But player 2 can replace this automaton at $t_1$ with another automaton which leaves out this state, and instead use the information provided by the other player who actually changes strategy. Therefore the initial choice of automata cannot be an equilibrium. □

We may now approach the characterization of payoffs which occur in equilibria of the repeated Prisoners' dilemma game.

THEOREM 9 *The payoff vectors $d$ which can occur in equilibrium are either $d = (0,0)$ or*

$$d = \alpha(3, -1) + (1 - \alpha)(-1, 3)$$

*for some rational number $\alpha \in [0, 1]$ such that $d \geq (0, 0)$.*

PROOF: From Lemma 4 we have that in the equilibrium play, the strategy pairs in the cycle is either $(E, E)$, $(T, T)$ or alternatively $(E, T), (T, E)$ (players must change strategies simultaneously).

We first consider equilibria of the first type. Suppose that $(E, E)$ occurs in the cycle at time $t_1$ for some state $q_1^{t_1}$ of player 1. Consider the state $q_2^{t_1+1}$ that would occur in player 2's automaton if 1 chooses $T$ instead. Since each state occurs at some stage of the equilibrium play, there is some $t' > t_1$ such that the automaton of 2 is in the state $q_2^{t+1}$ at $t'$ when the equilibrium strategies are played. The average payoff from $t_1 + 1$ to $t'$ is bigger than average payoff over the whole cycle, since otherwise player 1 could redesign her automaton so as to choose $T$ and then proceed as after $t'$; this would give an increase in payoff for player 1, contradicting equilibrium.

Since the average payoff to player 1 in the sequence from $t_1 + 1$ to $t'$ when playing the equilibrium is greater than the average over the cycle, and only $(E, E)$ and $(T, T)$ are played, there must have been some instance of $(E, E)$ between $t_1 + 1$ and $t'$, say at $t_2 > t_1$.

We now use the same argument as above from $t_2$ to obtain some $t_3 > t_2$ such that the average payoff in the cycle from $t_1$ to $t_3$ is above equilibrium, and we then proceed to $t_4, t_5$ etc. Since the cycle is finite, we get a contradiction, showing that $(E, E)$ cannot occur in an equilibrium with only $(E, E)$ and $(T, T)$ as possible choices, consequently the outcome must be $(0, 0)$.

The other equilibrium pattern occurs when the automata selects $(T, E)$ $K_1$ times and $(E, T)$ $K_2$ times, where $K_1/(K_1 + K_2) = \alpha$. □

## 5   Problems

**1.** The Prisoners' dilemma game in Example 1.2 is given below:

|               | confess     | not confess |
|---------------|-------------|-------------|
| confess       | $(-9, -9)$  | $(0, -10)$  |
| not confess   | $(-10, 0)$  | $(-1, -1)$  |

Find the minimax payoffs $v_1, v_2$ and the set of payoff vectors which can be obtained as Nash equilibria in the $\lambda$-discounted repeated game, for $\lambda$ close to 0.

Suppose now that the repeated game is modified as follows: Before each repetition, a coin is tossed, and the game terminates if tails and proceeds to a new instance of the one-shot game if heads. Does a counterpart of the Folk theorem hold for this type of repeated game?

**2.** Consider the following game $\Gamma$:

|       | $L$         | $R$         |
|-------|-------------|-------------|
| $T$   | $(-9, -9)$  | $(0, -10)$  |
| $B$   | $(-10, 0)$  | $(-1, -1)$  |

Describe the trigger strategy in $\Gamma^{\lambda}$. Check whether there are values of $\lambda$ for which the trigger strategies constitute a subgame perfect Nash equilibrium.

**3.** Consider the Cournot duopoly model in Chapter 5, Section 2.2. Suppose that the duopoly is repeated infinitely often, and the two contestants are guided by discounted present value of future profits, with a discount factor $\beta < 1$. Find the value of $\beta$ such that the equal division of the monopoly profit is sustained as a Nash equilibrium.

**4.** Suppose that in a repetition of the $\Gamma = (N, (S_i)_{i \in N}, (u_i)_{i \in N})$, the players can observe only the strategies played by the other players but not the player who chose this strategy. What are the implications for the Nash equilibrium payoffs of the repeated game? (For more about this type of situations, the reader is referred to Kaneko [1982].

**5.** (The one stage deviation property) Show that a strategy array in $\Gamma^{\lambda}$ is a subgame perfect Nash equilibrium if and only if no player can gain by deviating in a single period after any history.

Show by examples that there are Nash equilibria which do not satisfy the one stage deviation property.

# Chapter 9

# Selected Topics in Non-Cooperative Games

## 1 Introduction

The present chapter treats several topics, which are only very weakly connected to each other. They are collected here as an indication of the fertility of game-theoretical reasoning in many other fields of science. We consider first some applications in mathematical logic, which again point in several directions, as indicated by the work by Gale and Stewart [1953] on the one hand and the Ehrenfeucht-Fraïssé games on the other.

Following up on this, we give a brief treatment of combinatorial games, which formally may be considered as a subclass of the games that we have treated in previous chapters (two-person extensive form games with no chance moves, full information and simple payoff assignment) but which has an impressive mathematical theory going far beyond what is mentioned here. Finally, we give a short treatment of differential games.

The main objective in this chapter is to show that game theory has applications reaching out in many directions, some of them giving rise to spectacular further developments.

## 2 Games and logic

One of the fields where game theory has found widespread application is (mathematical) *logic*. One of the earliest instances where game theory appears in mathematical logic is Zermelo [1913], and the connection between games and logic were acknowledged in the early years of game theory with the contribution of Gale and Stewart [1953], which will be briefly outlined below, following the approach in Martin [1975].

Let $Y$ be a set of finite sequences, such that all initial segments of a sequence in $Y$ belongs to $Y$ (for technical reasons including also the empty segment), and such that each element of $Y$ is an initial segment of an element of $Y$ (for example, the set of all finite segments of the underlying set). We then let $\mathcal{F}(Y)$ be the set of all *infinite* sequences $(y_0, y_1, \ldots)$ such that the finite initial segments are in $Y$. For any subset $A$ of $\mathcal{F}(Y)$, define a two person game $\mathcal{G}(A, Y)$ as follows: Player 1 chooses an element $y_0$ (with the sequence $(y_0)$ belonging to $Y$, then player 2 chooses $y_1$ with $(y_0, y_1) \in Y$, player 1 chooses $y_2$ with $(y_0, y_1, y_2) \in Y$ etc. Player 1 *wins* if and only if the sequence $(y_0)_{i=0}^{\infty}$ belongs to $A$. Player 2 loses when player 1 wins and wins when player 1 loses.

A strategy for player 1 in the game $\mathcal{G}(A, Y)$ is a function $s$ defined on all elements (sequences) $(y_0, \ldots, y_{2n-1})$ of $Y$ of even length such that

$$(y_0, \ldots, y_{2n-1}, s(y_0, \ldots, y_{2n-1})) \in Y,$$

and a play of the game is a play according to the strategy $s$ if for each $n$, $y_{2n} = s(y_0, \ldots, y_{2n-1})$. Similarly, a strategy of player 2 takes sequences of odd length to new sequences in $Y$. A strategy is *winning* for player 1 (player 2) if every play according to $s$ results in player 1 (player 2) winning the game. The game is *determined* if one of the players has a winning strategy.

Consider now the family $\mathcal{B}$ of subsets of $\mathcal{F}(Y)$ of the form

$$B = \{(y_0', y_1', \ldots) \mid y_i' = y_i, i = 0, 1, \ldots, n\}$$

for some $(y_0, y_1, \ldots, y_n) \in Y$ (so that $B$ contains all sequences beginning with $(y_0, \ldots, y_n)$). If $B_1, \ldots, B_k$ are in $\mathcal{B}$, then $\cap_{i=1}^{k} B_i$ is either empty or belongs again to $\mathcal{B}$, so $\mathcal{B}$ is a base for a topology on $\mathcal{F}(Y)$. A game $\mathcal{G}(A, Y)$ is *open* if $A$ is an open set in this topology, and *closed* if $A$ is closed.

The following result is known as the Gale-Stewart theorem.

THEOREM 1 *Let* $\mathcal{G}(A, Y)$ *be a closed game. Then* $\mathcal{G}(A, Y)$ *is determined.*

PROOF: If player 2 has a winning strategy, then $\mathcal{G}(A, Y)$ is determined. Assume therefore that player 2 has no winning strategy for $\mathcal{G}(A, Y)$. Then there must be a first move $y_0$ by player 1 so that player 2 has no winning strategy for the game $\mathcal{G}(A, Y)$ with starting position $(y_0)$. Also, if player 2 has no winning strategy for $\mathcal{G}(A, Y)$ starting at $(y_0, \ldots, y_{2n})$, then for every $y_{2n+1}$ there is $y_{2n+2}$ such that player 2 has no winning strategy for $\mathcal{G}(A, Y)$ starting at $(y_0, \ldots, y_{2n+1}, y_{2n+2})$. It follows that we can choose $y_0, y_2, \ldots, y_{n+2}, \ldots$ in such a way that we get a strategy $s$ for player 1.

Now we notice that if $A$ is a closed subset of $\mathcal{F}(X)$, then its complement is open. Using the definition of openness, we have that if $(y_0, y_1, \ldots, y_n, \ldots)$ is not in $A$, then there is an initial segment $(y_0, \ldots, y_n)$ such that for all $(y_{n+1}', y_{n+2}', \ldots)$, the sequence $(y_0, \ldots, y_n, y_{n+1}', y_{n+2}', \ldots)$ is not in $A$. But then $s$ is a winning strategy for player 1, so that $\mathcal{G}(A, Y)$ is determined. $\qquad\square$

## 3   Ehrenfeucht-Fraïssé games

Another application of the machinery of game theory can be found in the so-called Ehrenfeucht-Fraïssé games. These games, played by two persons over a finite number of rounds, have emerged as an important tool in model theory, where they are used to investigate questions of equivalence of mathematical structures. The background idea, formulated by Tarski (1930), was to consider two structures as equivalent if exactly the same statements, formulated in first-order logic, are true in them. Through the work of several people emerged what is now known as Ehrenfeucht-Fraïssé games or alternatively back-and-forth games.

Following the approach of the previous section, we begin with the game[1]. In an Ehrenfeucht-Fraïssé game, the two players are denoted Spoiler and Duplicator. The game works with two graphs (given by their vertices and edges), which we denote $A$ and $B$. There are $n$ rounds, and in each round, the following moves are admissible for the players:

- The Spoiler chooses a graph and picks a vertex from this graph,
- The Duplicator then chooses a vertex from the other graph.

We denote by $x_i$ the vertex from $A$ and by $y_i$ the vertex from $B$ in the $i$th round (no matter who has chosen it). Let $X = \{x_1, \ldots, x_n\}$ and $Y = \{y_1, \ldots, y_n\}$ be the vertices chosen in the $n$ rounds. The Duplicator wins the $n$-round game if the map $f : X \to Y$ defined by $f(x_i) = y_i$ satisfies

$$(x_i, x_j) \text{ is an edge in } A \Leftrightarrow (f(x_i), f(x_j)) \text{ is an edge in } B,$$
$$x_i = x_j \Leftrightarrow f(x_i) = f(x_j),$$

i.e. $f$ reflects adjacency and equality. In this case $f$ is said to be a *partial isomorphism* of the two graphs, written $A \sim_n B$. If the Duplicator does not win, the Spoiler wins.

As usual, a strategy of a player is a family $\sigma_1, \ldots, \sigma_n$, where $\sigma_1$ is an admissible move for the player in the first round and $\sigma_i$ for $i = 2, \ldots, n$ maps any history (of choices in previous rounds, and for the Duplicator, a choice of the Spoiler in the present round) to an admissible move for the player in the $i$th round. A strategy is *winning* for the Spoiler if he wins playing this strategy no matter what the Duplicator chooses, and similarly winning for the Duplicator if she wins no matter what the Spoiler does.

The following examples may illustrate the notions: Consider the game in two rounds on the graphs $A$ and $B$. The two graphs are very simple, since there are no edges, and $A$ has one node more than $B$.

$$
\begin{array}{ccc}
\circ & \circ & \quad\quad \circ \\
a_1 & a_2 & \quad\quad b_1
\end{array}
$$

$$
\begin{array}{cc}
A & \quad\quad B
\end{array}
$$

Here the Spoiler may choose three vertices in the first round. If the Spoiler chooses a vertex, say $a_1$ from $A$, then the Duplicator must choose the only vertex in

---

[1]The exposition in this section is inspired by and follows rather closely that of http://www.math.cornell.edu/~mec/Summer2009/Raluca

the other graph. But then the choice of $a_2$ by the Spoiler in the second round leads to a win, since the Duplicator had to choose the same vertex twice, whereas the Spoiler chooses different vertices. Thus, the Spoiler has a winning strategy.

Next, consider the graphs $A$ and $B$ below, keeping two rounds as above.

Now the Duplicator has a winning strategy, which can be described as follows:

In the first round, if the Spoiler chooses from $A$, then the Duplicator chooses $b_1$. If not, the Duplicator chooses $a_1$. In the second round, the Duplicator can adapt to all choices of the Spoiler: If the Spoiler picks something chosen in the first round (either by himself or by the Duplicator), then Duplicator replies choosing the vertex from $A$ picked in the first round. If he chooses a vertex in either graph not picked in the first round, the Duplicator replies with a non-picked vertex in the other graph. In any case, this will give a partial isomorphism, so the Duplicator has won.

Having thus introduced the Ehrenfeucht-Fraïssé games, we may prodeed to their applications. For this, we must introduce the notion of *first order logic*. Here we start with *atomic formula*, which (in the language of the structures we are dealing with) are the primitive statements which may be true and false. Thus, dealing with graphs as we are doing here, the atomic formula express adjacency of vertices, $ad(a_1, a_2)$ means that there is an edge from $a_1$ to $a_2$, something which is true in the graph $A$ and false in the graph $B$ below. From atomic formulas we can make *propositions* using conjunctions ($\wedge$), disjunctions ($\vee$) and negation ($\neg$). At this level we have the so-called *propositional logic*, and we then obtain first order logic by adding the quantifiers $\exists$ and $\forall$ together with suitably many symbols for variables, constants, relations and functions.

An example or a sentence in first order logic is $\exists x\, ad(a_1, x)$, which is true in the graph $A$ below. Another one is $\exists x_1 \forall x_2 \neg ad(x_1, x_2)$ which is false in $A$ but true in $B$, where $b_1$ is a vertex which is not adjacent to any other vertex.

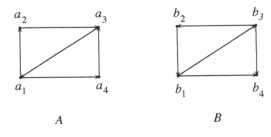

Now we need only one more notion: For $f$ a formula of first order logic, we define the *quantifier rank* of $f$, denoted $qr(f)$, as an expression of the number of quantifiers needed, in the following way:

(a) if $f$ is atomic, then $qr(f) = 0$,
(b) if $f$ has the form $\neg g$ for some first order formula $g$, then $qr(f) = qr(g)$,
(c) if $f$ has the form $g \wedge h$ or $g \vee h$, then $qr(f) = \max\{qr(g), qr(h)\}$,
(d) if $f$ has the form $\exists x g$ or $\forall x g$, then $qr(f) = qr(g) + 1$.

There is a close connection between the Spoiler-Duplicator games considered above and properties of first order logic. More specifically, a winning strategy for the Spoiler in an $n$-round game can be described as follows:

– There is a choice for the Spoiler (in first round) such that for all possible responses of the Duplicator,

– there is a choice for the Spoiler (in second round) such that for all possible moves of Duplicator,

– ...

– there is a choice for the Spoiler (in $n$th round) such that the Spoiler wins.

Translating this to notions referring to the graphs we get that

– There is a vertex in the graph $A$ (used by the Spoiler in his first move) such that for all vertices in $B$ (Duplicator's response)

– there is another vertex in $A$ or $B$ (chosen by the Spoiler) such that for all possible moves of Duplicator,

– ...

– some sentence with quantifier rank $n$ is true in $A$ and false in $B$ or vice versa.

We have now shown the following:

THEOREM 2 *Let $A$ and $B$ be graphs. Then the following statements are equivalent:*

(i) *$A$ and $B$ satisfy the same first-order sentences with quantifier rank $n$,*
(ii) *$A \sim_n B$, that is the Duplicator has a winning strategy in the n-round Ehrenfeucht-Fraïssé game on $A$ and $B$.*

Having come so far, one may guess that the results obtained so far have implications which do not mention games, meaning that the use of games makes it possible to derive results which are important and useful also outside the realm of game theory. Some of these results deal with the possibility of expressing properties in first order logic. Indeed, we can restate the result obtained in the following form, known as the Ehrenfeucht-Fraïssé theorem.

THEOREM 3 *Let $P$ be a property. There is no first order sentence that expresses $P$ if and only if for each n, there are two graphs $A_n$ and $B_n$ such that*

(i)  *property P is true on $A_n$,*

(ii)  *property P is false on $B_n$,*

(iii)  *$A \sim_n B$, that is the Duplicator has a winning strategy in the n-round game on A and B.*

A simple example of property of graph is that it has an even number of nodes. If $G_n$ is the graph

$$G_n$$

then for each $n$ one may choose graphs $A_n = G_{2n}$ and $B_n = G_{2n+1}$, and the Duplicator will win the $n$ round Ehrenfeucht-Fraïssé game. This means that this property is not expressable in first order logic.

Since our purpose with the treatment of Ehrenfeucht-Fraïssé games was to show that gametheoretical reasoning may be used in many different fields, but not to penetrate deeper into these fields, we shall not pursue the matter further. It should of course also be added, that results as the Ehrenfeucht-Fraïssé theorem could have been obtained also without the explicit use of games, but the Ehrenfeucht-Fraïssé games have proved themselves a useful tool which offers an intuitive treatment of the problems considered.

## 4   Combinatorial games

In the study of combinatorial games, we basically consider two-person extensive form games with some additional properties. It is however preferable to define combinatorial games in another way, to which we proceed in a moment.

The specific features of these games may be summarized as the following, staying as yet with the intuitive approach:

- Two players, usually called *Left* (*L*) and *Right* (*R*),
- No chance moves, so that only Left and Right move,
- Perfect information: no moves are hidden to the other players,
- Turn-based: the moves of Left and Right alternate,
- Winner: in every game, there is a winner, namely the first player to fulfill the winning condition below,
- Winning condition: the last player who makes a valid move wins.

An example of a combinatorial game with these properties – one which will play a prominent role in the sequel – is *Nim:* There are two players, and the game uses several piles (of matches or coins). In a move, the player must remove any number of items (but at least one) from one and only one of the piles. The player who removes the last item from the last pile has won.

It is seen that what matters for players in this game is the *position* of the game (how many items there are in each pile). This can be used in the formal approach to combinatorial games. Following Schleicher and Stoll [2006], we define a game recursively, using that a position in itself defines a new game, and the position itself is characterized by the options for $L$ and $R$.

DEFINITION 1 *The set of games is defined by the following two conditions:*

(1) *If L and R are two sets of games, then the ordered pair $G = (L, R)$ is a game.*
(2) *There is no infinite sequence of games $G^i = (L^i, R^i)$ with $G^{i+1} \in L^i \cup R^i$ for all $i \in \mathbb{N}$.*

The first condition in the definition tells us that new games are made from old ones in a particular way; the elements of $L$ and $R$ are called the *left* and *right options* of $G$. The second condition in the definition is called the Descending Game Condition, and it states that a game must come to an end eventually.

So far we have only a method for constructing games, but not an instance of a game. But we may create games starting with the empty set (of games): The simplest game is the zero game written as $0 = (\{\ \}, \{\ \})$, where no player has any move. The next simplest ones are then $1 = (\{0\}, \{\ \})$ with one move for L, $-1 = (\{\ \}, \{0\})$ with one move for R, and $* = (\{0\}, \{0\})$ with a free move for whoever has to take it. It is customary to use the notation $0 = \{\ |\ \}$, $1 = \{0|\ \}$, $-1 = \{\ |0\}$ and $* = \{0|0\}$.

**4.2. Ordering games.** The numbers $1$, $-1$ and $*$ used above were introduced as labels of specific games, but they have some further significance. To see this, we introduce an ordering of games which is related to the winning possibilities of Left and Right. If $G$ is a game, then we write

$G > 0$ if Left can enforce a win no matter who starts,
$G < 0$ if Right can enforce a win no matter who starts,
$G \parallel 0$ if First (the first player to move) can enforce a win, no matter who it is
(one says that '$G$ is fuzzy to zero'),
$G = 0$ if Second (player) can enforce a win.

As usually, $G \geq 0$ means $G > 0$ or $G = 0$ etc. The notation $G \rhd 0$ means $G > 0$ or $G \parallel 0$, i.e. Left can win as first player, and $G \lhd 0$ means that Right can win as first player. These notions can be made precise in accordance with Definition 1 if we define $\geq$ and $\leq$ by $G \geq 0$ unless there is a right option $G^R \leq 0$, and $G \leq 0$ unless there is a left option $G^L \geq 0$. Since the definition is again recursive, one needs to go back to other games to decide whether Left or Right wins, and this is fortunately possible due to (2) of Definition 1.

**4.3. Adding and subtracting games.** Let $G = \{G^L, \ldots \mid G^R, \ldots\}$ (where the notation indicates that the left and right options may consist of arbitrary sets of games) and $H = \{H^L, \ldots, \mid H^R, \ldots\}$ be games. Then we may define

$$G + H = \{G^L + H, G + H^L, \ldots \mid G^R + H, G + H^R, \ldots\},$$
$$-G = \{-G^R, \ldots \mid -G^L, \ldots\},$$
$$G - H = G + (-H).$$

**Example 9.1. Hackenbush.** A simple example of a combinatorial game is provided by *Hackenbush*: The two players have each a particular color (in the figure Left has Black and Right has White), and in a move they may remove ('hack') one stick of their color. If some stick becomes unconnected to the ground, then it cannot be used any more. The winner is the last player to have a valid move.

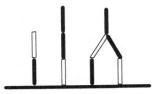

The Hackenbush version of the game 0 is a game with no sticks at all. The game $1 = \{0 \mid \}$ is one where Left has one move and Right has no moves, so it can be illustrated by a Hackenbush game with just one Black stick.

We may also illustrate the notions as $G > 0$ (where Left can win), $G > 0$ (Right has a win) and $G = 0$ where Second can win,

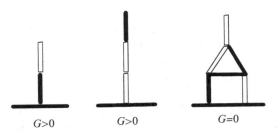

$G>0$                    $G>0$                    $G=0$

Again we have used a recursive definition, so that the sums $G^L + H$ etc. must be defined before we can find $G + H$. Notice that the left options of $G + H$ is the union of all the games $G^L + H$ for $G^L$ left options in $G$ together with the union of all $G + H^L$ for $H^L$ a left option of $H$, and similarly for the right options.

It is seen that $-1 = \{ \mid 0\}$ defined above is the negative of $1 = \{0 \mid \}$, so that our labeling makes sense so far. We also have that $* + * = \{* \mid *\}$ which is a zero element, in the sense that whoever begins will lose. It may be checked – using induction based on the recursive definition of games – that the operation $+$ as introduced above is associative and commutative with $0 = \{ \mid \}$ as zero element. One may then proceed defining $G = H$ if $G - H = 0$, $G > H$ if $G - H > 0$ etc. (the notion $=$ is justified since $G - G = 0$ for all games, it may be checked that $=$ is indeed an equivalence relation).

Using the ordering relations one obtains some methods for simplifying games. If $G$ and $H$ are games, and $H \lhd G$, then $H$ is a *left gift horse* for $G$. Similarly, if $H \rhd G$, then $H$ is a *right gift horse* for $G$. The notions are used to formulate the *gift horse principle*: If $H_L, \ldots$ are left gift horses and $H_R$ are right gift horses for some game $G = \{G^L, \ldots \mid H^R, \ldots\}$ then

$$\{H_L, \ldots, G^L, \ldots \mid H_R, \ldots, G^R, \ldots\} = G,$$

so that the gift horses can be added as options to the game without changing the value of the game – no player wants to use these options anyway. One can reverse the steps as well, removing options that will never be used: If $G$ is a game, then a left option $G^L$ is *dominated* by another left option $G^{L'}$ if $G^L \leq G^{L'}$, and a right option $G^R$ is dominated by another right option $G^{R'}$ if $G^R \geq G^{R'}$. If a game $G$ has some left option $G^L$ and right option $G^R$, then its value is unchanged if a left option dominated by $G^L$ is removed, or a right option dominated by $G^R$ is removed.

---

**Example 9.2. Adding Hackenbush games.** The addition of games turns out to be very simple when we look at Hackenbush games. If each game consists of a number of connected piles of sticks, then the sum game arises when all the piles of the summands are taken together. As an example, consider the game to the right in the figure below:

It has three piles, and consequently it can be considered as the sum of two copies of the one-pile game to the left, and one copy of the other one-pile game, which we can identify as the game $-1 = \{ \mid 0\}$. The other game can be written as $\{0 \mid 1\}$: Left has only the option of changing the game to 0, namely by removing the black stick. Right can remove the white stick, but then the game changes to 1.

Looking at the rightmost game, it can be seen that if Left is First, then Right wins, and if Right is First, then Left wins. In other words, this is the game 0. Using that it is the sum of two games and $-1$, we obtain that the game to the left is equivalent to $1/2$.

---

A left option $G^L$ of a game $G$ is *reversible* (through $G^{LR}$) if $G^L$ has a right option $G^{LR}$ so that $G^{LR} \leq G$, meaning that replacing $G^L$ with all the left options of $G^{LR}$ does not hurt Left. One may therefore replace the reversible left option $G^L$ with all the left options of $G^{LR}$ without changing the value of the game.

The operations of removing dominated options and replacing reversible options may be used to find a simplified version of a given game, belonging to the same equivalence class. For the subclass of *short* games, which are games with a finite number of positions, there is a unique simplest form of the game, which has no dominated and no reversible positions.

**4.4. Numbers.** In the process of classifying games, there are some classes which can be considered as particularly wellbehaved, since they give rather immediate answers to the basic question of who will win the game. We have already introduced games like $0, 1, -1, 2$, which are denoted by numbers, and which behave in some sense like numbers. The numbers measure the number of free movements left and they are therefore easy to compare.

---

**Box 9.3. Numbers and birthdays.** In Conway [1976] a system of *birthdays* of games were introduced. The game $0 = \{\,|\,\}$ was born on day 0, which then is its birthday.

The three games $1 = \{0\mid\}$, $-1 = \{\mid 0\}$ and $* = \{0\mid 0\}$ were all born on day 1, being defined by objects born on day 0. On day 2, we find $\{0\mid 1\}$, $\{0,1\mid\}$ and $\{-1,0\mid 1\}$. The games $0 = \{\,|\,\}$, $1 = \{0\mid\}$, $2 = \{1\mid\}$, $3 = \{2\mid\}$ are non-negative integers, and $-1 = \{\mid 0\}$, $-2 = \{\mid -1\}$, $-3 = \{\mid -2\}$ are negative integers.

It is clear that Left wins all the positive integers and Right wins all negative integers. The game $0 = \{\,|\,\}$ is won by Second.

Formally, the birthday of a game $G$, denoted $b(G)$ is defined by

$$b(G) = \{b(G^L), b(G^R)\mid\}.$$

It has the poperty that $b(G^L) < b(G)$ for all $G^L$ and $b(G^R) < b(G)$ for all $G^R$, and in addition, $b(-G) = b(G)$ for all $G$. The birthday may be used if one searchs for a bound for the game, since $-b(G) \le G \le b(G)$.

---

A game is a *number* if both left and right options are numbers, and no left option is greater than or equal to any right option.

The game $0 = \{\,|\,\}$ is a number, since if not, then some of its options would not be a number or some left option is $\ge$ some right option. But 0 has no options, so it is indeed a number.

If a number has the form $x = \{x^L\mid x^R\}$, then $x^L < x < x^R$. Indeed, if $x^L = x$, then *Second* can win in $x^L - x = x + \{-x^R\mid -x^L\}$, a contradiction since Right can take the sum to 0 if playing first. If $x^L > x$, that is $x^L + \{-x^R\mid -x^L\} > 0$, we get a similar contradiction. This shows that $x^L < x$, and $x < x^R$ is shown in the same way.

To turn the numbers into a field, a multiplication of numbers must be defined, and we define for $x = \{x^L, \ldots\mid x^R, \ldots\}$ and $y = \{y^L, \ldots\mid y^R, \ldots\}$ the product $x \cdot y$

(where the · will be omitted in the sequel) by

$$x \cdot y = \{x^L \cdot y + x \cdot y^L - x^L \cdot y^L, x^R \cdot y + x \cdot y^R - x^R \cdot y^R, \ldots \mid$$
$$x^L \cdot y + x \cdot y^R - x^L \cdot y^R, x^R \cdot y + x \cdot y^L - x^R \cdot y^L, \ldots\}, \tag{1}$$

where right and left options of $x \cdot y$ are all the games that can be formed from right and left options of $x$ and $y$ as indicated. The seemingly complicated formula is motivated by the demand that multiplication should satisfy $(x - x^L)(y - y^L) > 0$ or $xy > x^L y + xy^L - x^L y^L$, which points to the definition in (1). It can be shown that if $x$ and $y$ are numbers, then so is $xy$.

To find an inverse of a number $x$, we let $x > 0$ and define

$$y = \left\{0, \frac{1 + (x^R - x)y^L}{x^R}, \frac{1 + (x^L - x)y^R}{x^L} \,\middle|\, \frac{1 + (x^L - x)y^L}{x^L}, \frac{1 + (x^R - x)y^R}{x^R}\right\}.$$

where again the definition is recursive, and we use all options $x^L \neq 0$ and $x^R$ of $x$. Then $xy = 1$ (again the proof uses induction). We thus have that the numbers (or rather, the equivalence classes of numbers) is a field. One can show even more – the field is totally ordered, and it is algebraically closed (shown in Conway [1976]).

**4.5. Nim and nimbers.** We begin – for reasons to become clear later – at another place, namely by defining impartial games as games where Left and Right have the same possible moves in every position. Formally, an *impartial game* is a game for which the set of left and right options are the same, and all options are impartial games. If $G$ is impartial, then $G = -G$, so that $G + G = 0$, and this in its turn means that $G = 0$ or $G \parallel 0$. Since left and right options are always the same, we may write the game as $G = \{G_1, G_2, \ldots\}$, where the set on the right-hand side is the set of (left or right) options. Also, since Left and Right have exactly the same options, we are not interested in whether Left or Right wins, what matters is whether the game is a win of First or of Second.

Returning to the game Nim mentioned above, then Nim, which consists in removing an arbitrary number of elements (called 'stones') of any single heap as long as this is possible, is impartial, since at any stage the players have the same options. Conway [1976] introduced the term *nimber* for a Nim game with a single heap: $*n$ denotes such a game having a heap of $n$ elements. We then have that

$$*0 = 0, *1 = \{*0\}, \ldots, *n = \{*0, *1, \ldots, *(n-1)\}.$$

Notice that $*1 = \{0 \mid 0\}$ is *not* a number.

Other Nim games, with more than one heap, can be found by addition. Thus $*n + *k$ is the game with two heaps of size $n$ and $k$ respectively. It may be noticed that $*k = *n$ only if $n = k$: Since $*k = -*k$ we have that $*n = *k$ implies $*n + *k = 0$, but if $n \neq k$, then First wins in $*n + *k$ by removing stones from the larger heap so as to create two heaps of equal size. But nimbers behave in a different way in other contexts: a Nim heap of size $n > 0$ satisfies $*n \parallel 0$, *not* $*n > 0$. But the game $*n + *n$ is a win for Second, so $*n + *n = 0$.

There is a famous classification of impartial games, which relates to the nimbers, that is Nim games with only one heap. We begin by introducing an operation $\text{mex}(\cdot)$: If $G$ is a set of nimbers, then $\text{mex}(G)$ is the smallest nimber not contained in $G$. Now the following holds:

THEOREM 4 *Let $G$ be a set of nimbers. Then $G = \text{mex}(G)$.*

PROOF: The game $G$ is impartial as a set of options which are all nimbers and therefore impartial, and $\text{mex}(G)$ is a nimber and therefore impartial. We must show that $G - \text{mex}(G) = G + \text{mex}(G) = 0$. If we write $\text{mex}(G) = *n = \{*k \mid k < n\}$ for some nonnegative integer $n$, and $G = *n \cup G'$, where $G'$ is the set of options of $G$ exceeding $*n$.

The first player has three types of moves, namely (1) move in $\text{mex}(G)$ to some $*k < *n$, (2) move in $G$ to an option $*k < *n$, or (3) move in $G$ to an option $*k > *n$. Move (1) leads to $G + *k$ and is countered by the move in $*n$ to $*k + *k = 0$, move (2) leads to $*k + *n$ and is countered by the move in $*n$ to $*k + *k = 0$. Finally, move (3) leads to $*k + *n$ (with $k > n$) and it is countered by a move in $*k$ to $*n + *n = 0$. In all cases, Second moves to 0 and wins.                                                    □

The theorem shows that the game $G$ is reduced to a single heap Nim game. We can however go beyond the Nim games to get a general result about impartial games, the celebrated Sprague-Grundy theorem shown independently by Sprague [1936] and Grundy [1939].

THEOREM 5 *Every impartial game is equal to a nimber.*

PROOF: This is shown by induction: If $G = \{G_1, G_2, \ldots\}$ is a game given by all its options, then by the induction hypothesis, every option of $G$ is equal to a nimber and using Theorem 4, we get that $G$ is equal to the mex of all its options, which is a nimber.                                                                                    □

The fact that any impartial game can be reduced to a Nim game with a single heap means that if we know how to win such Nim games (and we do, cf. Example 9.4, then we can find winning strategies for all impartial games by transforming to nimber and then using the winning strategy for this game.

## 5   Differential games

There are several ways in which one can approach the field of differential games. One may follow the historical development, where differential games can be seen as an outgrowth of the research on control theory combined with game theory, which at that time was mainly the theory of two-person zero-sum games. Alternatively, one may look upon differential games as a subclass of the dynamic games which by now takes up a considerable space in the body of mainstream game theory.

**Example 9.4. Winning strategies in Nim and nimber addition.** There is a simple procedure for finding winning strategies in Nim (and for deciding whether First or Second has a win), due to Bouton [1905]. We show it by an example, namely a Nim game with 3 heaps of size 3, 14 and 7.

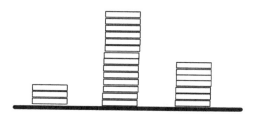

It turns out to be useful to write the sizes of the heaps (in the rows) as sums of powers of 2:

| | $2^3$ | $2^2$ | $2^1$ | $2^0$ |
|---|---|---|---|---|
| 3 | 0 | 0 | 1 | 1 |
| 14 | 1 | 1 | 1 | 0 |
| 7 | 0 | 1 | 1 | 1 |
| | 1 | 0 | 1 | 0 |

What matters now is whether there is an odd or even number of 1's in the columns. If we sum the columns modulo 2, as done in the table, we get 0 when there is an even and 1 when there is an odd number.

If there is at least one odd number, then First has a winning strategy, which consists in restoring 0's in all columns. In our example, this can be done for example by removing 10 of the stones in the middle heap, so that only 4 are left. The new table has the form shown below:

| | $2^3$ | $2^2$ | $2^1$ | $2^0$ |
|---|---|---|---|---|
| 3 | 0 | 0 | 1 | 1 |
| 4 | 0 | 1 | 0 | 0 |
| 7 | 0 | 1 | 1 | 1 |
| | 0 | 0 | 0 | 0 |

Now we have a game with even number of 1's in all columns. In such a game Second always has a winning strategy, since no matter what First does, the game will change into one with an odd number of 1's.

What was done in this analysis can be seen as performing an addition of nimbers. The result is the number the dyadic expression of which is given in the last line, and we get that $*(1 \cdot 2^3 + 1 \cdot 2^1) = *10$. We conclude again that First has a winning strategy.

Since we have already chosen to single out two-person zero-sum games for special treatment, the first approach seems more logical for us, and therefore this is where we begin. But this should not make us forget that in several respects differential games are only special cases of a more general theory of dynamic games.

**5.1. Defining the game and its value.** A differential game is a conflict which is based on a choice of state over time (the precise nature of this 'state' will depend on the applications), where the change of state is modelled as a dynamical system. Letting $\mathbb{R}^m$ be the space of possible states $x = (x_1, \ldots, x_m)$, the system is given by the vector differential equations

$$\frac{dx}{dt} = f(t, x, u_1, u_2), \ x(t_0) = x_0, \tag{2}$$

for $t \in [t_0, T_0]$, $0 \leq t_0 < T_0$, where $f : \mathbb{R} \times \mathbb{R}^m \times \mathbb{R}^p \times \mathbb{R}^q \to \mathbb{R}^m$ is a continuous function. We assume that $u_1 \in U_1$ and $u_2 \in U_2$ are chosen as functions of $t$,

$$u_1 = u_1(t), \ u_2 = u_2(t),$$

where $U_1$ and $u_2$ are compact subsets of $\mathbb{R}^p$ and $\mathbb{R}^q$, respectively. Functions $u_1$ and $u_2$ are called *controls* (of player 1 and 2, respectively), and the sets $U_1$ and $U_2$ are called the control sets of the two players. For given control functions, under suitable assumptions on the dynamical system in (2), it has a unique solution $x(t)$. The *payoff* is then defined as

$$P(u_1, u_2) = g(x(T)) + \int_{t_0}^{T} h(t, x(t), u_1(t), u_2(t)) \, dt \tag{3}$$

for some fixed $T \in [t_0, T_0]$, where $g$ and $h$ are continuous functions. We let the first player choose the control function such as to maximize payoff whereas the second player wants to minimize it.

We are close to having the full description of the differential game, but we still need to define suitable strategy spaces. Here we need that players can make their choice depend on the history of play as observed by the player. We let $M_i(t)$ be the set of all control functions defined on $[t, T]$, for $t \in [t_0, T_0]$, $i = 1, 2$, and formalize history-dependent control by defining strategies of player $i$ as mappings $\sigma_i : M_j(t_0) \to M_i(t_0)$, $j \neq i$, such that for all $s \in [t_0, T]$ and all $u_j, u'_j \in M_j(t_0)$,

$$\left[ u_j(\tau) = u'_j(\tau), \tau \leq s \right] \Rightarrow \left[ \sigma_i(u_j)(\tau) = \sigma_i(u_j)(\tau) \text{ for almost all } \tau \in [t_0, s] \right],$$

i.e. such that the control chosen according to the strategy $\sigma_i$ at time $\tau$ depends only on the controls chosen by the other player at time $\leq \tau$.

We have most of the ingredients of a two-person zero-sum game: The strategy sets are $\Sigma_i$, $i = 1, 2$, consisting of all the strategies $\sigma_i$ defined above, and whenever we have control functions $u_1$ and $u_2$, payoff to player 1 is $P(u_1, u_2)$ and payoff to player 2 is $-P(u_1, u_2)$. However, we still miss a way of assigning a payoff to any pair $(\sigma_1, \sigma_2)$ of strategies of the two players, so formally we hvae not quite defined a game yet.

Fortunately, we can get a long way towards general results even so. The main general result about two-person zero-sum games is the existence of a value, which

**Example 9.5.** Many of the early applications of differentiable games were *pursuer-evader games*. The following is a simple example of such a situation, modelling a situation where a rocket tries to hit an aircraft. The underlying dynamical system is

$$\dot{x} = u_1, \; \dot{y} = u_2, \; x(0) = x_0, y(0) = y_0,$$

where $x, y, u_1, u_2$ (rocket and aircraft positions, rocket and aircraft velocities) are all three-dimensional variables. The state space consists of all values of $(x, y)$, and control sets are suitable bounded sets contaning 0, we assume that $\|u_1\| \leq \alpha, \|u_2\| \leq \beta$ with $\alpha > \beta$. The distance between rocket and at time $t$ aircraft is

$$r(t) = \|x(t) - y(t)\|,$$

and the game terminates at $(t, x, y)$ for which either $t > T$ and $r(t) = 0$ or $t = T$. The pursuer wants to minimize time before arrival at the terminal set, and the evader wants to mazimize this time.

We notice that

$$\dot{r}(t) = \frac{x(t) - y(t)}{r(t)} u_1(t) - \frac{x(t) - y(t)}{r(t)} u_2(t),$$

so that the controls act in an additively separable way on the distance. This may be exploited to suggest suitable control functions, namely

$$u_1^* = \frac{\alpha}{r}(y - x)$$

for the pursuer, which means that the resulting change in distance is $-\alpha$, meaning that the pursuer goes in the direction of the aircraft at maximal speed, and

$$u_2^* = \frac{\beta}{r}(x - y)$$

for the evader who goes in the opposite direction at maximal speed.

A direct argument will show that these controls, which depend only on the state variables, constitute an equilibrium: Inserting in the expression for $\dot{r}$, we have that if the pursuer uses $u_1^*$, then

$$\dot{r} = -\alpha - \frac{x - y}{r} u_2 \leq -(\alpha - \beta)$$

since the change in $r$ caused by the evader cannot be greater than $\beta$ in the opposite direction. It follows that $r(t) \leq \|x_0 - y_0\| - (\alpha - \beta)t$, so that minimal capture time $t^0$ must satisfy

$$t^0 \leq \frac{\|x_0 - y_0\|}{\alpha - \beta}.$$

On the other hand, when the evader chooses $u_2^*$, then we get:

---

**Example 9.5, continued**

$$\dot{r} \geq \frac{x-y}{r}u_1 + \beta = -(\alpha - \beta).$$

It follows therefore that

$$t^0 \geq \frac{\|x_0 - y_0\|}{\alpha - \beta}.$$

Consequently, the pair $(u_1^*, u_2^*)$ constitutes a saddle point in the sense that capture time at any $(u_1, u_2^*)$ is at least $t_0$ which again is $\leq$ capture time at any $(u_1^*, u_2)$. We can therefore consider $t_0$ as the *value* of this game.

---

arises as maximin of player 1 and minimax of player 2 and is also the equilibrium payoff of player 1 in any equilibrium. We therefore introduce the *lower value* $V^-(t_0, x_0)$ of the differential game as

$$V^-(t_0, x_0) = \sup_{u_1 \in M_1(t_0)} \inf_{\sigma_2 \in \Sigma_2(t_0)} P(u_1, \sigma_2(u_1)),$$

so that $V^-(t_0, x_0)$ is the worst that can happen to player 2 when she chooses best response to any control function chosen by player 1, and similarly, the *upper value* $V^+(t_0, x_0)$ as

$$V^+(t_0, x_0) = \inf_{u_2 \in M_2(t_0)} \sup_{\sigma_1 \in \Sigma_1(t_0)} P(\sigma_1(u_2), u_2)$$

as the worst that can occur for player 1 choosing always best responses. As always, we have

$$V^+(t_0, x_0) \geq V^-(t_0, x_0).$$

In the case that the two are equal, we define the *value* of the differential game as

$$V(t_0, x_0) = V^-(t_0, x_0) = V^+(t_0, x_0).$$

Games that have a value are of course particularly interesting, and the value gives us useful information about the possible gains and losses in the conflict. We would however also like to know how to achieve this value or at least how to approach it.

**5.2. The Hamilton-Jacobi equation and viscosity solutions.** To get somewhat closer to actually finding values of differential games we introduce some further concepts, known from the theory of optimal control.

Let $H : [0, T] \times \mathbb{R}^m \times \mathbb{R}^m \to \mathbb{R}$ be continuous and let $g : \mathbb{R}^m \to \mathbb{R}$ be bounded and continuous. We are interested in solutions to the system of partial differential equations

$$v_t + H(t, x, \nabla_x v) = 0, \quad v(x, T) = g(x),$$

where $v_t = \dfrac{\partial v}{\partial t}$ and $\nabla_x v$ is the vector of partial derivatives of $v$ wrt. $x_i$, $i = 1, \ldots, m$. In general, there will be many solutions to this system, and therefore we strengthen our demands slightly as follows: A function $v : [0,1] \times \mathbb{R}^m \to \mathbb{R}$ is a *viscosity solution* to the system

$$v_t + H(t, x, Dv) = 0, \ (t, x) \in (0, T) \times \mathbb{R}^m, \ u(T, x) = g(x), \ x \in \mathbb{R}^m, \tag{4}$$

called the *Hamilton-Jacobi equations*, if for each $C^1$ function $\phi : (0, 1) \times \mathbb{R}^m \to \mathbb{R}$ the following conditions are satisfied:

(a) if $v - \phi$ attains a local maximum at $(t_0, x_0) \in (0, T) \times \mathbb{R}^m$, then

$$\phi_t(x_0, T_0) + H(t_0, x_0, D\phi(t_0, x_0)) \geq 0,$$

(b) if $v - \phi$ attains a local minimum at $(t_0, x_0) \in (0, T) \times \mathbb{R}^m$, then

$$\phi_t(x_0, T_0) + H(t_0, x_0, D\phi(t_0, x_0)) \leq 0.$$

The main advantage of a viscosity solution (which is easily seen to be a solution in the ordinary sense) is that it is uniquely determined (at least when $H$ satisfies suitable conditions (of Lipschitz type). We now specify the function $H$, or rather, we introduce two different versions, namely the *upper Hamiltonian*

$$H^+(t, x, p) = \min_{u_2 \in U_2} \max_{u_1 \in U_1} \{ f(t, x, u_1, u_2) \cdot p + h(t, x, u_1, u_2) \},$$

and the *lower Hamiltonian*

$$H^-(t, x, p) = \max_{u_1 \in U_1} \min_{u_2 \in U_2} \{ f(t, x, u_1, u_2) \cdot p + h(t, x, u_1, u_2) \}.$$

This gives us *two* systems of equations (the Isaacs equations)

$$V_t^+ + H^+(t, x, DV^+) = 0, \ V^+(T, x) = g(x), \ t \in [0, T], \ x \in \mathbb{R}^m, \tag{5}$$

$$V_t^- + H^-(t, x, DV^-) = 0, \ V^-(T, x) = g(x), \ t \in [0, T], \ x \in \mathbb{R}^m. \tag{6}$$

As was to be expected, there is a connection between viscosity solutions to the Isaacs equations and the upper and lower values as defined above. This is established by the following result due to Evans and Souganidis [1984].

THEOREM 6 *The lower value $U^-$ is a viscosity solution of (5), and the upper value $U^+$ is a viscosity solution of (6). If*

$$H^+(t, x, p) = H^-(t, x, p), \ \text{all} \ (t, x, p) \in [0, T] \times \mathbb{R}^m \times \mathbb{R}^m,$$

*then $V^+ = V^-$ and the differential game has a value $V$.*


**5.3. Saddle points and feed-back control.** For games with a value we want also to find optimal strategies realizing the value. First we introduce the notion of *feedback* *control* in our context, namely continuous functions $u_i^0 : [0,T] \times \mathbb{R}^m \times \mathbb{R}^m \to U_i$, $i = 1, 2$ such that

$$\min_{u_2 \in U_2} \max_{u_1 \in U_1} \{f(t,x,u_1,u_2) \cdot p + h(t,x,u_1,u_2)\}$$

$$= \min_{u_2 \in U_2} \left\{f(t,x,u_1^0(t,x,p),u_2) \cdot p + h(t,x,u_1^0(t,x,p),u_2)\right\}$$

$$= \max_{u_1 \in U_1} \left\{f(t,x,u_1,u_2^0(t,x,p)) \cdot p + h(t,x,u_1,u_2^0(t,x,p))\right\}.$$

With this notation, the Hamilton-Jacobi equations become

$$V_t + f(t,x,u_1^0(t,x,\nabla_x V), u_2^0(t,x,\nabla_x V)) \cdot \nabla_x V$$
$$+ h(t,x,u_1^0(t,x,\nabla_x V), u_2^0(t,x,\nabla_x V)) = 0, \ (t,x) \in (0,T) \times \mathbb{R}^m, \quad (7)$$

with $V(T,x) = g(x)$ for $x \in \mathbb{R}^m$.

THEOREM 7  *If $V$ is $C^1$ and solves (7), and the two functions $u_1^*, u_2^*$ defined by*

$$u_1^*(t,x) = u_1^0(t,x,\nabla_x V), \ u_2^*(t,x) = u_2^0(t,x,\nabla_x V)$$

*are $C^1$, then $(u_1^*, u_2^*)$ is a $C^1$ saddlepoint.*

In order to find the functions $u_1^*$ and $u_2^*$, we use that the partial differential equation (7) may be solved by finding a solution to the system of *ordinary* differential equations

$$\frac{dx_k}{dt} = f_k(t,x,u_1^0(t,x,\nabla_x U), u_2^0(t,x,\nabla_x U)), \ k = 1,\ldots,m,$$

$$\frac{dV_k}{dt} = -\sum_{i=1}^m V_i \frac{\partial f_i}{\partial x_k}(t,x,u_1^0(t,x,\nabla_x U))$$

$$+ \frac{\partial h}{\partial x_k}(t,x,u_1^0(t,x,\nabla_x U), u_2^0(t,x,\nabla_x U)), \ k = 1,\ldots,m,$$

together with the terminal conditions $V_k = \dfrac{\partial g}{\partial x_k}$ for $t = T$, to be solved for functions $x_1,\ldots,x_m, V_1,\ldots,V_m$.

A solution to this system of equations gives us functions $x_k(u,t)$, $V_k(u,t)$ for $k = 1,\ldots,m$. Solving for $u$ in the first $m$ equations gives us functions $u = u(x,t)$ which inserted into the $V_k$s gives

$$V_k = V_k(u(x,t),t) = \tilde{V}_k(x,t) \text{ for } k = 1,\ldots,m.$$

Now

$$V(t,x) = \int_T^t \Big(\sum_{k=1}^m \tilde{V}_k f_k(t,x,u_1^0(t,x,\nabla_x \tilde{V}), u_2^0(t,x,\nabla_x \tilde{V}))$$

$$- h(t,x,u_1^0(t,x,\nabla_x \tilde{V}), u_2^0(t,x,\nabla_x \tilde{V}))\Big) dt + g(x)$$

is a solution to (7).

While we have a general theory of two-person zero-sum differential games, such as outlined in the previous section, the situation is somewhat different when we turn to general, that is not necessarily two-person and not necessarily zero-sum, differential games. This is similar to what we have already encountered when dealing with static games, but there is an additional complication in the differential case: For one-person differential games (control problems) and two-person zero-sum games, one can rely on the Hamilton-Jacobi theory, and in particular, the Hamilton-Jacobi equation can be transformed to a system of ordinary differential equations which are easier to solve than the original partial differential equation. For general games, one may derive a Hamilton-Jacobi equation as before, but it is less useful since the transformation does not work any more. Therefore, successful studies of general differential games deal with particular classes of games, such as e.g. *linear quadratic differential games,* where the underlying dynamical system is linear and payoff functions quadratic, see e.g. Engwerda [2005].

## 6 Problems

**1.** Consider the two graphs $A$ and $B$ below.

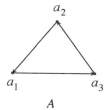 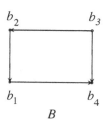

Show that the Duplicator has a winning strategy for the 3 rounds Ehrenfeucht-Fraïssé game.

Show that the Spoiler has a winning strategy for the 4 rounds game.

**2.** A graph is Eulerian if it has a cycle which uses each edge exactly once. Show that the property of being Eulerian cannot be expressed in first order logic.

Hint: Consider the graphs $E_n$, and check that they are Eulerian if and only if $n$ is even.

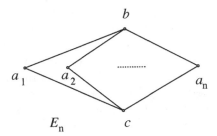

Let $A_n = E_{2n}$, $B_n = E_{2n+1}$, and show that the Duplicator wins the $n$ round Ehrenfeucht-Fraïssé game on $A_n$ and $B_n$.

**3.** In the impartial version of Hackenbush, the sticks are all of the same color, and each player moves one stick from a figure which is connected to the ground. The winner is the last player who has a valid move.

Who wins in the situations illustrated above?

**4.** In the game Road Trip (from Gartner [2007]), two players must move two cars, initially at $A$ and $E$ to Home following the edges in the graph below.

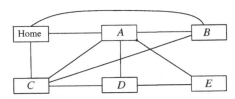

Only moves to cities with a letter which is earlier in the alphabet and to Home (provided there are suitable edges) are allowed. The game ends when both cars have returned and winner is the player taking the last car to Home. Determine who wins the game.

(Hint: Assign values to towns as follows: Home is 0 (a car at home has no available movements). Then $A$ gets value mex($\{0\}$) = $*1$, $B$ gets mex($\{0, *1\}$) = $*2$ etc. After the full assignment the game is shown to be equal to $*2 + *2$).

**5.** (Numbers) Show that the game $G = \{-1 \mid 5\}$ has value 0, and that the value of $\left\{\frac{1}{4} \mid 1\right\}$ is $\frac{1}{2}$.

**6.** Construct an addition table for the nimbers from 0 to $*9$ using mex($\cdot$).

# PART 2
# Cooperative Game Theory

# Chapter 10

# Introduction to Cooperative Games

## 1 Introduction

Cooperative games are conflict situations where players can enter into binding agreements about coordinated behavior. The theory of cooperative games deals with solutions to such conflicts, and it describes these solutions in detail – existence, uniqueness etc. The cooperative game theory has always constituted an important part of game theory as a whole. In von Neumann and Morgenstern [1944], the chapters 5 –12 (out of 12) deal with cooperative games, and rather much of what was added in the course of the 50s and 60s belonged to this field.

The important new aspect of cooperative as compared to non-cooperative game theory is, as already mentioned, the possibility of coalition building. The significance of such coalitions depends of course on what they can do for their members. In the present introductory chapter we introduce different representations of cooperative games, where we look for as simple a presentation as possible, given that the basic aspect, namely the presence of coalitions with their specific possibilities of action, should be kept.

The simplest way to begin the discussion is to use the basic description of a game, used also in the non-cooperative game theory, and then introduce the cooperative aspect in the description of the relevant solution or equilibrium concepts. This is indeed what we shall do in the sections 2 – 4. After this we discuss other forms of representing cooperative games, namely the effectivity functions, and their use.

## 2 Cooperative solution concepts for normal form games

**2.1. Strong Nash equilibrium.** We consider a game in normal form

$$\Gamma = (N, (S_i)_{i \in N}, u),$$

where, as usual, $N = \{1, \ldots, n\}$ is the set of players, and each $S_i$ is a nonempty set of strategies for player $i$, each $i \in N$, and $u : S_1 \times \cdots \times S_n \to \mathbb{R}^N$ is the payoff function of the game, which to each array $s = (s_1, \ldots, s_n)$ of strategies (one for each player) assigns a payoff $u_i(s)$ to each of the players.

This is of course the basic description of a game, no matter whether one is interested in cooperative or non-cooperative games. It is therefore reasonable to begin a discussion of cooperative games here – in the same way as one did for non-cooperative games.

In the game $\Gamma$ there is a possibility of forming *coalitions*; by a coalition we understand a nonempty subset $S$ of the player set $N$. $N$ itself is a coalition (but $\emptyset$ is not). Given that the members of a coalition $S$ can coordinate their choices of strategies, they have a certain possibility of influencing the course of the game; at the very least they can intuitively prevent mutual competition harming everyone. However, they cannot necessarily fully control the outcome of the game, which still depends on the strategies chosen in the complement of the coalition.

This straightforward consideration will be followed up in later sections, since it is at the heart of the derivation of simpler representations of a cooperative game. In the present introductory phase we shall have a closer look at some ways in which the coalitional aspects may be introduced without going into detailed considerations of what the players outside the coalition will do, something which in its turn will take us to a formalization of the power structure inherent in a given game. Instead, we shall use a stability argument: A strategy choice is a solution if no individual or coalition subsequently (that is given the choices made by the others) are unsatisfied with their choices.

In the non-cooperative theory, this approach took us to the *Nash equilibrium*, and we had much to say as well for as against it. The most appealing feature of this approach to solutions is its simplicity. It was extended to cooperative game theory in Aumann [1959]:

**DEFINITION 1** *Let* $\Gamma = (N, (S_i)_{i \in N}, (u_i)_{i \in N})$ *be a game. A strategy array* $s^0 = (s_1^0, \ldots, s_n^0)$ *is a* strong Nash equilibrium *if no coalition $S$ has an improvement of $s^0$, that is a coalitional strategy* $t_S = (t_i)_{i \in S}$ *such that*

$$u_i(t_S, s_{N \setminus S}^0) > u_i(s^0)$$

*for all* $i \in S$.

In the above expression we have used the notational convention $s_{N \setminus S}^0$ for $(s_i^0)_{i \in N \setminus S}$, and we identify the strategy arrays $s$ and $(s_S, s_{N \setminus S})$ for arbitrary coalitions $S$. This notational convention will be used also in the future, as it will simplify complicated expressions.

The strong Nash equilibrium inherits several properties of the Nash equilibrium. This is not surprising, since in particular we have that a strong Nash equilibrium is a Nash equilibrium in each two person game which is obtained by keeping all strategies except those of two players fixed (at the equilibrium choices).

So far, we have considered only equilibria in pure strategies. Introducing mixed strategies for the individual players offers no complications, and indeed the strategy sets $S_i$ may as well be the set of probability distributions on an underlying

**Example 10.1.** The following game involves 3 players, who each have two strategies, in the representation below player 1 chooses row, player 2 column and player 3 matrix.

|   | L | R |
|---|---|---|
| T | $(1,1,0)$ | $(1,0,1)$ |
| B | $(0,3,1)$ | $(0,0,0)$ |

|   | L | R |
|---|---|---|
| T | $(0,3,0)$ | $(1,2,1)$ |
| B | $(1,2,0)$ | $(1,1,1)$ |

I                                    II

This game has no strong Nash equilibrium: If we fix the choice of player 3 at the first matrix (I) and look at what can be chosen by player 1 and 2, then only $(T,L)$ is a Nash equilibrium in the resulting (non-cooperative) two person game. If similarly we keep player 3 fixed at matrix II, then only $(B,L)$ is a Nash equilibrium in the two person game between players 1 and 2. This last strategy combination is however not stable against reactions from player 3, who will prefer to change strategy to I given the choices of the others. Therefore only the strategy array $(T,L,I)$ remains, but also this strategy array does not work, since the coalition consisting of players 2 and 3 can improve by changing to $(R,II)$.

set of pure strategies. However, when the choice of strategies is coordinated inside a coalition, one might also consider the possibility of randomizing over such coalitional strategies. For the time being we assume that such coalitional randomization is not an option; we return to this later.

We shall use the notation $\Sigma_S$ for the set of coordinated strategy choices of the coalition $S$ in *both* cases; whether we allow for mixed strategies or not will be clear from the context.

At a first sight it seems discouraging that the proposed equilibrium concept fails to exist in such a simple case as that of Example 10.1. On the other hand, the strong Nash equilibria are interesting in many contexts, and we shall return to them later. However, the intuitive message of the example, which is that the conditions for a strong Nash equilibrium ("no coalition can improve given the strategy choices of the remaining players") are rather restrictive, and as a consequence one needs rather strong assumptions on the game to be sure that it has a strong Nash equilibrium.

While the existence of (non-cooperative) Nash equilibria can be assured under rather acceptable and easily interpreted assumptions on the underlying game, the situation is somewhat different for strong Nash equilibria. The matter is complicated in two different ways: First of all one needs more assumptions, so that apart from the conditions which are needed to show existence of non-cooperative Nash equilibria, it is necessary to add further conditions which guarantee the existence of strategy arrays that cannot be improved by any coalition. But this leads to the next complication, namely that these additional assumptions have a form which is rather difficult to interpret, a complication which has its root in the new feature that coalitions having possibility of improvement may overlap.

**2.2. Coalition proof Nash equilibrium.** The intuition behind the strong Nash equilibria – if a coalition can improve the payoffs of its members by changing to a new coordinated strategy choice, then it will do it – can be criticized from several different viewpoints.

The more fundamental of these, having to do with the basic idea of a Nash equilibrium, namely that the other players stay passive, will not be discussed here; instead we consider the possible objection related to the notion of an improvement. Is it clear that the availability of an improvement, as defined above, necesarily upsets the original strategy array?

Example 10.2 indicates a fundamental problem with coordinated strategy choices. If what is agreed in the coalition cannt be realized, then it is not a real possibility for the coalition, as it lacks credibility as an alternative to the status quo. One may of course refer to a possible existence of a legal framework which guarantees that agreements can be enforced, and indeed this was how game theorists used to sidestep the problem. But it seems preferable to address the fundamental problem of coalitional stability.

A more satisfactory way of treating the problem is to restrict attention to such coalitional improvement where no subset of the coalition has incentive to make additional improvements. In this way the external authority enforcing coalitional agreements is avoided, since no one will benefit from deviating from the agreement. The corresponding solution concept was introduced in Bernheim et al. [1987], who gave it the name *coalition proof Nash equilibrium*. The definition must be made inductively.

DEFINITION 2 *Let* $\Gamma = (N, (S_i)_{i \in N}, (u_i)_{i \in N})$ *be a game, and* $s^0 = (s_1^0, \ldots, s_n^0)$ *a strategy array in* $\Gamma$. *An internally consistent improvement (ICI) of* $s^0$ *for the coalition S is an S-strategy* $t_S \in \times_{i \in S} S_i$ *with the following properties*

 (i) *if* $|S| = 1$, *i.e.* $S = \{i\}$ *for* $i \in N$, *then* $t_S$ *is an improvement of* $s^0$,
 (ii) *Asume that ICIs have been defined for all coalitions with k members, $k \geq 1$, and that S is a coalition with* $|S| = k + 1$; *then*

  (a) $t_S$ *is an improvement of* $s^0$ *for S, and*
  (b) *no* $S' \subset S$ *has an ICI of* $(t_S, s_{N \setminus S}^0)$.

That a coalitional strategy is internally consistent means that no subcoalition has incentive to make subsequent corrections in the coalitional choice without mentioning this to the remaining members of the coalition (since such a correction would typically result in an inferior outcome for some of the coalition members, or it would itself be subject of further corrections by a smaller subcoalition).

**Example 10.2.** Consider the following 3 person game, where player 1 and 2 both have the strategy sets $\{a, b\} \times \{1, 2\}$, while player 3 only has $\{1, 2\}$. If players 1 and 2 both choose $a$, then the game takes the form

|   | 1 | 2 |
|---|---|---|
| 1 | $(3, 3, 0)$ | $(0, 5, 0)$ |
| 2 | $(5, 0, 0)$ | $(1, 1, 0)$ |

where the choice of row and column is determined by the second component in the strategies of player 1 and 2 (player 3 cannot influence the outcome); if one of the players have chosen $b$, then the game has the form

|   | 1 | 2 |
|---|---|---|
| 1 | $(2, 2, 2)$ | $(0, 0, 2)$ |
| 2 | $(0, 0, 2)$ | $(0, 0, 2)$ |

where again it is player 1 and 2, who choose row and column, while 3 is passive.

It is easy to see that the strategy choice $((b, 1), (b, 1), 1)$, which results in the payoff vector $(2, 2, 2)$, is *not* a strong Nash equilibrium. The coalition $\{1, 2\}$ can improve by changing their $b$ to $a$, whereby both get 3.

Closer inspection of the example shows however, that it can be risky for the two players to decide upon a change of the game matrix to the first mentioned, since this is the payoff matrix of a prisoners' dilemma game. Given that the coalition $\{1, 2\}$ has decided to change strategy to $((a, 1), (a, 1))$, both members will be tempted to change the second part of their strategy coordinate from 1 to 2. In other words, the coordinated improvement is vulnerable under a subsequent giving up of the coordinated choice by the members of the coalition.

With the concept of an internally consistent improvement we get the following counterpart of the definition of a strong Nash equilibrium.

DEFINITION 3 *Let* $\Gamma = (N, (S_i)_{i \in N}, (u_i)_{i \in N})$ *be a game. A strategy array* $s^0 = (s_1^0, \ldots, s_n^0)$ *is a coalition proof Nash equilibrium in* $\Gamma$ *if no coalition has an ICI of* $\sigma^0$.

A strong Nash equilibrium is a coalition proof Nash equilibrium, but the converse inclusion does not necessarily hold (this was one of the points of introducing the concept).

There are few general results on existence of coalition proof Nash equilibria.

**Example 10.3.** Consider the 2-person game with matrix

$$\begin{pmatrix} (1,1) & (0,0) \\ (0,0) & (0,0) \end{pmatrix}$$

Here the strategy choice 1st row/1st column is a coalition proof Nash equilibrium. It is also an ordinary Nash equilibrium and a strong Nash equilibrium. The combination 2nd row/2nd column is a Nash equilibrium, but it is neither strong or coalition proof Nash.

In the 2-person game with payoff-matrix

$$\begin{pmatrix} (1,1) & (3,0) & (1,0) \\ (0,0) & (2,2) & (0,0) \end{pmatrix}$$

is 1st row/1st column coalition proof Nash equilibrium, but it is not a strong Nash equilibrium (the coalition with both players can improve by changing to 2nd row/2nd column, but then player 1 can improve.

The exception is the class of 2-person games, where there is only a single coalition apart from the singletons, namely the grand coalition. The following holds:

THEOREM 1 *Let* $\Gamma = (\{1,2\}, S_1, S_2, u)$ *be a 2-person game, such that*
  *(i)* $S_i$ *is a compact metric space,* $i = 1, 2$,
  *(ii)* $u : S_1 \times S_2 \to \mathbb{R}^2$ *is continuous.*
*If* $\Gamma$ *has a Nash equilibrium, then it has a coalition proof Nash equilibrium.*

PROOF: Let $E$ be the set of all Nash equilibria in $\Gamma$. We claim that $E$ is closed. Indeed, let $(s_N^k)$ converge to $s_N^0$. We then need to show that $s_N^0 \in E$. Suppose on the contrary that there is $\tau_1 \in S_1$ so that $u_1(\tau_1, s_2^0) > u_1(s_1^0, s_2^0)$. Since $u_1$ is continuous, we have $u_1(s_1^k, s_2^k) \to u_1(s_1^0, s_2^0)$, and $u_1(\tau_1, s_2^k) \to u_1(\tau_1, s_2^0)$. Therefore there must be $m$ such that $u_1(\tau_1, s_2^m) > u_1(s_1^m, s_2^m)$, a contradiction which proves our claim.

Since $S^N$ is compact, then also $E$ is compact. Suppose that $E$ is nonempty, and let $s_N^*$ be a point where $u_1 + u_2$ attains its maximum on $E$. Then $s_N^*$ is a coalition proof Nash equilibrium. □

COROLLARY 1 *Let* $\Gamma = (\{1,2\}, S_1, S_2, u)$ *be a 2-person game. If* $|S_i| < \infty$, $i = 1, 2$, *and* $\Gamma$ *has a Nash equilibrium, then* $\Gamma$ *has a coalition proof Nash equilibrium.* □

## 3   Characteristic function

The solution concepts discussed in the preceding sections are directly related to the game as formulated through strategy sets, payoff etc., that is to the normal

**Example 10.4.** The following example shows that there are 3-person games without coalition proof Nash equilibria. Let $\Gamma = (S_1, S_2, S_3, u)$, where $S_i = \{s_i^1, s_i^2\}$, $i = 1, 2, 3$, og $u_i$, $i = 1, 2, 3$, is given by the following pair of matrices:

|       | $s_2^1$ | $s_2^2$ |
|-------|---------|---------|
| $s_1^1$ | $(0,0,0)$ | $(-1,-1,0)$ |
| $s_1^2$ | $(-1,-1,0)$ | $(1,1,0)$ |

$$s_3^1$$

|       | $s_2^1$ | $s_2^2$ |
|-------|---------|---------|
| $s_1^1$ | $(1,1,-k)$ | $(0,0,-k)$ |
| $s_1^2$ | $(0,0,-k)$ | $(-1,-1,1)$ |

$$s_3^2$$

(Here player 1 chooses row, player 2 column, and player 3 matrix).

Let $\Gamma^*$ be the mixed extension of $\Gamma$ (wellknown from the non-cooperative theory, here the players choose probability distributions over the above strategies, and the payoffs are found as the mean values). We claim that if $0 < k < 1/8$, then $s_N = (s_1^1, s_2^1, s_3^1)$ is the only Nash equilibrium of $\Gamma^*$. Let $p_i$ be the probability for $s_i^1$ in a mixed strategy for $i$. Then $s_3^2$ is best reply at $(p_1, p_2)$ if and only if

$$(1 - p_1)(1 - p_2) \geq \frac{k}{k+1}. \tag{1}$$

Further we have that the Nash equilibria in the mixed extension of the 2-person game

|       | $s_2^1$ | $s_2^2$ |
|-------|---------|---------|
| $s_1^1$ | $(0,0)$ | $(-1,-1)$ |
| $s_1^2$ | $(-1,-1)$ | $(1,1)$ |

are $(s_1^1, s_2^1)$, $(s_1^2, s_2^2)$, and $(p_1, p_2) = (2/3, 2/3)$.

Let $(p_1, p_2, p_3)$ be a Nash equilibrium in $\Gamma^*$. Since $k > 0$ we have $p_3 > 0$. If $p_3 = 1$ then from

$$\left(1 - \frac{2}{3}\right)\left(1 - \frac{2}{3}\right) = \frac{1}{9} = \frac{\frac{1}{8}}{1 + \frac{1}{8}}$$

and $k < 1/8$ we have that $p_1 = q_1 = 1$. In other words, if $(p_1, p_2, p_3) \neq (1,1,1)$, then the following inequalities hold:

$$(1 - p_1)(1 - p_2) = \frac{k}{k+1} < \frac{1}{9}, \quad 0 < p_3 < 1. \tag{2}$$

We may asume that $p_1 > 2/3$. Then it follows that

$$u_2(p_1, s_2^1, p_3) > -\frac{1}{3}p_3 + \frac{2}{3}(1 - p_3),$$

$$u_2(p_1, s_2^2, p_3) < -\frac{2}{3}p_3 + \frac{1}{3}p_3 = -\frac{1}{3}p_3.$$

Therefore we must have $p_2 = 1$ contradicting (2).

It is obvious that $s_N^1 = (s_1^1, s_2^1, s_3^1)$ is not a coalition proof Nash equilibrium, since $(s_1^2, s_2^2)$ is an ICI for $\{1, 2\}$ against $s_N^1$.

form of the game, and as such they may be considered as most directly applicable. The cooperative aspects of the theory emerge through the very formulation of the solution concepts and the considerations lying behind. At this level, there is no distinction between cooperative and non-cooperative *games*, the distinction comes only when the requirements of the solutions are considered.

Even though this approach may be considered as satisfactory from a purely logical point of view, it has some practical drawbacks, since the solution concepts formulated in this way may be lacking in transparency. In the general setup there is rather too many details that may be relevant for the solution, and it may be difficult to see which part of it matters and what is less important.

To getter a better understanding of the peculiarities arising through the formation of coalitions and the strength of the coalitions thus formed, one has to concentrate on some aspects of the game and then neglect all the other aspects. This leads to special representations of cooperative games. The most widely used of these is the *characteristic function*. The characteristic function is – at least at our present level of discussion – a correspondence or a multimap, which to each coalition assigns a set of admissible payoffs to its members, but the tradition from von Neumann and Morgenstern [1944] has been decisive for the terminology, and in the version which they introduced, the characteristic function was actually a function. The characteristic function can be defined in several different ways, not resulting in the same object. We begin with this historical approach to the characteristic function, which has by now been largely superseded by others, since it works only when the game is zero-sum.

Before we proceed, we return briefly to the problem of randomizing strategies. While the previous approaches presumed that no coordinated randomization takes place, there are no such limitations from now on, actually we need the possibility of coalition randomization for the first approach to be considered now. In this case the set of strategies open to a coalition $S$ is $\triangle (\times_{i \in S} S_i)$, the set of probability distributions over $\times_{i \in S} S_i$. The payoff function $u$ extends to $\triangle (\times_{i \in N} S_i)$ by linearity.

Let $\Gamma = (N, (S_i)_{i \in N}, (u_i)_{i \in N})$ be *zero-sum* in the sense that $\sum_{i \in N} u_i(\sigma) = 0$ for all $s \in \Sigma^N$. The zero-sum games played a central role in the initial development of game theory, and although they are nowadays largely abandoned from the point of view that few conflicts in real life have this property, they remain an important subclass of games, since they can be considered as a very simple approximation to more realistic games, and since they enjoy many attractive properties which make them useful for initial considerations of concepts and structures which may turn up again in the general family of games, even if in a somewhat different version.

The advantage of restricting attention to zero-sum games can be seen when we want to express what should be understood by the *strength* of a coalition. Assume that the coalition $S$ contemplates a coordinated choice of strategy. What can the coalition obtain by this? Clearly this depends on what the remaining players, that is the players in the complement $N \setminus S$ do. Implicitly in our discussion in the previous sections we assumed that they remained passive, but we cannot in general be sure that this will be the case. Unfortunately, the results obtained by the coalition depends in a crucial way on the behavior of the remaining players.

In the considerations about the behavior of the complement we will be helped by the zero-sum assumption. What the coalition $S$ can achieve for its members (measured as $\sum_{i \in S} u_i$, where we assume that payoff can be measured in dollars or some other unit that can be transferred among players, so that summation of individual payoffs is meaningful), must be paid by the complement, so it seems reasonable to assume that the complement will have a hostile attitude towards possible coordinated strategy choices in $S$. Let us further assume, that also the complement coordinates its behavior; then we are left with a *2-person zero sum game*

$$\Gamma_S = (\{S, N\backslash S\}, \Sigma_S, \Sigma_{N\backslash S}, (u_S, u_{N\backslash S})),$$

where $u_T = \sum_{i \in T} u_i$ for every coalition $T$. It does not immediately look as a simplification to replace the game $\Gamma$ by the family of games $(\Gamma_S)_{S \subset N}$, but it is: We may now use the fundamental result (also due to von Neumann), that every 2-person zero-sum game has Nash equilibria in mixed strategies, and the payoff to each player is the same in all Nash equilibria, namely the *value* $v(\Gamma_S)$ of the game $\Gamma_S$, cf. Definition 6 in Chapter **2**.

Now we have a characteristic function of the game $\Gamma$, given as $v : 2^N \backslash \{\emptyset\} \to \mathbb{R}$ defined by

$$v(S) = v(\Gamma_S).$$

The characteristic function indicates what the coalition can assure for its members for subsequent distribution among the latter, no matter what the complement might do.

We have thus introduced the von Neumann-Morgensterns characteristic function (ch.VI, section 25.1.2-3 i von Neumann and Morgenstern [1944]). It has the obvious limitation of being defined only for games which are zero-sum (or constant-sum, what is basically the same). We need some additional considerations to obtain something similar for general games. This does not mean that we shall not see this particular characteristic function in the sequel. It appears in some applications, and furthermore, much of the classical cooperative game theory works for any characteristic function, no matter where it came from.

In the generalization of the above construction of the characteristic function a further complication occurs. When considering zero-sum games, it made sense to use sums of payoffs of the individual players. This is fine as long as payoff is an amount of money, but unfortunately this is not always the case, and in many connections it is important that payoff is a purely subjective utility gain of the player, so that summation is no longer meaningful.

The differentiation between situations where summation is meaningful, and where it is not, is fundamental and has give rise to specific terminology. Traditionally, what is considered here are *side payments*, money (or utility) transfers between players. If summation of payoffs is acceptable (in the sense that the coalition subsequently can distribute the payoff sum among members in any way desired), the game was said to be a *game with side payments* (or a sidepayment game, and if this is not the case, we have a *game without side payments* (a non-sidepayment game). In recent years the terminology has shifted, so that one speaks of TU (transferable utility) games in the first case and NTU (non-transferable utility) games in the second.

In the general NTU game one cannot perform summation of the payoffs of individual players. Consequently, track must be kept of the payoff vectors that a coalition can achieve for its members. The value of the characteristic function for the coalition $S$ becomes a subset of $\mathbb{R}^S$, consisting of vectors $(z_i)_{i\in S}$ of (utility-)payoffs achievable by coordinating strategies inside $S$.

We have still some ambiguity left in determining what it should mean that $S$ can obtain a certain payoff vector. This cannot be removed in a simple and unique way, since there are several possibilities, depending on the applications, and the rather convincing approach taken in the zero-sum case is not generalizable. Here are some traditional ways of defining characteristic functions.

DEFINITION 4 *Let* $\Gamma = (N, (S_i)_{i\in N}, (u_i)_{i\in N})$ *be a game in normal form. The $\alpha$-characteristic function for $\Gamma$ is defined by*

$$V_\alpha(S) = \left\{ (z_i)_{i\in S} \mid \exists \sigma_S \in \Sigma_S \forall \tau_{N\setminus S} \in \Sigma_{N\setminus S} : u_i(\sigma_S \tau_{N\setminus S}) \geq z_i, \text{ all } i \in S \right\},$$

*and the $\beta$-characteristic function of $\Gamma$ is given by*

$$V_\beta(S) = \left\{ (z_i)_{i\in S} \mid \forall \tau_{N\setminus S} \in \Sigma_{N\setminus S} \exists \sigma_S \in \Sigma_S : u_i(\sigma_S \tau_{N\setminus S}) \geq z_i, \text{ all } i \in S \right\}.$$

The two types of characteristic function differ, as it is seen, in the order of the quantifiers. In the $\alpha$-characteristic function the coalition can achieve a payoff vector only if it has a coordinated strategy choice which will give its members at least this payoff *no matter* what the complement does. For the $\beta$-characteristic function the demand is somewhat weaker: The coalition must have a reply on each of the choices of the complement which gives its member at least this payoff vector.

It is clear that $V_\alpha(S) \subset V_\beta(S)$ for every coalition $S$. Also each of the characteristic functions has the property

$$V(S) - \mathbb{R}^S_+ \subset V(S)$$

for each coalition $S$. This property, known in economics as "free disposal", is called *comprehensiveness* in game theory.

Apart from this, not much can be said in general about characteristic functions. Otherwise formulated, any $V$ satisfying comprehensiveness and some regularity condition, is candidate for being the characteristic function of a suitable normal form game. Consequently, it is more appropriate to abstract from the precise way in which the characteristic function was derived from the normal form, defining a characteristic function quite independently.

DEFINITION 5 *A cooperative (NTU) game is a pair* $(N, V)$, *where* $N = \{1, \ldots, n\}$ *is a finite set of players and* $V$ *a map, which to each coalition* $S$ *assigns a subset* $V(S)$ *of* $\mathbb{R}^S$, *such that the following conditions are satisfied:*

   *(i)* $V(S)$ *is closed for each coalition* $S$,

  *(ii)* $V(S) - \mathbb{R}^S_+ \subset V(S)$ *for all* $S$ *(comprehensiveness),*

 *(iii)* $V(N)$ *is upper bounded, i.e. there is* $b \in \mathbb{R}^N$ *such that* $z \leq b$ *for all* $z \in V(N)$ *(vector inequalities are assumed satisfied coordinatewise).*

After the introductory considerations, which by now can be seen mainly as motivating the above definition, we need no longer care whether the characteristic function is of the $\alpha$- or the $\beta$-type or possibly something different from both – at least as long as we develop the formal aspects of the theory. When the results are assessed and interpreted, we shall occasionally have to return to the underlying background.

Having introduced general NTU games we can also define TU games:

DEFINITION 6 *A cooperative TU game is a cooperative game* $(N, V)$, *where for each coalition* $S$

$$V(S) = \left\{ z \in \mathbb{R}^S \; \middle| \; \sum_{i \in S} z_i \leq v(S) \right\}$$

*for a number* $v(S) \in \mathbb{R}$. *The function* $S \mapsto v(S)$ *is called the characteristic function of the TU game, which is written as* $(N, v)$.

## 4   Simple games

**4.1. Definition of a simple game.** At this stage we have introduced the characteristic function, which is the point of departure for most of the theory of cooperative games. Before we turn to the latter, it may however be useful to look at a special case, introduced – as so many other parts of game theory – by von Neumann and Morgenstern [1944] (chapter X deals with these games).

Let us – in the spirit of von Neumann and Morgenstern assume that we are dealing with a TU game $(N, v)$, and further we assume that the characteristic function takes its values in $\{0, 1\}$. In this case we are dealing with a so-called *simple game*. Coalitions can be characterized as *winning* (if $v(S) = 1$) or *losing* ($v(S) = 0$). Formally we shall use this characterization in the definition of a simple game.

DEFINITION 7 *A simple game is a* $(N, \mathcal{W})$, *where* $N = \{1, \ldots, n\}$ *is a finite set, and* $\mathcal{W}$ *is a family of subsets of* $N$, *satisfying the conditions*

   *(i)* $\mathcal{W} \neq \emptyset$, $\emptyset \notin \mathcal{W}$,

  *(ii)* $[C \in \mathcal{W}, C \subset D] \Rightarrow D \in \mathcal{W}$ *(monotonicity).*

*If* $(N, \mathcal{W})$ *also satisfies*

*(iii)* $C \in \mathcal{W} \Rightarrow N \backslash C \notin \mathcal{W}$,

then it is said to be a *proper simple game*.

A simple game is alternatively called a family of majorities, and this latter name may be useful since it points to the main application of the concept. Coalitions $C$ belonging to the family $\mathcal{W}$ are called winning (as mentioned before), and coalitions which are not winning, can also be of interest; if the complement of a coalition is not winning, the coalition is said to be *blocking*.

A prominent class of simple games is that of *weighted majority games*. Such a game is defined using an $n$-vector $(q_1, \ldots, q_n) \in \mathbb{Q}_+^N$ with $\sum_{i \in N} q_i = 1$ by

$$\mathcal{W} = \left\{ C \subseteq N \,\middle|\, \sum_{i \in C} q_i \geq \frac{1}{2} \right\}.$$

Here the winning coalitions are exactly those which obtain a majority when adding the weights of their members. Another basic type of simple games is the *unanimity* game, defined by

$$\mathcal{W} = \{ C \subseteq N \mid T \subseteq C \}$$

for some designated coalition $T$, so that only the supersets of $T$ are winning. Unanimity games will play an important role as fundamental building blocks, also for constructions reaching far beyond the simple games, in the chapters to follow.

**4.2. The Nakamura number of a simple game.** In order to indicate some further applications of simple games, we shall look at a solution concept for simple games, or rather, for realizations of simple games. One may look at simple games in another way that our usual; rather than seeing the simple game as a description of a single, specific conflict situation (where players fight for a dollar which some of the coalitions can get, others not), we may look at it as a description of the *power structure* in society. If a problem occurs which needs solution in the form of a joint decision, the solution will depend on how the parties assess the possible alternatives and of the power structure.

Define a *realization* of a simple game $(N, \mathcal{W})$ as a pair $(A, u)$, where $A$ is a finite set of alternatives, and $u = (u_1, \ldots, u_n)$ denotes an array of utility functions $u_i : A \rightarrow \mathbf{R}$, one for each individual $i \in N$ (i.e. individual $i$ finds the alternative $a$ to be better than alternative $a'$ if and only if $u_i(a) > u_i(a')$); in accordance with the tradition in social choice theory we call such an array a *profile*. We want now to determine which alternative will be chosen by the society as represented by $(N, \mathcal{W})$. This power structure indicated that if a coalition in $\mathcal{W}$ can agree to choose an alternative $a^0$, then this will be the choice of society. In this situation we can say something about the final choice for any profile: If all members of a coalition $C \in \mathcal{W}$ considers an alternative $a$ to be better than another alternative $b$, then $b$ will not be chosen, since the coalition $C$ can pretend that it has agreed that the outcome should be $a$. This argument leads to the following definition:

**Example 10.5.** Most applications of simple games deal with voting:

Consider a group $N$ of individuals having to make a collective decision. The most obvious to do in this case would be to put the matter to a voting. We assume for simplicity that the case is one which can be decided by "Yes " or "No". A simple majority counts the number of Yes and No. If the there are more "Yes" than "No", the collective result will be "Yes", otherwise it is "No".

In this connection it may be of interest to see which pressure groups in society have the power to set their will through (given that they agree). Clearly we get all the majority coalitions, i.e. such $C$ for which $|C| > n/2$. Notice that $\mathcal{W} = \{C|\,|C| > n/2\}$ is a proper simple game.

As is well known, one may use other methods of voting than just simple majority, and the above can be immediately generalized to weighted majority. Here every individual has a weight $q_i \geq 0$, so that $\sum_{i \in C} q_i = 1$, and a coalition has a majority if $\sum_{i \in C} q_i > \frac{1}{2}$ (simple majority is the special case with $q_i = \frac{1}{n}$ for all $i$). Again we have a simple game $\mathcal{W} = \{C|\sum_{i \in C} q_i > \frac{1}{2}\}$.

A somewhat more special example, indicating that simple games turn up in many contexts, we may consider parliaments with several chambers. To have a transparent example, we shall assume that there are three chambers (this is unusual, but two-chamber parliaments are analytically less interesting since a majority of chambers means both of them). Assume that there are $n$ represented in all three chambers of the parliament. The representation of the parties may not be the same in each; we let party $i$ have the fraction $q_i^s$ of the seats in chamber $s$, $s = 1, 2, 3$.

The procedure for legislation is assumed to be that it should pass (with simple majority) in a majority of the three chambers. In order for a coalition $C$ to have its legislation pass, and to be the support of a government in the parliament, we must have

$$\left| \left\{ s |\, \textstyle\sum_{i \in C} q_i^s > \tfrac{1}{2} \right\} \right| \geq 2$$

or

$$\sum_{\{s|\sum_{i \in C} q_i^s > 1/2\}} \frac{1}{3} > \frac{1}{2},$$

since each of the chambers has the weight $\frac{1}{3}$ in the majority game between chambers.

It may be checked that the set of all possible parliamentary supports $C$ defined as above is a proper simple game. Moreover, it can be shown (Keiding [Ke]) that every simple game can be obtained as a hierarchically weighted majority game as in the example, but possibly with more hierarchies that the two we have used here. The minimal number of hierarchies used in such a representation of the simple game is called the *height* of the simle game. This height is of magnitude smaller than $n$, the number of individuals in $N$.

An alternative $a$ dominates an alternative $b$ (via the coalition $C$ in the profile $u$), if $C \in \mathcal{W}$ and $u_i(a) > u_i(b)$ for all $i \in C$. Further we define the *core* of $(N, \mathcal{W})$ over $A$ wrt. the profile $u$, written $\text{Core}(\mathcal{W}, A, u)$, as the set of alternatives which are not dominated (by some alternative via some coalition) in the profile $u$.

**Example 10.6.** It may be expected that the Nakamura number itself depends on the number $n$ of players, but the relationship is rather complicated. If we look at simple majorty games and assume that $n$ is odd, so that the winning coalitions are all those having at least $(n + 1)/2$ members, we see that as soon as $n \geq 3$, we can find 3 coalitions with empty intersection, so that the Nakamura number is 3 independent of number of players.

If conversely $\mathcal{W}$ is such that a coalition must contain at least $n - 1$ to be winning, then one needs $n$ coalitions to have empty intersection. Small majorities result in small Nakamura numbers, which fits well with intuition: If it is easy to collect a majority against each alternative, it is difficult to make a decision.

The core as defined above depends on the profile. For the applications, where the profile may not be known or observed, it is important to find properties of the underlying simple game $(N\mathcal{W})$, which ensures that the core is never empty. The simple game $(N, \mathcal{W})$ is said to be *stable* over $A$ if $\mathrm{Core}(\mathcal{W}, A, u) \neq \emptyset$ for every profile $u$.

It turns out that stability of a simple game may be determined from a combinatorial property of its winning coalitions. Define the Nakamura-number for $\mathcal{W}$ as

$$\nu(\mathcal{W}) = \min \left\{ r \,\middle|\, \exists C_1, \dots, C_r \in \mathcal{W}, \cap_{j=1}^r C_j = \emptyset \right\}$$

i.e. as the smallest number of coalitions from $\mathcal{W}$ with empty intersection. Note that if $(N, \mathcal{W})$ is proper, then $\nu(\mathcal{W}) \geq 2$. The following result is due to Nakamura [Na]:

THEOREM 2 *Let* $(N, \mathcal{W})$ *be a simple game and $A$ a set of alternatives. Then* $(N, \mathcal{W})$ *is stable over $A$ if and only if* $\nu(\mathcal{W}) > |A|$.

PROOF: Assume that $|A| \geq \nu(\mathcal{W}) \geq r$, and choose coalitions $C_1, \dots, C_r \in \mathcal{W}$ with $\cap_{j=1}^r C_j = \emptyset$ together with $r$ different alternatives $a_1, \dots, a_r$ from $A$. For each $i \in N$ we can choose $j = j(i)$ such that $i \notin C_j$. Let $u_i$ be such that

$$u_i(a_j) > u_i(a_{j+1}) > \cdots > u_i(a_r) > u_i(a_1) > \cdots > u_i(a_{j-1})$$

og $u_i(a) > u_i(b)$ for $a \in \{a_1, \dots, a_r\}$. Finally, let $u$ be the profile consisting of $u_1, \dots, u_n$. For all $i \in C_j$ we now have $u_i(a_{j-1}) > u_i(a_j)$. Consequently $a_j$ is dominated by $a_{j-1}$, $j = 1, \dots, r$.

Clearly, we have also that $b \notin \{a_1, \dots, a_r\}$ is dominated. It follows that $\mathrm{Core}(\mathcal{W}, A, u) = \emptyset$, so that $(N, \mathcal{W})$ is not stable over $A$.

Conversely, assume that $\mathrm{Core}(\mathcal{W}, A, u) = \emptyset$ for a certain profile $u$. Then every $a \in A$ is dominated. Choose $a_1 \in A$ arbitrarily, and let $C_1 \in \mathcal{W}$ be such that $u_1(a_2) > u_1(a_1)$, alle $i \in C_1$ for some alternative $a_2 \in A$. Proceeding similarly with $a_2, a_3, \dots$, we will at some stage (since $A$ is finite) get an alternative that we had already. If $a_{k+1}$ is such an alternative, we then know about the sets $C_1, \ldots, C_k$, that $\cap_{j=1}^{k} C_j = \emptyset$: if this was not the case we would have $i \in N$ and $k' < k$ such that $u_i(a_k) > u_i(a_{k-1}) > \cdots > u_i(a'_k)$. We have thus shown that $v(W) \geq k \geq |A|$. □

The result can be used to assess whether a given voting procedure, formulated as a simple game, can be used for decision making in cases where there are many alternatives. This depends, as it is seen, on the Nakamura number. If this number is small, then even for a small set $A$ there are profiles such that all the alternatives are dominated.

## 5  Problems

**1.** Let $\Gamma = (N, (S_i)_{i \in N}, (u_i)_{i \in N})$ be a 2-person zero-sum game, so that $u_1(s) = -u_2(s)$ for all $s \in \Sigma_N$. Show that the mixed extension of $\Gamma$ has a strong Nash equilibrium.

Does this result hold for zero-sum games with more than 3 players?

**2.** Construct an example of a 3-person game with no coalition proof Nash equilibrium.

**3.** Let $\Gamma = (N, (S_i)_{i \in N}, (u_i)_{i \in N})$ be a game. A *Pareto optimal* Nash equilbrium in $\Gamma$ is a Nash equiliibrium $s^0 = (s_1^0, \ldots, s_n^0)$ which is Pareto optimal in the sense that there is no other strategy array $s' = (s'_1, \ldots, s'_n)$ such that $u_i(s') \geq u_i(s^0)$ for all $i \in N$ and $u_j(s') > u_j(s^0)$ for some $j \in N$.

Show that the 3-person game with payoffs

|   | L | R |
|---|---|---|
| T | (1, 1, −5) | (−5, −5, 0) |
| B | (−5, −5, 0) | (0, 2, 7) |

$M_1$

|   | L | R |
|---|---|---|
| T | (1, 1, −6) | (−5, −5, 0) |
| B | (−5, −5, 0) | (−2, −2, 0) |

$M_2$

where player 1 chooses row, player 2 column and player 3 matrix, has two Pareto Optimal Nash equilibria in $\Gamma$. Are they also strong Nash equilibria?

**4.** Let $\Gamma = (N, (S_i)_{i \in N}, (u_i)_{i \in N})$ be a game. For any strategy array $s = (s_1, \ldots, s_n)$, define that map $V_s$ by

$$V_s(S) = \left\{ (z_i)_{i \in S} \mid \exists t_S = (t_i)_{i \in S} : z_i \leq u_i(t_S, s_{N \setminus S}), \text{ all } i \in S \right\}$$

for every coalition $S$.

Show that $(N, V_s)$ is a cooperative NTU game.

If $(N, V)$ is a cooperative game, then a payoff vector $u = (u_1, \ldots, u_n) \in V(N)$ belongs to the *core* of $(N, V)$ if $u$ is Pareto optimal in $V(N)$ (there is no vector $u' \neq u$ in $V(N)$ with $u'_i \geq u_i$ for all $i$) and there is no coalition $S$ such that $V(S)$ contains a vector $u^S$ with $u_i^S > u_i$ for all $i \in S$.

Show that $s$ is a strong Nash equilibrium of $\Gamma$ if and only if $u(s)$ belongs to the core of $(N, V_s)$.

Show that if $s$ is a strong Nash equilibrium of $\Gamma$ and $u(s)$ is Pareto optimal in $V(N)$, then $u(s)$ belongs to the core of $(N, V_\beta)$ (also called the $\beta$-core of $\Gamma$).

**5.** Find the Nakamura-number of the simple game $(N, \mathcal{W}^k)$, where

$$\mathcal{W}^k = \{C \subseteq N \mid |C| \geq k\}$$

for $1 \geq k \geq n$.

**6.** Let $(N, \mathcal{W}_k)_{k \in K}$ be a finite family of simple games with the same player set, and let $(K, \mathcal{U})$ be a simple game on the set $K$. The *inner compound* of $(N, \mathcal{W}_k)_{k \in K}$ with quotient $(K, \mathcal{U})$ is the simple game $(N, \mathcal{W})$ with

$$\mathcal{W} = \{C \subseteq N \mid \{k \in K \mid C \in \mathcal{W}_k\} \in \mathcal{U}\}.$$

Let $\mathfrak{W}^{(1)}$ be the set of weighted majority games, and define for $m > 1$ $\mathfrak{W}^{(m)}$ as the set of simple games obtained as inner compound of simple games in $\mathfrak{W}^{(m-1)}$ with a quotient which is a weighted majority game (on some index set).

Show that if $(N, \mathcal{W})$ is a unanimity game with a unique minimal (for inclusion) winning coalition $C$ of cardinality $|C| = s < n$, then $(N, \mathcal{W})$ belongs to $(N, \mathcal{W})^{(m)}$ for $m = \lceil \log_2 s \rceil$ (the smallest integer $\geq \log_2 s$).

# Chapter 11

# Bargaining

## 1 Introduction

We begin our treatment of cooperative games with a discussion of a class of games which is particularly simple in the sense that only the grand coalition consisting of all players can achieve anything. Proper coalitions play no role whatsoever.

The underlying conflict situation can therefore be described as one where the players have the choice between finding a common solution or getting nothing at all. Since the interesting of such conflicts are those where each player has some potential gain from participating in finding the common solution, the problem boils down to deciding how much of the common gain that should be allocated to each player. This may be expected to depend on the "bargaining position" of the player, and the task before us is to embed this concept into a precise model of bargaining.

It may be noticed that since proper coalitions are irrelevant (since they can do nothing for their members), this bargaining strength is something which concerns the player in an isolated sense (not being influenced by membership of alternative coalitions). One may therefore with some justification consider the bargaining situation as a *non*-cooperative game which does not really belong to the present discussion. On the other hand, from a formal point of view it is more convenient to treat it as a cooperative game, which is what we are doing.

In a "realistic" bargaining situation the solution of the conflict will probably have been preceded by a phase of argumentations, where the individuals will have pointed out both their bargaining strength (whatever that may be) and some considerations of fairness or equality. These latter considerations will be made more precise in the axiomatic approach to bargaining problems, which may be seen as a normative theory of solving conflicts; though it is our first instance of the axiomatic approach, it will not be the last one, since axiomatic characterizations have found widespread use in cooperative game theory. We proceed to possible formalizations of the bargaining process, and finally we consider briefly the case of bargaining under uncertainty and incomplete information.

## 2    Axiomatic bargaining theory

The bargaining situations to be treated in the following are characterized by a set of possible results of the bargaining process together with a specific result, which is in the literature is referred to either as *status quo* or the *disagreement outcome*; this is what the players get if they cannot reach an agreement in the course of the bargaining process.

DEFINITION 1 *An n-person bargaining problem is a pair $(B, a)$, where B is a subset of $\mathbb{R}^n_+$ and $a \in \mathbb{R}^n$, satisfying*

(i) *B is nonempty, convex and compact,*
(ii) $B \subset \{x \in \mathbb{R}^N \mid x_i \geq a_i, i = 1, \ldots, n\}$,
(iii) *If $x \in B$ and $x' \in \mathbb{R}^n$ satisfies $a_i \leq x'_i \leq x_i$ for all i, then $x' \in B$,*
(iv) *B contains some x with $x_i > a_i$, all $i \in N$.*

The vector $a$ is the above mentioned disagreement outcome. We will interpret the vectors $x \in B$ as specifying a utility level for each player. With our usual assumptions about the underlying preferences of the players the bargaining problem is essentially unchanged if we add a (positive or negative) constant to the utility function of each player (and it may even vary between players). We may therefore assume – as we indeed shall do in the sequel – that the disagreement payoff $a$ is the zero vector $0 \in \mathbb{R}^n_+$.

It can be seen that a bargaining problem $(B, 0)$ or just $B$ can be considered as a cooperative game in characteristic function form $(N, V)$, where

$$V(S) = \begin{cases} \{x' \in \mathbb{R}^n \mid \exists x \in B, x'_i \leq x_i, i \in N\} & \text{for } i \in N, \\ \{x' \in \mathbb{R}^S \mid x'_i \leq a_i, i \in S\} & \text{for } i \in S, \end{cases}$$

that is as a game where only the grand coalition has any strategical impact on the outcome differing from what the players can do as individuals. The condition (iv) in Definition 1 secures that there is a potential gain for each player by participating in the bargaining for a joint outcome.

Condition (i) is partly a standard assumption about $V(N)$ in a cooperative game, partly an extension, in particular with respect to the convexity assumption. Usually the latter is explained by having lotteries as some of the outcomes of the bargaining process (together with the assumption that players' utility functions satisfy the expected-utility hypothesis).

We denote the class of bargaining problems (for a given player set $N$) by $\mathcal{B}$. As mentioned above it is everywhere assumed that the disagreement outcome is 0.

DEFINITION 2 *A bargaining solution is a function $\Phi$ which to every bargaining problem $B \in \mathcal{B}$ assigns a payoff vector $\Phi(B) \in B$.*

A bargaining solution can be seen as a rule for solving bargaining problems: To every conceivable problem the solution indicates what the result of the bargaining should be. It is straightforward that there exist bargaining solutions, for example the trivial solution $\Phi_T$ with $\Phi_T(B) = 0$ for each $B \in \mathcal{B}$, but we are here interested in bargaining solutions which give a "reasonable" or "fair" outcome of the bargaining.

To identify such solutions we must specify our requirements to them. Some properties of reasonable bargaining solutions are rather straightforward.

AXIOM 1 (Pareto optimality) *For each $B \in \mathcal{B}$ the payoff $\Phi(B)$ is Pareto optimal: If $\Phi(B) = x \in B$ and $y \in B$ is such that $y_i \geq x_i$ for all $i$, then $y = x$.*

This property seems natural as a demand for efficiency in bargaining. Another, similarly uncontroversial, requirement is the following.

AXIOM 2 (Relative invariance under scale transformations) *For every n-tuple $c = (c_1, \ldots, c_n)$ of positive numbers and every $B \in \mathcal{B}$, if the bargaining problem $cB$ is defined by*

$$cB = \{(c_1 x_1, \ldots, c_n x_n) \mid x = (x_1, \ldots, x_n) \in B\},$$

*then $\Phi(cB) = (c_1 \Phi_1(B), \ldots, c_n \Phi_n(B))$.*

Axiom 2 may be considered as a property of the utility representations of players' preferences over outcomes rather than a property of the bargaining solution. Since we consider lotteries as possible outcomes and consequently assume the utility functions to be consistent with expected utility maximization, we have that affine transformations of utilities are acceptable. Having already used addition of scalars in determining the value 0 of the disagreement payoff, we have only the scale transformations left over. In this situation Axiom 2 is a consistency requirement – scale transformations, which leave the fundamental situation as well as the underlying preferences unchanged, should not influence the outcome of the bargaining.

The next axiom may also seem intuitively reasonable, and in different disguises it will turn up in many other situations. It might however be less readily acceptable in many applications than the previous two.

AXIOM 3 (Symmetry) *If $B \in \mathcal{B}$ is symmetric (in the sense that for every permutation $\sigma$ of $N = \{1, \ldots, n\}$, if $x \in B$ then so is $(x_{\sigma(1)}, \ldots, x_{\sigma(n)})$, then $\Phi(B)$ belongs to the diagonal $\{x \in \mathbb{R}^n \mid x_1 = \cdots = x_n\}$.*

Symmetry means here that the players are not in any way favorized based on considerations which are external to the bargaining problem. It should be noticed that the axiom is formulated in such a way that it refers only to bargaining problems which are themselves symmetric, so that every individual plays exactly the same role in the problem.

The axioms 1 – 3 restrict the set of bargaining solutions ($\Phi_T$ is no longer a member of this set), but there are still rather many bargaining solutions which satisfy all of these requirements. In order to have a characterization by axioms of a single bargaining solution we need to add further axioms, which now may be somewhat debatable. Among these, the most prominent is the following.

Axiom 4 (Independence of irrelevant alternatives) *If $B, B' \in \mathcal{B}$, $B \subset B'$ and $\Phi(B') \in B$, then $\Phi(B) = \Phi(B')$.*

Also this axiom (or related versions) is one which will show up in many different contexts. At a first sight, it seems rather obvious: If what is best (or "most reasonable" or "most fair") in the larger set of alternatives actually belongs to the smaller set, then it should also be best in this smaller set. In particular, this is the case whenever the selected element maximizes some given function, since what maximizes in the large set must also maximize in the smaller one. But looking closer one sees that the intuitiveness hinges on the idea that "best" can be measured on a scale or determined using a complete preorder. We shall return to the more problematic aspects of Axiom 4 later.

Whatever may be our future reservations about this axiom, it delivers the goods in the sense that we get a unique bargaining solution on the basis of Axioms 1 – 4.

THEOREM 1 *There is one and only one bargaining solution $\Phi_N$ which satisfies Axioms 1 – 4, and for $B \in \mathcal{B}$, this solution is the solution of*

$$\max x_1 x_2 \cdots x_n$$

*subject to*

$$x = (x_1, \ldots, x_n) \in B.$$

PROOF: First of all we note that $\Phi_N$ is welldefined. This follows easily since each $B \in \mathcal{B}$ is nonempty, convex and compact, and the function which takes each $(x_1, \ldots, x_n)$ to the product of its coordinates, is continuous and strictly quasiconcave, so that it attains a unique maximum on $B$.

Next we show, that $\Phi_N$ satisfies the Axioms 1 – 4. Of these, Axiom 1, 3 and 4 are trivially satisfied. For Axiom 2, we notice that if

$$x_1^0 \cdots x_n^0 \geq x_1 \cdots x_n, \text{ all } x \in B,$$

then also

$$(c_1 x_1^0) \cdots (c_n x_n^0) \geq (c_1 x_1) \cdots (c_n x_n)$$

for all $x \in B$, from which it is seen that $(c_1 x_1^0, \ldots, c_n x_n^0)$ maximizes the product of the coordinates over all vectors in $cB$, so that Axiom 2 is satisfied.

Suppose now that $\Phi$ is a bargaining solution which satisfies Axioms 1 – 4. Choose an arbitrary $B \in \mathcal{B}$. We show that $\Phi(B) = \Phi_N(B)$.

Using Axiom 2 we may assume that $\Phi_N(B) = (1, \ldots, 1)$, since if it was not the case, we would just change the scale by $1/x_i^0$ for each $i$.

We now construct a new bargaining problem $\widehat{B}$ defined by

$$\widehat{B} = \text{conv}(\{B^\sigma \mid \sigma \text{ is a permutation of } N\}),$$

where $B^\sigma = \{(x_{\sigma(1)}, \ldots, x_{\sigma(n)}) \mid x \in B\}$. Then $\widehat{B}$ is symmetric by definition, so by Axiom 3, $\Phi(\widehat{B}) = \lambda(1, \ldots, 1)$ for some $\lambda > 0$. Since $\Phi_N(B) = (1, \ldots, 1)$, we have that

$B \subset \{x \in \mathbb{R}^n_+ \mid \sum_{i=1}^n x_i \leq n\}$ (the hyperplane defined by $\sum_{i=1}^n x_i = n$ is a tangency plane to the surface defined by the maximal attainable product of coordinates), and then $\widehat{B}$ must be contained in this set. From Axiom 1 we have that $\Phi(\widehat{B}) = (1, \ldots, 1)$.

Since $B \subset \widehat{B}$ and $\Phi(\widehat{B}) = 1$, we have from Axiom 4 that $\Phi(B) = (1, \ldots, 1) = \Phi_N(B)$ as contended. Thus, $\Phi_N$ is the only bargaining solution which satisfies Axioms $1-4$. $\qquad\square$

The above result is due to Nash [1953], and $\Phi_N$ is called *Nash's bargaining solution.*

Though the Nash bargaining solution appears as a reasonable recipe for solving bargaining problem – and for many years it remained the only one discussed in the literature – it is not the only possible one. Below we take a look at another solution, which may be considered as equally plausible. To simplify the exposition we restrict attention to the case $n = 2$.

The point of departure is Axiom 4, which as already mentioned is the weak spot of the Nash solution. Consider the two bargaining problems (as shown in Fig. 1)

However, it seems intuitively unreasonable that player 2 gets as much as player 1 in the problem $B$, actually she will then get the maximal payoff in any payoff feasible in the problem $B$. That it is so is of course a consequence of Axiom 4, which we consequently will replace by something else.

For the formulation of the alternative axiom we use some notation: Let $a(B)$ be the vector with coordinates

$$a_i(B) = \max\{x_i \mid x \in B\}, \ i = 1, 2.$$

The coordinates of the vector $a(B)$ is called the *aspiration levels* for players 1 and 2.

Further, define the function $g_B : [0, a_1(B)] \to [0, a_2(B)]$ by

$$g_B(x_1) = \begin{cases} x_2 & \text{if } (x_1, x_2) \text{ is Pareto optimal,} \\ a_2(B) & \text{if there is no such } x. \end{cases}$$

The function $g_B$ assigns to every $x_1$ the greatest payoff which can be given to player 2 if player 1 should have $x_1$.

We may now formulate the new axiom:

AXIOM 5 (Monotonicity) *If $B, B' \in \mathcal{B}, a_1(B) = a_1(B')$, and $g_B \leq g_{B'}$, then $\Phi_2(B) \leq \Phi_2(B')$.*

The following theorem presents an alternative to the Nash bargaining solution.

THEOREM 2 *There is a unique bargaining solution $\Phi_{KS}$ which satisfies Axioms $1-3$ and 5, and for $B \in \mathcal{B}$, $\Phi_{KS}(B)$ is the vector which maximizes*

$$\min\left\{\frac{x_1}{a_1(B)}, \frac{x_2}{a_2(B)}\right\}.$$

PROOF: It is easily seen that $\Phi_{KS}$ is well-defined and satisfies the axioms $1-3$ and 5.

Assume that $\Phi \neq \Phi_{KS}$ satisfies the properties of Axioms $1-3$ and 5. We may assume (after application of Axiom 3 if necessary) that $(a_1(B), a_2(B)) = (1, 1)$.

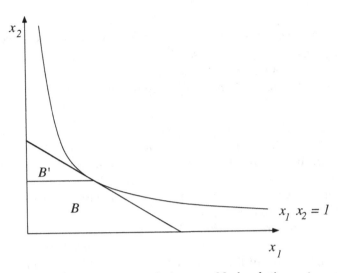

**Fig. 1.** Two bargaining problems with the same Nash solution outcome.

Clearly $(0, 1) \in B$ and $(1, 0) \in B$. Let

$$\tilde{B} = \mathrm{conv}(\{(0, 1), (1, 0), \Phi_{KS}(B)\}).$$

Then $\tilde{B} \in \mathcal{B}$, $B$ is symmetric, and $\tilde{B} \subset B$. From axiom 2 we have that

$$\Phi(B) \in \{\lambda(1, 1) \mid \lambda \geq 0\},$$

and from Axiom 1 we get that $\Phi(\tilde{B}) = \Phi_{KS}(B)$. Applying Axiom 5, we have that $\Phi_2(\tilde{B}) \leq \Phi_2(B)$, from which we get that also $\Phi_1(\tilde{B}) \leq \Phi_1(B)$. Summing up, we have that $\Phi(\tilde{B}) = \Phi(B)$, which together with $\Phi_{KS}(B) = \Phi(\tilde{B})$ and the arbitrariness of $B$ gives that $\Phi = \Phi_{KS}$.                                    □

The solution $\Phi_{KS}$ is called the Kalai-Smorodinsky solution and is due to Kalai and Smorodinsky [1975].

## 3   Bargaining solutions as result of a process

In this section we look at a bargaining solution which differs from the two considered above. For simplicity we again restrict attention to the two-player case.

Consider a bargaining problem as that shown in Fig. 1 as $B$. Assume that the actual bargaining goes as follows: Initially, the players state their maximal demands, and then they reduce their claims successively. So, at the outset player 1 proposes the payoff vector $(a_1(B), 0)$, and player 2 proposes $(0, a_2(B))$. A concession from player 1 is then an accept of some $a_1$ smaller than $a_1(B)$ against a guarantee from player 2 that she will never get less than $s_1 \geq 0$, and similarly one can define a concession from player 2.

As it can be seen, the concessions will transform the original bargaining problem to a new one (the coordinate system should have its origin in $(s_1, s_2)$, with aspiration levels $(a_1 - s_1, a_2 - s_2)$. For each player, we can determine the *concession rates*

$$\frac{s_1}{a_1(B) - a_1}, \text{ respectively } \frac{s_2}{a_2(B) - a_2},$$

which gives the relation between what the player gets (in terms of increased security against bad results in the future of the bargaining process) and what she gives up (possibility of advantageous results).

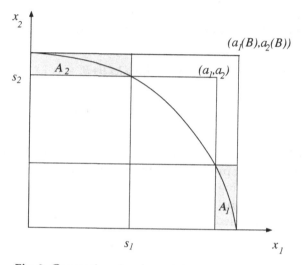

**Fig. 2.** Concessions in a bargaining process.

The basic idea of the bargaining solution that we consider now is that every step in the process should be *fair* in the sense that the concession rates should be equal for the two players. Geometrically this means that the areas $A_1$ and $A_2$ in Fig. 2 are identical.

The *Maschler-Perles solution* [Maschler and Perles, 1981] to the bargaining problem $B$ is the final result of a step of concessions of this type. The solution occurs when the players agree after having made sufficiently many concessions.

To make this precise we must work with small (actually: infinitesimally small) bargaining steps. Formally, we introduce the functions $\theta_i : B \to \mathbb{R}^2$ given by

$$\theta_i(x) = \max\{x_i' \mid x' \in B, x \le x'\}, \ i = 1, 2.$$

We assume that the boundary of $B$ is such that $\theta_i$ is differentiable. A vector $c \in \mathbb{R}^2$ with positive coordinates is a *direction of equal concession rates* if

$$\frac{\theta_1'(x; c)}{c_1} = \frac{\theta_2'(x; c)}{c_2},$$

where $\theta_i'(x; c)$ is the directional derivative of $\theta_i$ at $x$ in the direction $c$. It can be shown that this equation has a unique (except for scalar multiplication) solution, which we will denote by $c^*(x)$. We then have a differential equation

$$\frac{dx}{dt} = c^*(x(t)),$$

which describes the bargaining process. This differential equation has a unique solution $x^0$ with initial condition $x^0(0) = 0$, and the point where the solution curve $\{x^0(t) \mid t \in \mathbb{R}\}$ intersects the boundary of $B$ is the *Maschler-Perles* solution at $B$.

As it can be seen, the definition of the Maschler-Perles solution is rather complex. Nevertheless, the solution has some intuitive appeal as mimicking a "real" bargaining. In addition, it has nice properties: It satisfies the Axioms 1 – 3 as well as the following:

AXIOM 6. (Superadditivity) *Let $B, B' \in \mathcal{B}$ and let $B + B' = \{x \mid x = x_1 + x_2, x_1 \in B, x_2 \in B'\}$. Then*

$$\Phi(B + B') \geq \Phi(B) + \Phi(B').$$

Superadditivity is a useful property, in particular when we take into consideration that the bargaining problem is subject to uncertainty. Suppose that the players know that the bargaining problem will be either $B$ or $B'$, and that each of these problems have the probability $1/2$. If the players postpone bargaining untill the actual problem is revealed (by "nature"), then they will get the expected result

$$\frac{1}{2}\Phi(B) + \frac{1}{2}\Phi(B').$$

On the other hand, they could consider the situation as one single bargaining problem $\hat{B}$ with elements

$$x = \frac{1}{2}x + \frac{1}{2}x',$$

where $x \in B$, $x' \in B'$, and use the bargaining solution $\Phi$ on $\hat{B}$, corresponding to performing the bargaining before nature has chosen. If $\Phi$ satisfies Axiom 6 (and Axiom 2), then it is at least as good for both to bargain now as to postpone bargaining.

## 4   The Nash program: Bargaining as a non-cooperative game

In our discussion above the bargaining situation has been treated as a cooperative game – indeed this was the obvious approach given the context. The bargaining problem was considered as one where a common decision was necessary, since the disagreement payoff was clearly inferior to all. Nevertheless, there is also a non-cooperative aspect, which we have neglected so far but which should be faced in some way or another: How can the players obtain an agreement about collective action? It may be argued that the bargaining theory treated so far, with the possible exception of the Maschler-Perles solution, should be considered as a *normative* theory of bargaining, but in this case it should be supplemented with a descriptive theory about the way in which a solution is reached in practice. Such a theory will typically rely on non-cooperative game theory, and in this sense we are making an excursion back to Part 1 of this book.

We shall take a closer look at some approaches to a description of a bargaining procedure. The first of these is the bargaining model proposed in Rubinstein [1982]. Contrasting with the broad generality of our previous approach, we shall be much more specific now about the object over which bargaining takes place: We assume that what should be negotiated is a division of a certain bundle of goods between the involved players, and for simplicity we take it as one homogeneous good available in the quantity 1. We look only at 2-person bargaining problems, so that the final result of the bargaining can be stated as a number $s$ between 0 and 1, indicating player 1's share of the good. The share of player 2 is then $1 - s$.

A further simplification is achieved by adding an assumption about players' utilities. Since bargaining takes time, we assume, quite realistically in fact, that the utility of a certain share obtained quickly is greater than the utility of the same share obtained after a lengthy negotiation. More specifically it is assumed that player 1's utility of the share $s$ obtained at time $t$ is

$$u_1(s, t) = s - c_1 t,$$

and player 2's utility correspondingly

$$u_2(s, t) = 1 - s - c_2 t$$

(where $s$ is the share going to player 1 after a negotiation ending at $t$).

The principle of the bargaining procedure which we model here is the following: at even dates $t = 0, 2, 4, \ldots$ player 1 proposes a division (or rather, a share to player 1) $s$ of the good, and player 2 responds with either 'Yes' or 'No'. In the case of a 'Yes' bargaining has ended, and the players get $s$ and $1 - s$ at that date. If player 2 responds 'No' the game proceeds to the next round (so that $t$ is odd), and player 2 similarly proposes a share $s$ (to player 1) which may be accepted or rejected.

Formally, we have a game where the strategies for player 1 are sequences $(f^t)_{t=0}^{\infty}$, where $f^0$ is an element of $[0, 1]$, and $f^t$ for $t > 0$ even is a function from $[0, 1]^t$ (the set of all previously suggested divisions, that is the history of the play before $t$) to $[0, 1]$, and for $t$ odd a function from $[0, 1]^t$ to {Yes, No}. Correspondingly player 2's strategies are sequences $(g^t)_{t=0}^{\infty}$, where $g^t$ for $t$ odd is a function from $[0, 1]^t$ to $[0, 1]$, $g^0$ an element of {Yes, No}, and for $t > 0$ even a function from $[0, 1]^t$ to {Yes, No}. The procedure described is the simplest possible way of formulating a negotiation where each player can respond to the former proposals of the other player.

It is seen that with the utility functions already defined we obtain a game in normal form: A pair of strategies $\left((f^t)_{t=0}^{\infty}, (g^t)_{t=0}^{\infty}\right)$ give rise to a *play*: Player 1 proposes $f^0$, which is a particular share $s^0$ of the good, player 2 responds with $g^0(s^0)$, if the game does not terminate, player 2 proposes $g^1(s^0)$, player 1 responds with $f^1(s^0, g^1(s^0))$, either the game ends or she proposes $f^2(s^0, g^1(s^0))$ etc.

Let $t$ be the first date at which a player chooses 'Yes' (if there is such a date); then the result of the bargaining is the associated proposal for a division of the good, and the negotiation time is $t$, from which the utility payoff of the two players can be determined. If the game does not stop (both players reject all the time), the result is defined as $(0, \infty)$ for both players, and this result is assumed inferior to any other result for both players.

We are interested in the non-cooperative equilibria of this game. However, it turns out that Nash equilibria are in general not very interesting.

**THEOREM 3** *Every division s can occur as the outcome of a Nash equilibrium.*

PROOF: Let $s \in [0, 1]$ be given and define $f = (f^t)_{t=0}^{\infty}$, $g = (g^t)_{t=0}^{\infty}$ by

$$f^t = s, \quad g^t(s^1, \ldots, s^t) = \begin{cases} \text{Yes} & \text{hvis } s^t \leq s, \\ \text{No} & \text{hvis } s^t > s, \end{cases}$$

for $t$ even, og

$$g^t = s, \quad f^t(s^1, \ldots, s^t) = \begin{cases} \text{Yes} & \text{hvis } s^t \geq s, \\ \text{No} & \text{hvis } s^t < s, \end{cases}$$

for $t$ odd. Then $(f, g)$ is a Nash equilibrium, and the play stops at $t = 0$, that is immediately, with the share $s$ to player 1.                                    □

Since there are so many Nash equilibria and many of them may be trivial, established using unrealistic threats, it seems reasonable to look for subgame perfect Nash equilibria.

Writing down the equilibrium conditions is tedious, and a less detailed description suffices for our purposes: $(f, g)$ is a subgame perfect Nash equilibrium if for every $T$ and every sequence $s^1, \ldots, s^T$ of previous suggestions for a division with associated accepts or rejections it holds that $f|_{s^1, \ldots, s^T}$ and $g|_{s^1, \ldots, s^T}$ are Nash equilibrium strategies in the game starting at $T + 1$. Here $f|_{s^1, \ldots, s^T}$ is the sequence of functions $(f(s^1, \ldots, s^T, r^{T+1}, \ldots, r^t))_{t=T+1}^{\infty}$ with fixed $s^1, \ldots, s^T$, and $g|_{s^1, \ldots, s^T}$ is defined similarly.

To describe the divisions that can be obtained in subgame perfect Nash equilibria, we begin with the following result.

**LEMMA 1** *Let $(x, y) \in [0, 1]^2$ be such that*

$$y \text{ is the smallest number with } x - c_1 \leq y,$$
$$x \text{ is the largest number with } y + c_2 \geq x.$$

*Then there is a subgame perfect Nash equilibrium which results in $x$, and in the opposite game, where player 2 has the first move, there is a subgame perfect Nash equilibrium resulting in $y$.*

It may be noticed that $y$ is defined in such a way that player 1 is indifferent between the division $y$ proposed by player 2 and the division $x$ obtained after another round, so $y$ is the maximal amount that player 2 can obtain if player 1 can obtain $x$ when choosing. Similarly $1 - x$ is exactly as good for player 2 as $1 - y$ after another round.

PROOF OF LEMMA 1: Define strategies $(f^t)_{t=0}^\infty, (g^t)_{t=0}^\infty$ by

$$f^t = x, \quad g^t(s^1, \ldots, s^t) = \begin{cases} \text{No} & \text{if } x > s^t, \\ \text{Yes} & \text{if } x \leq s^t, \end{cases}$$

for $t$ even, and

$$g^t = y, \quad f^t(s^1, \ldots, s^t) = \begin{cases} \text{No} & \text{if } y < s^t, \\ \text{Yes} & \text{if } y \geq s^t, \end{cases}$$

for $t$ odd. It is easily verified that $((f^t)_{t=1}^\infty, (g^t)_{t=1}^\infty)$ gives a subgame perfect equilibrium; a play of the game will stop after first step with the division defined by $x$. If the game started only in the second step the play would correspondingly stop after this step with the division given by $y$. □

I can be seen that pairs $(x, y)$ satisfying the conditions in Lemma 1, are solutions to the system of equations

$$\begin{aligned} y &= \max\{x - c_1, 0\} =: d_1(x), \\ x &= \min\{y + c_2, 1\} =: d_2(y). \end{aligned} \tag{1}$$

The composite $d_2 \circ d_1$ of the two functions $d_1$ and $d_2$ is a continuous map from $[0, 1]$ to itself, so it has a fixed point $x^0$ (the graph of $d_2 \circ d_1$ must intersect the diagonal), and then $(x^0, d_1(x^0))$ satisfies the conditions in Lemma 1. The mappings $d_1$ and $d_2$ as well as the set of pairs $(x, y)$ satisfying (1) are shown in Figure 3 in each of the cases $c_1 > c_2$, $c_1 = c_2$, and $c_1 < c_2$. The solutions are found where the two graphs intersect, and it is uniquely determined in the cases $c_1 > c_2$ and $c_1 < c_2$.

The last statement, pertaining to uniqueness of equilibrium divisions, does not follow from our argumentation so far. We have shown uniqueness of solutions to (1), but there might be other equilibrium divisions. However, this can be shown not to be the case. For a detailed proof the reader is referred to Rubinstein [1982]; here we sketch the argument in one of the cases, namely $c_1 > c_2$:

Suppose that there is a division $a > c_2$ which is also sustained by a subgame perfect Nash equilibrium. Assume that $a$ is the largest of all such divisions, and choose a number $b$ sufficiently close to $a - c_2$, so that $b + c_2 < a$ but $b + c_1 > a$. Then $(1 - b) - c_2 > 1 - a$, and $b$ can be seen as a better result for player 2 than $a$ and a potential offer from 2 if she rejects $a$. But as $a$ is an equilibrium offer, player 1 must have had a reason for rejecting it, which means that some equilibrium result $c$ is better than $b$ even after a further round of the game, that is $c - c_1 \geq b$ or $c \geq b + c_1$. But then $c > a$ contradicting that $a$ is the largest equilibrium result.

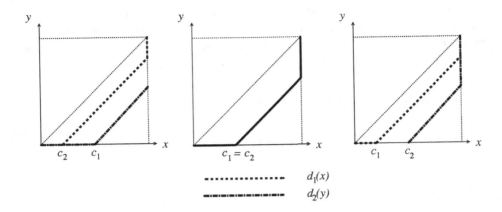

Fig. 3. Solutions to the bargaining game.

The uniqueness result sketched here is closely related to the choice of very simple utility functions of players. The parameters $c_1$ and $c_2$ can be interpreted as negotiation costs, and the player with lower cost gets the best result.

## 5  Bargaining with incomplete information

In our discussion of the property of superadditivity for bargaining solutions (cf. Section 3) we touched upon the problems of uncertainty in the bargaining situation. In our context the uncertainty was connected with the sets of attainable bargaining outcomes, while the players were assumed to know all other relevant data, among these the utility functions of the other participants.

This may not be a particularly realistic way of treating uncertainty and incomplete information. It seems more often to be the case that the alternatives to be obtained are known to all while the preferences, that is the utility payoffs, of the involved parties are subject to uncertainty.

Following Harsanyi and Selten [1972] such a situation can be formalized as follows (where we restrict ourselves to the case of two-person bargaining):

We assume that each of the two players may have a finite number of possible utility functions, so that in any given outcome the utility of player 1 is chosen from the set $\{u_1^1, \ldots, u_1^K\}$, and that of player 2 from the set $\{u_2^1, \ldots, u_2^M\}$. We say that player 1 has type $k$ (player 2 has type $m$), if she has the utility function $u_1^k$ ($u_2^m$). Here we follow the approach to games with incomplete information already outlined in Chapter 6. In the same vein, we introduce a subjective probability distribution for player 1 over the types $\{1, \ldots, M\}$ of player 2, and correspondingly a subjective probability distribution for player 2 over the set of player 1's types. The two subjective distributions are *consistent* in the sense that they can be represented by a matrix

$$r = \begin{pmatrix} r_{11} & \cdots & r_{1M} \\ \vdots & & \vdots \\ r_{k1} & \cdots & r_{KM} \end{pmatrix}$$

satisfying $r_{km} \geq 0$ for all $k$ and $m$, and

$$\sum_{k=1}^{K} \sum_{m=1}^{M} r_{km} = 1.$$

Thus, $r$ defines the *common prior* probability distribution over combinations of types, and the marginal probabilities $p_k = \sum_{m=1}^{M} r_{km}$ and $q_m = \sum_{k=1}^{K} r_{km}$ give the probability that player 1 has type $k$ and player 2 type $m$.

We assume that there is a given finite set $D$ of possible outcomes of the bargaining process, and that $D$ contains a designated element $d^0 \in D$, interpreted as status quo or the disagreement outcome. We can then state the *bargaining problem with incomplete information* as a triple $(D, r, d^0)$, where $D$ is a set of bargaining outcomes, $r$ a $(K \times M)$-matrix of subjective probabilities over type assignments, and $d^0$ is the disagreement outcome. We assume (as in previous sections) that all utility functions are normalized in such a way that $u_1^k(d^0) = 0$ og $u_2^m(d^0) = 0$ for all $k$ and $m$.

What is to be understood by a final result of the bargaining problem with incomplete information? Following the tradition (which was established in Chapter 6) we look for a rule specifying how the outcome should depend on the types of the individuals. We therefore define a *mechanism* for the bargaining problem as a function $\mu : T \to \Delta D$, where $T$ consists of all pairs $(k, m)$ of player types and $\Delta D$ is the set of all probability distributions on $D$; $\mu$ assigns to each $d$ and each pair of types $(k, m)$ the probability $\mu(k, m)(d)$ that outcome $d$ is chosen given that types are $(k, m)$.

Given a mechanism $\mu$, let $U_1(\mu|r, k)$ be expected utility of player 1 in state $k$, that is

$$U_1(\mu|r, k) = \sum_{m=1}^{M} \sum_{d \in D} \frac{r_{km}}{p_k} \mu(k, m)(d) u_1^k(d),$$

where $r_{km}/p_k$ is the conditional probability of player 2 being in state $m$ given that player 1 is in state $k$. Expected utility of player 2 in state $m$ is defined similarly as

$$U_2(\mu|r, m) = \sum_{k=1}^{K} \sum_{d \in D} \frac{r_{km}}{p_m} \mu(k, m)(d)(u_2^m(d).$$

We have now – at least apparently – extended the formalism for dealing with usual (complete information) bargaining problems to our case of bargaining with incomplete information. However, a new phenomenon occurs when players have private information about their type. Choosing any mechanism $\mu$, the subsequent application of the rule means that agents must send a message about their type, and this message may be true or it may not be, depending on what in the given situation is best for the player. Obviously, the possibility of opportunistic behavior makes it difficult to ascertain that the decision finally made reflect the interests of society, since the types may have been distorted on the way. Therefore it seems natural to restrict attention to mechanisms where such a misrepresentation does not occur.

Let

$$U_1(\mu, k'|r, k) = \sum_{m=1}^{M} \sum_{d \in D} \frac{r_{km}}{p_k} \mu(k', m)(d) u_1^k(d)$$

be expected utility of player 1 if the true type is $k$ but $k'$ is signaled for the final decision, and define $U_2(\mu, m'|r, m)$ similarly. The mechanism $\mu$ is *incentive compatible* if

$$U_1(\mu|r, k) \geq U_1(\mu, k'|r, k), \text{ all } k, k',$$
$$U_2(\mu|r, m) \geq U_2(\mu, m'|r, m), \text{ all } m, m'. \tag{2}$$

Thus, if $\mu$ is incentive compatible, then no player can benefit from misrepresenting the type. We add another condition known from mechanism theory, namely *individual rationality*, which can be formulated as

$$U_1(\mu|r, k) \geq 0, \text{ all } k, \ U_2(\mu|r, m) \geq 0, \text{ all } m. \tag{3}$$

We have now reached the final formulation of the bargaining problem with imperfect information, which amounts to searching for a mechanism satisfying the two conditions (2) and (3) – and the additional axioms which one might want to introduce to characterize the choice. It can be shown that there is a counterpart of the Nash solution for this case, namely the mechanism which maximizes

$$\left( \prod_{k=1}^{K} U_1(\mu|r, k) \right)^{p_k} \left( \prod_{m=1}^{M} U_2(\mu|r, m) \right)^{q_m}$$

over all mechanisms $\mu$ satisfying the two conditions (2) and (3). Again this solution has a characterization by axioms which are extensions of the Axioms 1-4 used in Section 2. For details we refer to Harsanyi and Selten [1972], Myerson [1984].

## 6   Problems

**1.** Consider the 2-person normal form game with payoff matrix

|   | E | T |
|---|---|---|
| E | (6, 3) | (2, 4) |
| T | (3,2) | (7, 0) |

Find the set of feasible utility vectors and the associated bargaining problem $B$. Find the Nash bargaining solution to this bargaining problem.

**2.** The *egalitarian* solution $\Phi_E$ assigns to each bargaining problem $B$ the point $\lambda_B(1,\ldots,1)$, where $\lambda_B$ is the maximal number $\lambda$ such that $\lambda(1,\ldots,1) \in B$. Check that $\Phi_E$ satisfies Pareto optimality, symmetry and strong monotonicity: If $B' \subseteq B$, then $\Phi^E(B') \leq \Phi^E(B)$.

More generally, for $\lambda = (\lambda_1,\ldots,\lambda_n)$ an array of positive numbers, a solution $\Phi$ is $\lambda$-egalitarian if solution $\Phi(\lambda \cdot)$ is egalitarian. Which of the axioms $1-6$ are violated by such a bargaining solution?

**3.** A *utilitarian* solution $\Phi$ selects for each bargaining problem $B$ an element $x$ of $B$ which maximizes $x_1 + \cdots + x_n$ on $B$, and $\Phi$ is $\lambda$-*utilitarian* for $\lambda = (\lambda_1,\ldots,\lambda)$ if $\Phi(\lambda \cdot)$ is utilitarian. Check the Axioms $1-6$ for a $\lambda$-utilitarian bargaining solution.

Show for that if $B$ is a 2-person bargaining problem, then there is $\lambda = (\lambda_1, \lambda_2)$ such that $\Phi_N(B)$ coincides with both $\Phi_E(\lambda B)$ (see Problem 2) above and $\Phi(B)$ for some $\lambda$-utilitarian solution at $B$.

**4.** The (discrete) Raiffa solution is the bargaining solution $\Phi_R$, where for any $B \in \mathcal{B}$, $\Phi_R(B)$ is defined as the limit of a sequence $(z^t)_{t \in \mathbb{N}}$ with

$$z^1 = \frac{1}{N} \sum_{i \in N} a_i(B),$$

$$z^2 = \frac{1}{N} \sum_{i \in N} a_i(B^1), B^1 = \{x \in B \mid x_i \geq z_i^1, i \in N\},$$

$$\cdots$$

$$z^t = \frac{1}{N} \sum_{i \in N} a_i(B^{t-1}), B^{t-1} = \{x \in B \mid x_i \geq z_i^{t-1}, i \in N\},$$

Check which of the Axioms $1-6$ are satisfied by the Raiffa solution.

It is straightforward that the Raiffa solution satisfies *Midpoint Domination*: $\Phi_R(B)_i \geq \frac{1}{N} \sum_{i \in N} a_i(B)$ for all $B \in \mathcal{B}$.

**5.** Prove the following result due to Moulin [1983]: The Nash bargaining solution is the only solution which satisfies Midpoint Domination (see Exercise 4) and Independence of Irrelevant Alternatives.

**6.** For the set $\mathcal{B}_2$ of 2-person bargaining problems, the Equal Area bargaining solution selects a point $x$ on the boundary of $B$ such that the ray through $x$ divides $B$ into two sets of equal area.

Check the Axioms $1-6$ for the Equal Area bargaining solution.

# Chapter 12

# TU Games: Classical Solutions

## 1 Introduction

In Chapter 10 we introduced the TU games, also known as cooperative games with side payments, as a special case of the cooperative games defined by a characteristic function. In this and the following chapters we have a closer look at this particularly simple class of games. From the viewpoint of economic theory it may seem strange that the bulk of cooperative game theory was and still is to find in the analysis of this type of conflicts, which may be considered as occurring less often than those corresponding to NTU games. This has a historical explanation – for many decades all the attention was concentrated on TU games, and game theory as such had only few contacts to other fields such as economics. But there is also a more reasonable explanation, namely that knowledge of what happens in the relatively simple TU games makes it easier to treat the more complicated NTU games. It is therefore by no means a superfluous activity to investigate TU games as close as we are doing.

As previously notes we have a given set $N$ of players, and a family of coalitions which may be formed. For simplicity we shall assume that the set $S$ of admissible coalitions consists of all nonempty subsets of $N$. We now repeat the definition of a TU game.

DEFINITION 1 *A TU game is a pair $(N, v)$, where $N$ is a set of players and $v : S \to \mathbb{R}$ a real function defined on the set $S$ of all nonempty subsets of $N$ (coalitions) . The function $v$ is called the characteristic function.*

A particularly simple example of a characteristic function is obtained when for given numbers $x_1, \ldots, x_n$ we define $v : S \to \mathbb{R}$ by

$$v(S) = \sum_{i \in S} x_i$$

for $S \in S$. Such an *additive* characteristic function is *trivial* in our game-theoretical context. In the interpretation of the values of the characteristic function as amounts of money that the coalition can secure for its members, we have that the coalitions in an additive game have no other possibilities (or "strength") than the sum of what its members bring with them.

Two TU games (in the following referred to as games) $(N, v)$ and $(N, u)$ with the same player set are equivalent if the game $(N, v - u)$, where the characteristic function $v - u : S \to \mathbb{R}$ is given by

$$(v - u)(S) = v(S) - u(S), S \in S$$

is trivial. Clearly this is an equivalence relation on the set of TU games with player set $N$.

LEMMA 1 *Every game is equivalent to exactly one game in 0-normalized form, i.e. a game* $(N, u)$ *where* $u(\{i\}) = 0$ *for all* $i \in N$.

PROOF: Uniqueness: Assume that the given game $(N, v)$ is equivalent to 0-normalized games $(N, u_1)$ and $(N, u_2)$. Then both $v - u_1$ and $v - u_2$ are additive. An additive set function is fully determined by its values at the sets $\{i\}$, for $i \in N$, and since

$$(v - u_1)(\{i\}) = v(\{i\}) = (v - u_2)(\{i\}), \text{ all } i \in N,$$

we have $v - u_1 = v - u_2$, so $u_1 = u_2$.

Existence: Let $w : S \to \mathbb{R}$ be the additive set function defined by $w(\{i\}) = v(\{i\})$. Then $u = v - w$ is 0-normalized and equivalent to $v$, since $v - u = w$ is additive. $\square$

In an arbitrary game the coalition $S$ is said to be *inessential* if there is a partition of $S$ in two sets $R$ and $T$ such that

$$v(S) = v(R) + v(T);$$

in the opposite case $S$ is said to be *essential*. The game $(N, v)$ is inessential if all coalitions are inessential. An inessential game is equivalent to the zero game $(N, 0)$, whose characteristic function is identically 0.

A *dummy* for the game $(N, v)$ is a player $i \in N$, for whom

$$v(S \cup \{i\}) - v(S) = v(\{i\}), \text{all } S \subset N \backslash \{i\},$$

so that joining a coalition, no matter which, the player cannot increase its power. A 0-dummy is a dummy with $v(\{i\}) = 0$. It is easily seen that a player is a dummy if and only if she does not belong to any essential coalition.

A game is *constant sum* if

$$v(S) + v(N \backslash S) = v(N) \text{ for alle } S \in S.$$

A constant sum game is equivalent to a zero sum game, that is a game where

$$v(S) + v(N \backslash S) = 0 \text{ for alle } S \in S$$

There is a standard procedure for constructing a constant sum game, the so-called *constant extension*, from a given game $(N, v)$: Let $N' = N \cup \{j\}$, where $j \notin N$, and define the characteristic function $v' : S' \to \mathbb{R}$, where $S'$ is the set of nonempty subsets of $N'$, by

$$v'(S) = \begin{cases} v(S) & j \notin S, \\ v(N) - v(N \backslash S) & j \in S; \end{cases}$$

then $(N', v')$ is constant-sum. Intuitively, the adjoined player $j$ captures any surplus or deficit as compared to the constant sum $v(N)$.

In the following chapters we shall investigate solution concepts for classical TU games. Here we need an initial consideration of what should be understood by a solution. In our interpretation the game must eventually result in a downpayment of a payoff (positive or negative) to each player. We therefore introduce the following concepts.

DEFINITION 2 *A payoff vector for the game $(N, v)$ is a vector $(\alpha_1, \ldots, \alpha_n) \in \mathbb{R}^N$. An imputation is a payoff vector $(\alpha_1, \ldots, \alpha_n)$ which satisfies*

(i) *individual rationality:* $\alpha_i \geq v(\{i\})$, *all $i \in N$,*
(ii) *Pareto optimality:* $\sum_{i \in N} \alpha_i = v(N)$.

The set of imputations in the game $(N, v)$ is denoted $A(N, v)$ or simply $A$.

Initially we notice that a trivial game has exactly one imputation. Therefore, finding solutions for such games does not present any difficulty. But these games are uninteresting and present no challenges to theory. On the other hand, we shall see that the idea behind a given solution concept will be to find a trivial game which is in some respect close to the given game, and then take the unique imputation of this approximating game.

## 2   The von Neumann-Morgenstern solution

**2.1. Dominance and stability.** In order to formulate the historically first solution concept we introduce a *dominance relation* on the set $A$ of imputations in the given game.

Let $X$ be an arbitrary set and Dom a binary relation on $X$, i.e. Dom $\subseteq X \times X$. We write $x$ Dom $y$ for $(x, y) \in$ Dom. Let

$$\text{Dom}(x) = \{y \in X \mid x \, \text{Dom} \, y\}$$

be the set of all $y$ such that $x$ is Dom-better than $y$. We express this by saying that all such $y$ are *dominated* by $x$, and similarly, let

$$\text{Dom}^{-1}(x)$$

be all the elements of $X$, which dominate $x$. Correspondingly we shall write Dom($U$) for $\cup_{x \in U}\text{Dom}(x)$, all the elements $y$ which are dominated by some $x \in U$.

We may now define the Dom-maximal elements of $X$ as $\{x \in X \mid \text{Dom}^{-1}(x) = \emptyset\}$, that is the elements of $X$, which are not dominated by any other element of $X$. This set of maximal elements is known as the *core* (of $X$ with respect to Dom). We shall not discuss the core further for the moment, but we shall return to it again and again in what follows.

At present we shall be interested in a related concept, namely the so-called *stable sets*. In general, a Dom-stable subset of $X$ is a set $V$, which satisfies the two conditions

$$V \cup \mathrm{Dom}(V) = X \quad \text{(external stability)}$$

that every element of $X$ not in $V$ is dominated by something in $V$, and

$$V \cap \mathrm{Dom}(V) = \emptyset \quad \text{(internal stability)}$$

saying that no element of $V$ is dominated by another element of $V$.

Since the above definition of $V$ is given implicitly by two conditions that must be satisfied, it is not obvious that such a set exists, (and indeed it turns out that in many instances it doesn't). We shall see in the sequel that stable sets is a complicated study object, where relatively few good results can be obtained.

Before getting so far, we must specify the underlying set and the definition of the relation Dom in our context of cooperative games. The first step is straightforward, we are of course interested in choosing among all imputations belonging to a given game $(N, v)$; we denote this set by $I(N, v)$ or shorthand $I$. The definition of dominance may be slightly less self-evident; for notational reasons it is convenient to begin with the notion of "dominance via $S$", denoted $\mathrm{Dom}_S$, for an arbitrary coalition $S$, and then to let Dom be the union $\cup_S \mathrm{Dom}_S$ (where the relations are considered as subsets of $I \times I$).

The relation $\mathrm{Dom}_S$ is defined by $x \, \mathrm{Dom}_S \, y$ if

$$\sum_{i \in S} x_i \leq v(S)$$

$$x_i > y_i, \text{ all } i \in S$$

i.e. the coalition $S$ can give its members the payoff specified by the imputation $x$, and this payoff is considered as better than $y$ by all in $S$.

We have now introduced all the relevant concepts, and a *von Neumann-Morgenstern solution*, (shorthand a vNM-solution), also known as a *stable set* for the game $(N, v)$, can now be defined as a stable subset of $I$ for the relation Dom defined above.

From the preceding discussion it can be seen that the vNM-solution is conceptually more complex than other solutions as e.g. the core. It may therefore seem strange that this solution concept was introduced before the core, and that much of the research in cooperative games in the 1950s was related to stable sets.

There is however a simple explanation of this, namely that von Neumann and Morgenstern, as mentioned earlier, developed the representation of cooperative games by the characteristic function for zero-sum games, or, what in this context amounts to the same, constant-sum games. And for such games the core is empty, as we shall see below.

A cooperative game is constant sum, if

$$v(N) = V(S) + v(N \backslash S)$$

for all coalitions $S$. This corresponds to a situation where a given sum of money is at stake, and if a coalition is effective for some part of it, the complement will get the rest.

**Example 12.1.** Below we give an example of a 3-person game and consider its vNM-solutions. In such games the set of imputations can be illustrated in a 2-simplex which shows all the distributions of $v(N)$ among the players.

Let $(N, v)$ be given by $N = \{1, 2, 3\}$,

$$v(\{1, 2\}) = v(\{2, 3\}) = v(\{1, 3\}) = 1,$$

and $(N, v)$ is 0-normalized and constant-sum. Then

$$v^0 = \left\{\left(0, \frac{1}{2}, \frac{1}{2}\right), \left(\frac{1}{2}, 0, \frac{1}{2}\right), \left(\frac{1}{2}, \frac{1}{2}, 0\right)\right\}$$

is an *NM*-solution, but it is not the only one. For every $a$ with $0 \le a \le \frac{1}{2}$ each of the sets

$$V_{1,a} = \{\alpha \in A | \alpha_1 = a, \alpha_2 + \alpha_3 = 1 - a\}$$
$$V_{2,a} = \{\alpha \in A | \alpha_2 = a, \alpha_1 + \alpha_3 = 1 - a\}$$
$$V_{3,a} = \{\alpha \in A | \alpha_3 = a, \alpha_1 + \alpha_2 = 1 - a\}$$

is also a vNM-solution. This actually gives all vNM-solutions of the game. We notice that there can be infinitely many vNM-solutions (and of course a vNM-solution can consist of infinitely many imputations).

THEOREM 1 *Let $(N, v)$ be constant sum and essential in the sense that*

$$v(N) > v(\{i\}) + \cdots + v(\{n\}).$$

*Then the core is empty.*

PROOF: Let $x$ be an arbitrary imputation. We then have that

$$\sum_{i \in N} x_i = v(N), \ x_i \ge v(\{i\}), \ i \in N;$$

since the game is essential, there must be a player $i$ such that $x_i > v(\{i\})$. But since $(N, v)$ is constant sum, we may write

$$x_i + \sum_{j \ne i} x_j = v(N) = v(\{i\}) + v(N \setminus \{i\}),$$

and we get that

$$\sum_{j \in N \setminus \{i\}} x_j < v(N \setminus \{i\}).$$

But this means that there is $y \in I$ so that $y \operatorname{Dom}_{N \setminus \{i\}} x$, and consequently $x$ is not in the core. As $x$ was chosen arbitrarily, the core must be empty. $\qquad \square$

---

**Example 12.2. The Lucas game.** Here is an example of a 10-person game having no vNM-solutions.

$$v(N) = 5, \; v(\{1,3,5,7,9\}) = 4, v(\{3,5,7,9\}) = v(\{1,5,7,9\}) = v(\{1,,3,7,9\}) = 3,$$
$$v(\{1,2\}) = v(\{3,4\}) = v(\{5,6\}) = v(\{7,8\}) = v(\{9,10\}) = 1,$$
$$v(\{3,5,7\}) = v(\{1,5,7\}) = v(\{1,3,7\}) = v(\{3,5,9\}) = v(\{1,5,9\}) = v(\{1,3,9\}) = 2,$$
$$v(\{1,4,7,9\}) = v(\{3,6,7,9\}) = v(\{5,2,7,9\}) = 2$$

and $v(S) = 0$ for all other $S$.

It is not quite straightforward to show that there are no vNM-solutions, so we refer to Lucas [1969] for details.

---

There may be vNM-solutions even if the core is empty. Actually there are always vNM-solutions, also in constant-sum games, as long as there are only 3 or 4 players (and von Neumann and Morgenstern were mainly concerned with such games). These solutions may however have a rather complicated structure.

Most of the hopes for nice results about the vNM-solutions had subsequently to be abandoned. A striking expression of this is the result in Shapley [1959]: For every closed and bounded set $B$ of dimension $d$ (which in our context means that $B$ can be mapped continuously and bijectively on a subset of $\mathbb{R}^n$) there is a $(d + 3)$-person game $(N, V)$ (i.e. $|N| = d + 3$) such that $B$ is a subset of a vNM-solution of $V$ (and such that $V$ and $V \subset A$ can be separated by open sets. With this result one may give up the search for a nice structure of the vNM-solutions.

It turns out that also other conjectures about the structure of the solutions can be disproved by suitable examples. Even the most basic property, existence, is in trouble.

## 3   The core

**3.1. Definition of the core.** At the present stage of our discussion the core comes as no surprise. It has already played a role in previous chapters, so it is high time to give it a more detailed investigation.

Let $(N, v)$ be a game. The *core* of $(N, v)$, denoted Core $(N, v)$, is the set of imputations $x \in I$ for which

$$\sum_{i \in S} x_i \geq v(S), \; \text{all } S \in \mathcal{S}.$$

In other words, the core is the set of imputations that are coalitionally rational. Another characterization of the core uses the *excess function* $e : \mathcal{S} \times I$ defined by

$$e(S, x) = v(S) - \sum_{i \in S} x_i,$$

so that

$$\text{Core}(N, v) = \{x \in I \mid e(S, x) \geq 0, \text{ alle } S \in \mathcal{S}\}.$$

If an imputation $y$ is not in the core, then there must be a coalition $S$, such that

$$V(S) > \sum_{i \in S} y_i.$$

This coalition is said to *block*, or to have an improvement of, $y$.

The core – considered as a set of imputations – is uniquely defined, but there may be many elements in the core. Also, the core may be empty. For the trivial games the core equals the set of imputations, which is nonempty and unique.

The following simple connection holds in general between the core and the vNM- solutions.

THEOREM 2 *Let $(N, v)$ be a TU game. Then $\text{Core}(N, v) \subset V$ for every vNM solution $V$.*

PROOF: If $y \notin V$, there must be $x \in V$, so that $x$ dominates $y$. Since every $y \in \text{Core}(N, v)$ is undominated, $y$ cannot be outside $V$. □

It follows from the theorem that if $\text{Core}(N, v) \neq \emptyset$, then the vNM solutions must have nonempty intersection. As the vNM solutions of Example 12.1 did not have this property, we may conclude that the core of this game is empty. This followed by the way already from the fact that the game is constant sum.

If $\text{Core}(N, v) = V$ for some vNM-solution, then this solution is unique, as different vNM solutions cannot be contained in each other. Such games are said to have the CV property.

**3.2. Convex games.** A class of games with many nice properties (among which the CV property) is that of convex games. The conditions defining a convex game, can be seen as an extension of superadditivity, which we have already seen: A game $(N, v)$ is *superadditive* if

$$v(S) + v(T) \leq v(S \cup T), \text{ alle } S, T \in \mathcal{S}, S \cap T = \emptyset.$$

In the superadditive games the coalitions are at least as strong as the members taken together and typically stronger (there are synergy effects). In the interpretations this is usually a reasonable property; in the case it should not hold, then the coalition might just agree that its members should play the game individually.

The following property is however more restrictive: A game $(N, v)$ is *convex* if

$$v(S) + v(T) \leq v(S \cup T) + v(S \cap T) \text{ for all } S, T \in \mathcal{S}.$$

The reason for the term 'convex' in connection with set functions has to do with the difference operator $\triangle_R$ for $R \in \mathcal{S}$, which sends $v$ to $\triangle_R v$ defined by

$$(\triangle_R v)(P) = v(P \cup R) - v(P \cap R) \text{ for } P \in \mathcal{S}.$$

The second difference $\triangle_Q(\triangle_R)$ or $\triangle_{QR}^2$ is correspondingly given by

$$(\triangle_{QR}^2 v)(P) = (\triangle_Q)((\triangle_R v)(P))$$
$$= v(P \cup R \cup Q) - v((P \cup R)\backslash Q) - v((P\backslash R) \cup Q) + v((P\backslash R)\backslash Q)).$$

If we set $S = (P\backslash R) \cup Q, T = (P \cup R)\backslash Q$ in the above expression, we get that the second difference is $\geq 0$, corresponding to the nonnegativity of the second derivative of a convex function.

An interpretation of the convexity condition can be obtained by rewriting the definition as

$$v(S\backslash\{i\}) - v(S) \leq v(T\backslash\{i\}) - v(T) \tag{1}$$

for all $i \in N$ and all $S, T \in \mathcal{S}$, such that $S \subset T \subset N\backslash\{i\}$. This condition expresses a kind of increasing marginal product for coalitions. A subclass of the convex games is that of so-called convex measure games $(N, v)$ defined as

$$v(S) = f(\mu(S)) \text{ for } S \in \mathcal{S}$$

with $f : R \to \mathbb{R}$ a convex function and $\mu : \mathcal{S} \to \mathbb{R}$ an additive set function. There are however convex games which do not belong to this class.

Let $(N, v)$ be a convex game. By $\sigma$ we denote a permutation of the player set $N$, i.e. a bijective map from $N$ to $N$. Every $\sigma$ corresponds to a certain ordering of the players, and $\sigma(i)$ is the number of player $i$ in this ordering. For $\sigma$ a permutation of $N$ define the coalitions

$$S_{\sigma,k} = \{i \in N | \sigma(i) \leq k\}, \; k = 1, \ldots, n,$$

which are all the initial segments of players in the ordering determined by $\sigma$. Next, define the $n$-vector $x^\sigma$ by

$$x_i^\sigma = v(S_{\sigma,\sigma(i)}) - v(S_{\sigma,\sigma(i)-1}), \; i = 1, \ldots, n,$$

as an expression of what player $i$ brings with her to the coalition when the formation of coalitions is done in the order determined by $\sigma$ (intuitively, the players enter a room in the order given by $\sigma$, and the players in the room is the coalition formed at any stage).

It is easily seen that

$$\sum_{i \in N} x_i^\sigma = v(N) \; (= v(S_{\sigma,n})),$$

and that $x_i^\sigma \geq v(\{i\})$; for the last inequality we use the expression in (1). Consequently, $x^\sigma$ is an imputation.

The imputations $x^\sigma$, where $\sigma$ is a permutation of $N$, will be very useful in the sequel.

The first result that we get, is the non-emptiness of the core of convex games.

THEOREM 3 *Let $(N, v)$ be convex. Then $x^\sigma \in$ Core for $\sigma$ any permutation of $N$. In particular, Core$(N, v) \neq \emptyset$.*

PROOF: Let $x^\sigma$ be an imputation of this type, for $\sigma$ a permutation of $N$. Choose $T \in S$ arbitrarily, and let $j$ be the first (in the ordering determined by $\sigma$) player in $N \backslash T$. We then have

$$T \cup S_{\sigma,\sigma(j)} = T \cup \{j\}$$
$$T \cap S_{\sigma,\sigma(j)} = S_{\sigma,\sigma(j)-1},$$

and from convexity of $v$ we get

$$v(T) + v(S_{\sigma,\sigma(j)}) \le v(T \cup \{j\}) + v(S_{\sigma,\sigma(j)-1}),$$

or

$$x_j^\sigma \le v(T \cup \{j\}) - v(T).$$

Adding $x^\sigma(T) = \sum_{i \in T} x_i^\sigma$ on both sides and reshuffling terms we get that

$$x^\sigma(T) - v(T) \ge x^\sigma(T \cup \{j\}) - v(T \cup \{j\}). \tag{2}$$

Since (2) holds for any coalition $T$ when $j$ is the $\sigma$-first player not in $T$, we may proceed with the $\sigma$-second player $k$ not in $T$, which is the $\sigma$-first player not in $T \cup \{j\}$, to get

$$x^\sigma(T) - v(T) \ge x^\sigma(T \cup \{j\}) - v(T \cup \{j\}) \ge x^\sigma(T \cup \{j,k\}) - v(T \cup \{j,k\}),$$

and proceeding in this way with all the players not in $T$ we eventually get that

$$x^\sigma(T) - v(T) \ge x^\sigma(N) - v(N) = 0.$$

Since $T$ was arbitrary, we have shown that $x^\sigma \in \text{Core}(N,v)$, which therefore is $\ne \emptyset$. $\square$

We can get even more from this type of reasoning.

THEOREM 4 *Convex games have the CV property.*

PROOF: Suppose that $(N,v)$ is convex and that $y \notin \text{Core}\ (N,v)$. Then there must be $T \in S$, such that

$$v(T) > \sum_{i \in T} y_i.$$

Choose a permutation $\sigma$ such that $T = S_{\sigma,k}$ for some $k$, and let $x^\sigma$ be the imputation associated with $\sigma$. Then we have from the construction that

$$\sum_{i \in T} x_i^\sigma = \sum_{i \in S_{\sigma,k}} x_i^\sigma = v(S_{\sigma,k}) = v(T),$$

from which it is seen that $x^\sigma$ dominates $y$. By Theorem 3 $x^\sigma \in \text{Core}(N,v)$.

We have shown that $\text{Core}(N,v)$ is externally stable. Since $\text{Core}(N,v)$ is internally stable, it is a vNM-solution, so $(N,v)$ has the CV property. $\square$

We return to the situation of Theorem 3 which gave us a rather precise description of some of the elements in the core. The question is whether there are other elements. At the outset there is a straightforward positive answer to this, using the following simple fact:

LEMMA 2 Core$(N, v)$ *is a convex subset of* $I(N, v)$.

PROOF: The core is defined as the intersection of finitely many halfspaces of the form $\{x \in \mathbb{R}^N | \sum_{i \in S} x_i = v(S)\}$ and the hyperplane $\{x \in \mathbb{R}^N | \sum_{i \in N} x_i = v(N)\}$. □

We thus have that the core contains the convex hull of all the imputations $x^\sigma$. Actually the core coincides with this set. This follows once we see that these imputations are all the extreme points of the core.

THEOREM 5 *Let* $(N, v)$ *be convex, and let* $x \in$ Core$(N, v)$ *be an extreme point of the core. Then* $x$ *has the form* $x^\sigma$ *for some permutation* $\sigma$ *of* $N$.

PROOF: By induction in $n$, the number of players. For $n = 1$ and $n = 2$ the theorem is trivial.

Assume that the theorem has been shown to be true for all games with less than n players. Let $x$ be an extreme point in Core$(N, v)$. Then there must be coalitions $S \neq N$, such that $\sum_{i \in S} x_i = v(S)$; let $C$ be the set of all such coalitions. Since $v$ is convex, we have that

$$\sum_{i \in S} x_i + \sum_{i \in T} x_i = v(S) + v(T) \leq v(S \cup T) + v(S \cap T)$$

for all $S, T \in C$. But we also have

$$\sum_{i \in S} x_i + \sum_{i \in T} x_i = \sum_{i \in S \cup T} x_i + \sum_{i \in S \cap T} x_i.$$

Since $x \in$ Core$(N, v)$, it may be concluded that $\sum_{i \in W} x_i \geq v(W)$ for all coalitioner $W$, so that $S \cup T$ and $S \cap T$ also belong to $C$.

Suppose that $S \in C$ is minimal (for inclusion). If $|S| > 1$, then $x_i < v(\{i\})$ for all $i \in S$. Now choose all imputations $x^\sigma$, where $\sigma$ is a permutation sending the elements of $S$ to the set of numbers $\{1, \ldots, |S|\}$. For each of these imputations $x^\sigma$ we construct an imputation $y^\sigma$ by

$$y_i^\sigma = \begin{cases} x_i^\sigma & i \in S, \\ x_i & \text{otherwise.} \end{cases}$$

From the induction hypothesis we have that $y^\sigma$ are the extreme points of the set $\{y \in$ Core$(N, v) \mid y_i = x_i, i \notin S\}$, and it follows that $x$ can be written as a convex combination of the imputations $y^\sigma$, nontrivial since $x_i \neq v(\{i\})$ for all $i \in S$. We conclude that $S$ contains a one-element coalition, say $\{1\}$.

We now consider the game $(N\setminus\{1\}, v^{(1)})$ defined by

$$v^{(1)}(T) = v(T \cup \{1\}) - x_1$$

for all $T \subset N\setminus\{1\}$. It is easily seen that $v^{(1)}$ is convex: for arbitrary $S, T \subset N\setminus\{1\}$ we have that

$$v^{(1)}(S) + v^{(1)}(T) = v(S) - x_1 + v^{(1)}(T) - x_1$$
$$\leq v(S \cup T) + v(S \cap T) - 2x_1 = v^{(1)}(S \cup T) + v^{(1)}(S \cap T).$$

Consider now the imputation $(x_i)_{i \neq 1}$ in this game. From the induction hypothesis, we get that it is an extreme point in $\mathrm{Core}(N\backslash\{1\}, v^{(1)})$, and now it is easy to check that $x$ must be an extreme point in $\mathrm{Core}(N, v)$. $\qquad\square$

**3.3. Balanced games.** As it emerges from the results above, the convex games have many desirable properties, and we shall return to these games later. In relation to a determination of the class of games with nonempty core the class of convex games is however too narrow. We begin with an example.

Let $(N, v)$ be a game with $N = \{1, 2, 3\}$, such that $v(\{i\}) = 0$, $i = 1, 2, 3$, og $v(N) = 1$; we use the notation $v_{ij} = v(\{i, j\}), i, j = 1, 2, 3$. Then the core of this game is nonempty is equivalent to the existence of numbers $x_i \geq 0, i = 1, 2, 3$, such that

$$x_1 + x_2 \geq v_{12}, x_2 + x_3 \geq v_{23}, x_1 + x_3 \geq v_{13}, x_1 + x_2 + x_3 = 1.$$

For this to be possible, we must have that

$$v_{12} \leq 1, v_{23} \leq 1, v_{13} \leq 1, \frac{1}{2}\left(v_{12} + v_{23} + v_{13}\right) \leq 1. \tag{3}$$

We have thus derived necessary conditions for nonemptiness of the core.

In our present case they are also sufficient, since we may define $x$ by

$$x_1 = 1 - v_{23}, \ x_2 = \min\{v_{23}, 1 - v_{13}\}, \ x_3 = \max\{0, v_{13} + v_{23} - 1\},$$

and it is easily verified that $x \in \mathrm{Core}(N, v)$. In this situation we see that it is possible to formulate a condition which is both necessary and sufficient for the game to have a nonempty core. Except for the fact that the condition is about weighted sums of the $v(S)$, it may not be easy to see how it might generalize to other games. For this, we need to introduce the class of balanced games.

Let $(N, v)$ be a game. A *balanced family of coalitions* is an indexed family of numbers (the weights) $(c_S)_{S \in \mathcal{S}}$ satisfying $c_S \geq 0, S \in \mathcal{S}$, such that

$$\sum_{S \in \mathcal{S}: i \in S} c_S = 1 \text{ for all } i \in N.$$

The condition may also be expressed as follows: For each coalition $S$ we let $e_S$ be the vector with coordinates

$$(e_S)_i = \begin{cases} 1/|S| & i \in S, \\ 0 & \text{otherwise,} \end{cases}$$

i.e. the barycenter of the subsimplex of $\Delta = \{x \in \mathbb{R}_+^N | \sum_{i \in N} x_i = 1\}$ which arises if all coordinates outside $S$ are 0. A balanced family of coalitions is then a subset $C$ of $\mathcal{S}$, such that $e_N$ belongs to the convex hull of the vectors $e_S$ for $S \in C$,

$$e_N \in \mathrm{conv}(\{e_S \mid S \in C\}).$$

It is relatively easy to see that the two definitions give the same result; in the first approach the focus is on the weights, which is more relevant for our present problem.

The game $(N, v)$ is *balanced* if for each balanced family of coalitions $(c_S)_{S \in \mathcal{S}}$, we have that

$$\sum_{S \in \mathcal{S}} c_S v(S) \geq v(N).$$

It is not easy to find an intuitive interpretation of balanced games, the condition is technical and is interesting in view the results which it gives rise to. One way of looking at balanced coalitions is to see them as part-time engagements, whereby $c_S$ gives the fraction of time used in $S$ by its members (so to say the opening hours of coalition $S$). With this interpretation we get that the grand coalition is at least as powerful as a split of it into part-time coalitions, a strengthened version of superadditivity.

We have the following characerization of games with a nonempty core, first obtained by Bondareva [1962]:

THEOREM 6 *Let $(N, v)$ be a game. Then* $\mathrm{Core}(N, v) \neq \emptyset$ *if and only if $(N, v)$ is balanced.*

PROOF: We shall use the main theorem of linear programming. Let $A$ be a matrix with $n$ rows and with a column for each coalition $S$ in $\mathcal{S}$, so that the column corresponding to $S$ has a 1 in each row $i$ for which $i \in S$, and 0 otherwise. Let $b$ be the vector consisting of all $v(S)$ in the same ordering as the columns of $A$, and let $c$ be a vector with all coordinates equal to 1. Consider the LP problem

$$\min c \cdot u$$
$$\text{subject to}$$
$$uA \geq b,$$
$$u \geq 0.$$

(4)

The corresponding dual program is

$$\max b \cdot w$$
$$\text{subject to}$$
$$Aw \leq c$$
$$w \geq 0.$$

(5)

Suppose that $\mathrm{Core}(N, v) \neq \emptyset$ and let $x \in \mathrm{Core}(N, v)$. Then $u = x$ is an optimal solution to the primary problem (the constraints are satisfied since $\sum_{i \in S} x_i \geq v(S)$ for $S \in \mathcal{S}$, and the optimality follows from $\sum_{i \in N} \alpha_i = v(N)$). Consequently there must be both feasible and optimal solutions to the dual problem, and for each feasible solution $w$, the value of the objective function $b \cdot w$ is $\leq$ the optimal value of the primary problem, which is $v(N)$. If in particular $w$ satisfies the constraints with equality, then we get the balancedness condition.

Conversely, suppose that $(N, v)$ is balanced. Clearly, there are feasible solutions to the dual problem. Further we notice that if the array $(c'_S)_{S \in \mathcal{S}}$ satisfies

$$\sum_{S : i \in S} c'_S \leq 1 \text{ for all } i \in N,$$

then it can be rescaled to an array $(c_S)_{S \in S}$, for which equality holds; consequently

$$\sum_{S \in S} c'_S v(S) \leq v(N).$$

We thus have that the value of the objective function is $\leq v(N)$ for every feasible solution $w$ to the dual problem. Therefore the objective function has a maximum, so that the dual problem has an optimal solution. It follows that also the primary problem has an optimal solution $u^0$ with $c \cdot u^0 = v(N)$.

It is easily seen that the imputation $u^0$ is in $\text{Core}(N, v)$. □

We conclude – for the moment– our treatment of the core by a consideration of some extensions of the concept of the core, which may take into account the cost of coalition formation. For $\varepsilon$ a real number we define the $\varepsilon$-core $\text{Core}_\varepsilon(N, v)$ as the set of payoff vectors (note: not necessarily imputations) $x \in \mathbb{R}^N$, which satisfy

$$\sum_{i \in N} x_i = v(N)$$

and

$$\sum_{i \in S} x_i \geq v(S) - \varepsilon, \text{ all } S \in S$$

(often called the "strong" $\varepsilon$-core). The weak $\varepsilon$-core $\text{Core}_\varepsilon^w(N, v)$ is the set of all payoff vectors $x$, which satisfy $\sum_{i \in N} x_i = v(N)$ and

$$\sum_{i \in S} x_i \geq v(S) - \varepsilon|S|, \text{ all } S \in S$$

where $|S|$ as usually is the number of elements of $S$. In these definitions $\varepsilon$ has the role of a friction parameter, giving the amount which has to be put aside to cover the cost of coalition formation (either in total or per coalition member).

For $\varepsilon > 0$ we have

$$\text{Core}(N, v) \subset \text{Core}_{\varepsilon/n}^w(N, v) \subset \text{Core}_\varepsilon(N, v) \subset \text{Core}_\varepsilon^w(N, v).$$

Further we define the near-core n-$\text{Core}(N, v)$ as

$$\text{n-Core}(N, v) = \bigcap_{\varepsilon > 0} \left\{ \text{Core}_\varepsilon(N, v) \,\middle|\, \text{Core}_\varepsilon(N, v) \neq \emptyset \right\};$$

since we always have that $\varepsilon' < \varepsilon$ and $\text{Core}_{\varepsilon'}(N, v) \neq \emptyset$ imply $\text{Core}_{\varepsilon'}(N, v) \subset \text{Core}_\varepsilon(N, v)$, we get that

$$\text{n-Core}(N, v) = \text{Core}_{\varepsilon_0}(N, v),$$

where $\varepsilon_0$ is the smallest number $\varepsilon$, for which $\text{Core}_\varepsilon(N, v) \neq \emptyset$. If $\varepsilon_0 = 0$, then n-$\text{Core}(N, v) = \text{Core}(N, v)$.

Correspondingly we may define the weak near-core n-$\text{Core}^w(N, v)$ as

$$\bigcap_{\varepsilon > 0} \left\{ \text{Core}_\varepsilon^w(N, v) \,\middle|\, \text{Core}_\varepsilon^w(N, v) \neq \emptyset \right\}.$$

These specific versions of the core are however of interest mainly in the context of particular models where coalition formation is given particular attention.

## 4    The Shapley value

**4.1. Definition.** In the solution concepts that we have seen up to now (namely the vNM solution and the core), the argument has been linked to the dominance relation introduced at the outset. In our interpretation of this this relation dominance is an expression of strength, ability to manage the distribution of wealth better than it was done in the dominated situation.

Instead of this strength or efficiency argument we might as well have started with a fairness or equality argument. This would obviously have given rise to other solutions, but one cannot say that one argument is more natural or more fundamental than the other.

The Shapley value, which we consider below, may be considered as based on some kind of fairness consideration, and it thus represents the second type of argument. But there are other, more formal aspects, where the Shapley value differs from what we have seen so far. The most striking difference is that the Shapley value provides a single payoff or imputation (at least for sufficiently well-behaved games, as we shall see), where we have previously had to accept a set, or even a family of sets, as solutions. To this may be added that the Shapley value can be computed in a relatively straightforward way from the data of the game, and it is clear that this gives us a solution concept which is rather attractive. Indeed, the Shapley value has had more practical applications than most other solution concepts.

In the proof of nonemptiness of the core for convex games we introduced a procedure which has interest beyond above the concrete case: For each permutation of players, we let the coalitions be formed successively as the initial segments of players in the order given by this permutation. Consider such a permutation $\sigma$ of the players. For a player $i$ such that $\sigma(i) = k$,

$$v(S_{\sigma,\sigma(i)}) - v(S_{\sigma,\sigma(i)-1})$$

is an expression of what the coalition $S = S_{\sigma,\sigma(i)}$ loses if $i$ leaves the coalition from (and $S$ thus becomes $S_{\sigma,\sigma(i)-1}$) and thereby an assessment of what the coalition $S$ may ultimately be forced to pay in order to make $i$ stay. As such it is an expression of player $i$'s bargaining power against the coalition $S$ (given the permutation $\sigma$).

Now suppose that the actual sequence of players to emerge in the formation of coalitions is completely random, so that each permutation is considered equally likely. We can then find player $i$'s average bargaining power, which will be

$$\phi_i(v) = \frac{1}{n!} \sum_{\sigma} [v(S_{\sigma,\sigma(i)}) - v(S_{\sigma,\sigma(i)-1})]. \tag{6}$$

The sum in (6) can be written in a somewhat simpler way: Instead of summing over all permutation $\sigma$ we collect all cases where $S_{\sigma,\sigma(i)}$ is the same coalition $S$. The number of permutations of the player where this happens (for a given $S$ which contains $i$) is $(s-1)!(n-s)!$, where $s = |S|$, corresponding to all $(n-1)$-tuples with something from $S$ in the first $s-$ positions and something from $N \backslash S$ in the last $n - s$ places. In total we have

$$\phi_i(v) = \sum_{S:i\in S} \frac{(s-1)!(n-s)!}{n!} [v(S) - v(S \backslash \{i\})]. \tag{7}$$

The function $\phi$ that sends the game $(N, v)$ in the vector $\phi(v) = (\phi_1(V), \ldots, \phi_n(v))$, is called the *Shapley value*. Thus, in our context of solutions to TU games, a value is not a payoff vector but a function assigning a particular payoff vector to each game.

Using the expression (6) we may find $\sum_{i\in N} \phi_i(v)$ by first summing over $i$ for each $\sigma$ and then summing over $\sigma$, and we get

$$\sum_{i\in N} [v(S_{\sigma,\sigma(i)}) - v(S_{\sigma,\sigma(i)-1})] = v(S_{\sigma,1}) + [v(S_{\sigma,2}) - v(S_{\sigma,1})] +$$

$$+ \cdots + [v(S_{\sigma,n}) - v(S_{\sigma,n-1})] = v(S_{\sigma,n}) = v(N),$$

which shows that $\sum_{i\in N} \phi_i(v) = \sum n! \frac{1}{n!} v(N) = v(N)$. The payoff vector $(\phi_1(v), \ldots, \phi_n(v))$ thus sums to $v(N)$. However, it is not necessarily an imputation: Consider the game $v$ with $v(S) = 1$, all $S \in S$. Here $\phi(v) = (\frac{1}{n}, \ldots, \frac{1}{n})$, but $v(\{i\}) = 1 > \frac{1}{n}$ for all $i$. However, if the game $(N, v)$ is superadditive, then

$$v(S) \geq v(S \backslash \{i\}) + v(\{i\})$$

for all $i$ and $S \in S$, so that $\phi_i(v) \geq v(\{i\})$ for all $i$. We conclude that for superadditive games the Shapley value is an imputation and can be considered a solution for the game.

Having introduced the Shapley value, we proceed to a discussion of its properties. This discussion is rather closely connected with the axiomatic characterizations of this solution concepts, to which we turn now.

**4.2. The axiomatic approach to the Shapley value.** The Shapley value $\phi$ has several nice properties: First of all it has the purely technical advantage that $\phi(v)$ exists (as distinct from core and vNM solutions) for any game. Secondly – and for the interpretation maybe more importantly – it satisfies some reasonable demands to a payoff vector, which should reflect each player's power both individually and in coalitions.

A first illustration of this can be obtained by considering the so-called *unanimity games*. For $T \in S$ we let $e_T$ be the game given by

$$e_T(S) = \begin{cases} 1 & \text{if } T \subset S \\ 0 & \text{otherwise.} \end{cases}$$

In order to obtain positive payoffs in this game, at least the coalition $T$ must be formed. Players outside $T$ has little power in this game, while members of $T$ can be considered as being equally powerful. It seems therefore natural to view the imputation $x$ with $x_i = 1/|T|$ for $i \in T$ and $x_i = 0$ for $i \notin T$ as a fair or just (with respect to players' position) solution to this game. This payoff vector, which is an imputation since $e_T$ is superadditive, is precisely $\phi(e_T)$.

Before we proceed, we note that the unanimity games $e_T$ for $T \in S$ have a special position among all games. If we consider a game (with fixed set of players $N$ and set of coalitions $S$) as a vector with $|S|$ coordinates, where the coordinates corresponding to $S \in S$ is $v(S)$, then $e_T$ is the vector with coordinate 1 for on the place corresponding to $T$ and 0 otherwise, a kind of unit vector. Actually, the following holds.

LEMMA 3 *The set $\{e_T \mid T \in S\}$ is a basis for the vector space $\mathbb{R}^S$, i.e. any game $v$ can be written uniquely as a linear combination*

$$v = \sum_{T \in S} c_T e_T,$$

*where the coefficients $c_T, T \in S$, are given by*

$$c_T = \sum_{S \subseteq T} (-1)^{t-s} v(S).$$

PROOF: We show first that the vectors $e_T$, $T \in S$, are linearly independent. Suppose that

$$\sum_{T \in S} a_T e_T = 0$$

where 0 is the null vector in $\mathbb{R}^S$. For $T = \{i\}$ this means that $a_{\{i\}} = 0$, $i = 1, \ldots, n$. For $T = \{i, j\}$ we then get that $a_{\{i,j\}} = 0$, etc.. In total, we find that $a_T = 0$ for all $T$.

It now suffices to show that every $v$ has a representation as stated, since uniqueness follows from the linear independence of the $e_T$. Let $v_R$ be the coordinate corresponding to the coalition $R$ in the vector $\sum_{T \in S} a_T e_T$. We have then

$$v_R = \sum_{T \in S} \sum_{S \subseteq T} (-1)^{t-s} v(S) e_T(R) = \sum_{T \subseteq R} \sum_{S \subseteq T} (-1)^{t-s} v(S),$$

since $e_T(R) = 0$ for $T$ such that $T \not\subseteq R$. For every $T \subseteq R$, we must then sum over $S \subseteq T$ and then over $T$. This is equivalent to summation, for each $S \subseteq R$, over $T$ between $S$ and $R$ (and then over $S$), that is

$$v_R = \sum_{S \subseteq R} \sum_{T : S \subseteq T \subseteq R} (-1)^{t-s} v(S) = \sum_{S \subseteq R} v(S) \sum_{T : S \subseteq T \subseteq R} (-1)^{t-s}.$$

There are $\binom{r-s}{t-s}$ coalitions $T$ with $|T| = t$ between $S$ and $R$, so we have that

$$v_R = \sum_{S \subseteq R} v(S) \sum_{t-s}^{r} (-1)^{t-s} \binom{r-s}{t-s} = \sum_{S \subseteq R} v(S)(1-1)^{r-s} = v(R),$$

where we have used the binomial formula $(a+b)^n = \sum_{i=0}^n \binom{n}{i} a^i b^{n-i}$. $\qquad\square$

We can use this result to express the Shapley value of a game $(N, v)$ in another way.

LEMMA 4 *The Shapley value $\phi$ is a linear mapping from $\mathbb{R}^S$ to $\mathbb{R}^N$.*

PROOF: Let $v = v_1 + v_2$ (where games $(N, v)$ are regarded as elements in the vector space $\mathbb{R}^S$. Then for each $i \in N$ we have

$$\phi_i(v) = \sum_{S:i \in S} \frac{(s-1)!(n-s)!}{n!}[v(S) - v(S \setminus \{i\})] = \sum_{S:i \in S} \frac{(s-1)!(n-s)!}{n!}[v_1(S) - v_1(S \setminus \{i\})]$$

$$+ \sum_{S:i \in S} \frac{(s-1)!(n-s)!}{n!}[v_2(S) - v_2(S \setminus \{i\})] = \phi_i(v_1) + \phi_i(v_2).$$

Similarly, it is shown that $\phi_i(av) = a\phi_i(v)$ for $a \in \mathbb{R}$. □

Since a linear mapping, in this case $\phi$, is uniquely defined by its values at a basis, we get the following:

COROLLARY 1 *Let the game $v$ be given by the expression $v = \sum_{T \in S} c_T e_T$. Then $\Phi(v) = \sum_{T \in S} c_T \phi(e_T)$.* □

We can now elaborate on the comments above regarding the characterization of the Shapley value as a "reasonable" payoff vector in $v$. In the following, we consider games $(N, v)$ with fixed set of players $N$ and family of coalitions $S$. We introduce an abstract value of a game and then state some properties of this abstract value which characterize it as the Shapley value.

An (abstract) value is a mapping $\psi$ from games $v$ to payoff vectors $(x_1, \ldots, x_n)$. The Shapley value is an example of a value, but there may be many others (as indeed there are, as we shall see later).

Below we state some fairly obvious demands of reasonableness to such a value in the form of axioms.

AXIOM 1 (Linearity) $\psi(av_1 b v_2) = a\psi(v_1 b\psi(v_2)$ *for all $v_1, v_2$ and all real numbers $a, b \in \mathbb{R}$.*

AXIOM 2 (Pareto optimality) $\sum_{i \in N} \psi_i(v) = v(N)$ *for all $v$.*

AXIOM 3 (Symmetry) *Let $\sigma : N \to N$ be a permutation. If $\sigma v$ is the game defined by $(\sigma v)(S) = v(\sigma^{-1}(S))$ for $S \in S$ and $\sigma x$ is the payoff vector determined by $(\sigma x)_i = x_{\sigma^{-1}(i)}$ for $i \in N$, then*

$$\psi(\sigma v) = \sigma \psi(v).$$

AXIOM 4 (Dummy axiom) *If $i$ is such that $v(S) \mp v(S \setminus \{i\})$ for all $S \in S$, and $v(\{i\} = 0$, then $\psi_i(v) = 0$.*

We now have the following characterization of the Shapley value.

THEOREM 7 *There is exactly one value satisfying Axioms 1 to 4, and this value is the Shapley value.*

PROOF: First of all we show that $\phi$ satisfies Axioms 1 - 4. We have already shown Axiom 1 (Lemma 3) and 2. Axiom 4 follows immediately from the expression (7). Finally we obtain Axiom 3 from

$$\phi_i(\sigma v) = \sum_{S:i\in S} \frac{(s-1)!(n-s)!}{n!} \left[ v(\sigma^{-1}(S)) - v(\sigma^{-1}(S \setminus \{i\})) \right]$$

Instead of summing over all $S$ with $i \in S$ we may perform the summation over $S' = \sigma^{-1}(S)$ with $\sigma^{-1}(i) \in S'$, i.e.

$$\phi_i(\sigma v) = \sum_{S':\sigma^{-1}(i)\in S'} \frac{(s-1)!(n-s)!}{n!} \left[ v(S') - v(S' \setminus \{\sigma^{-1}(i)\}) \right]$$

$$= \phi_{\sigma^{-1}(i)} v.$$

In order to show that $\phi$ is the only value satisfying Axioms 1 to 4 we consider an arbitrary value $\psi$. Let $e_T$ for $T \in \mathcal{S}$ be a unanimity game. Then $\psi_i(e_T) = 0$ for $i \notin T$ (Axiom 4). According to Axiom 2 we must have that $\sum_{i\in T} \psi_i(e_T) = (e_T)(N) = 1$, and from Axiom 3 we get that $\psi_i(e_T) = \psi_j(e_T)$ for $i, j \in T$, that is

$$\psi_i(e_T) = \begin{cases} 1/|T| & \text{for } i \in T \\ 0 & \text{otherwise,} \end{cases}$$

or $\psi(e_T) = \phi(e_T)$.

Now let $v = \sum_T c_T e_T$ be an arbitrary game. We then have

$$\psi(v) = \psi(\sum_{T\in\mathcal{S}} c_T e_T) = \sum_{T\in\mathcal{S}} \psi(c_T e_T) = \sum_{T\in\mathcal{S}} c_T \psi(e_T)$$

$$= \sum_{T\in\mathcal{S}} c_T \phi(e_T) = \phi(\sum_{T\in\mathcal{S}} c_T e_T) = \phi(v)$$

where we have used the linearity axiom on both the abstract value $\psi$ and the Shapley value $\phi$. Summing up, we have $\psi(v) = \phi(v)$ for all $v$ or $\psi = \phi$.  □

**4.3. Properties of the Shapley value.** In general little can be said about the relation between the Shapley value of a game $(N, v)$ and the previously introduced solution concepts vNM-solutions and the core. However, the class of convex games again turns out to be well-behaved:

THEOREM 8 *Let $(N, v)$ be a convex game. Then $\phi(v) \in \text{Core}(N, v)$.*

PROOF: Writing the expression (6) in vector form we get that

$$\phi(v) = \frac{1}{n!} \sum_\sigma x^\sigma,$$

where we sum over all permutations of players, cf. Section 3.2. By Theorem 3 we have that $x^\sigma \in \text{Core}(N, v)$ for each $\sigma$. Since $\text{Core}(N, v)$ is a convex set, and $\phi(v)$ is a convex combination of imputation $x\sigma$, we get that $\phi(v) \in \text{Core}(N, v)$.  □

**4.4. Power indices.** In Chapter 10, Section 5 we introduced the simple games which are TU-games $v$ such that $v(S) \in \{0, 1\}$ for all coalitions $S$. The unanimity games $e_T$ from the last paragraph are simple games – but there are obviously many others.

In the analysis of simple games as a formalization of the power structures in a committee it turned out to be useful to obtain an index which would tell something about each player's position in the hierarchy of power, a number between 0 and 1 so that a player with no influence whatsoever gets the value 0 of the power index, and a dictator (whose wishes are always the committee's) has index 1. The specific numerical values of the index would then reflect the approach towards measuring influence.

We see from the previous sections that we already have a candidate for such an index, namely the Shapley value. If $(N, v)$ is a simple game (so that the family of winning coalitions is $\mathcal{W} = \{S \in \mathcal{S} \mid v(S) = 1\}$, we obtain the $i$th player's index value as

$$\phi_i = \sum_{S:i \in S} \left( \frac{(s-1)!(n-s)!}{n!} \right)[v(S) - v(S \setminus \{i\})].$$

This index is called the *Shapley-Shubik power index*. It has the properties of a power index suggested above: Since the Shapley value has the dummy property, we get that a dummy, that is a player who is not himself a winning coalition and does not change any non-winning coalition into a winning one by joining it, gets index value 0. Conversely, a veto player, who belongs to all the winning coalitions and is the only one of this kind, gets the index value 1.

Besides these properties, we obviously have the axioms for the Shapley value, where in particular the symmetry property is reasonable. But linearity no longer makes sense, since the sum of two simple games is not necessarily simple, consider for example two unanimity games $e_T$ and $e_S$ where $T \setminus S \neq \emptyset$, $S \setminus T \neq \emptyset$.

An axiomatic derivation of the Shapley-Shubik index therefore requires a different system of axioms, or at least replacing the linearity with something else. A proposal for such a replacement has been given by [Dubey [1975]] who suggested the following property of a general power index $\psi$ defined on simple games:

AXIOM 5 (Transfer) *If $v_1$ and $v_2$ are simple games on $N$, and $i \in N$, then*

$$\psi_i(v_1 \vee v_2)\psi_i(v_1 \wedge v_2) = \psi_i(v_1)\psi_i(v_2)$$

*where $v_1 \vee v_2$ and $v_1 \wedge v_2$ is defined by $(v_1 \vee v_2)(S) = \max\{v_1(S), v_2(S)\}$, $(v_1 \wedge v_2)(S) = \min\{v_1(S), v_2(S)\}$.*

Now the following holds:

THEOREM 9 *Let $\psi$ be a power index which satisfies Axioms 2 – 4 of the Shapley value and Axiom 5. Then $\psi$ is the Shapley-Shubik index.*

Another power index, which has had more applications than the Shapley-Shubik index, is the Banzhaf-Coleman index. Like the previous one it also makes sense in the larger class of all TU-games (only the axiomatization has to be adapted to the case under study), and we define therefore a Banzhaf-Coleman value $\psi$ by the formula

$$\psi_i(v) = \sum_{S \subset N \setminus \{i\}} \frac{1}{2^{n-1}}[v(S \cup \{i\}) - v(S)].$$

One may think of the Banzhaf-Coleman value as obtained by a consideration roughly corresponding to the random coalition formation behind the Shapley value. However, in the present setup each player contemplates all possible sets of future partners (and there are $2^{n-1}$ sets of potential partners to be joined by the coalition formation) and takes average over her marginal contribution to each of them.

## 5   Problems

**1.** A group of individuals have access to a technology with constant returns to scale, producing a single commodity using several input commodities, which have to be inserted in fixed proportions. Each individual has a certain amount of each of the input commodities

The individuals contemplate going together and then dividing the income from the sale of the output. Explain that this gives rise to a cooperative TU game (a production game), which is superadditive.

Show that the game is balanced.

A game $(N, v)$ is totally balanced if each subgame $(S, v)$ (with player set $S$ and the same value $v(T)$ of the characteristic function for subsets $T$ of $S$) is balanced. Explain that a production game is totally balanced.

**2.** A TU game $(N, v)$ is superadditive if for every two disjoint coalitions $S$ and $T$, the worth of their union $v(S \cup T)$ is the sum of the worths of S and T.

Check whether the core is nonempty for any superadditive game (if not, give an example of a superadditive game with empty core).

From any game $(N, v)$, superadditive or not, one can construct a superadditive game $(N, \hat{v})$, the superadditive cover of $(N, v)$, by letting $\hat{v}(S)$ be the largest number obtained by summing the worths of disjoint subcoalitions of $S$. Investigate the relation between the core of the game $(N, v)$ and the core of its superadditive cover.

**3.** In order to obtain uniqueness of core solutions, one has in general to select a particular element from among the imputations in the core. A possible way of doing this would be to select the maximizers of some function $U$, which may be considered as a social welfare function defined on the payoff vectors.

For $U : \mathbb{R}^N \to \mathbb{R}$ a function, define the *core selection* $\phi_U$ by

$$\phi_U(v) = \{z \in v(N) \mid z \in \mathrm{Core}(N, v), \text{ and } z' \in \mathrm{Core}(N, v) \text{ implies } U(z') \le U(z)\}.$$

Let $U(z) = \sum_{i \in N} z_i$ for all $z$. Let $S \in \mathcal{S}$ be an arbitrary coalition different from $N$. Show that the core selection $\phi_U$ is *S-monotonic* in the sense that if $(N, v)$ and

$(N, w)$ are two games such that $v(S) > w(S)$ but $v(T) = w(T)$ for all other coalitions $T$, and $z \in \phi_U(v)$, $z' \in \phi_U(w)$, then $z_i > z'_i$ for all $i \in S$. Give an interpretation of this condition.

Show that there are functions $U$ such that $\phi_U$ is not $S$-monotonic.

**4.** (Cost allocation) Let $N = \{1, \ldots, n\}$ be a set of projects or products that can be provided jointly or separately by some organization. Let $c(\{i\})$ be the cost of providing project $i$ alone, and let $c(S)$, for $S$ any subset of $N$, be the cost of providing the projects in $S$ jointly; we define $c(\emptyset) = 0$. The pair $(N, c)$ is a *cost-sharing game*, and $c$ is called its *cost function*.

A cost allocation method is a function $\phi$ which to each cost sharing game $(N, c)$ assigns a *cost allocation*, a vector $(x_1, \ldots, x_n)$ such that $\sum_{i \in N} x_i = c(N)$. It is usually assumed that the cost function $c$ is subadditive, i.e.

$$c(S \cup T) \le c(S) + c(T)$$

for all disjoint subsets $S$ and $T$ of $N$, and monotonic in the sense that $c(S) \le c(S')$ for $S \subseteq S'$.

Define the potential *cost saving* associated with $(N, v)$ as

$$v(S) = \sum_{i \in S} c(\{i\}) - c(S).$$

Show that $(N, v)$ is a superadditive TU game.

The *separable cost* of $i \in N$ is defined as $s_i = c(N) - c(N \setminus \{i\})$, the *alternate cost avoided* by $r_i = c(i) - s_i$. The *alternate cost avoided method* assigns cost by

$$x_i = s_i + \frac{r_i}{\sum_{j \in N} r_j} \left( c(N) - \sum_{j \in N} s_j \right).$$

The *semicore* of $(N, c)$ is the set of cost allocations $x$ such that $x_i \le c(\{i\})$ and $\sum_{j \ne i} x_j \le c(N \setminus \{i\})$. Define the associated set of imputations in the cost-savings game, and show that the alternatve cost avoided gives cost allocations in the semicore when the latter is nonempty.

**5.** For a TU game $(N, v)$, the *Weber set* consists of all the payoff vectors which can be found as convex combinations of the marginal vectors $x^\sigma$, defined by

$$x_i^\sigma = v(S_{\sigma, \sigma(i)}) - v(S_{\sigma, \sigma(i)-1}), \ i \in N,$$

where $S_{\sigma, \sigma(k)}$ consists of the initial $k$ elements of $s$ ordered according to the permutation $\sigma$.

Give an example of a game where the Weber set contains a payoff vector which does not belong to the core.

**6.** In a TU game $(N, v)$, one can define the (Harsanyi) dividends $\Delta_v(S)$, for $S$ any coalition, inductively (on the size of coalitions) by $\Delta_v(\{i\}) = v(\{i\})$ for each player $i$ and $\Delta_v(S) = v(S) - \sum_{T \subset S} \Delta_v(T)$ for coalitions with more than one player.

A *selector* is a function $\alpha$, which to each coalition $S$ chooses a player $\alpha(S)$ from the coalition $S$. The *selector value* belonging to $\alpha$ is the vector $m^\alpha(v)$ with $i$th coordinate

$$m_i^\alpha(v) = \sum_{S: i = \alpha(S)} \Delta_v(S).$$

Show that $m^\alpha(v)$ is a pre-imputation in the game $(N, v)$. Give an example of a game (with at least 4 essential players) where $m^\alpha(v)$ does not belong to the core for some selector $\alpha$.

**7.** Let $(N, v)$ be a game with $v(N) = 1$. A general power index may be defined as

$$p_j = \sum_{S \subseteq N: j \in S} P(S) \left[ v(S) - v(S \setminus \{j\}) \right].$$

If we normalize the index numbers so that they sum to 1, we have that $\sum_{S \subseteq N} P(S) = 1$. If we require that power depends on sizes of coalitions rather than identity of their members, we may assume that $P(S) = \lambda_i$, where $\lambda_i$ is a number depending only on $i = |S|$, the normalization condition takes the form

$$\sum_{i=1}^n \binom{n-1}{i-1} \lambda_i = 1$$

for numbers $\lambda_i \geq 0$.

Show that the Shapley-Shubik index and the Banzhaf-Coleman index have this form, and find the relevant $\lambda_i$'s.

# Chapter 13

# TU Games: Other Solutions

## 1 The bargaining set and the kernel

In the previous sections we have searched for solutions for a game $(N, v)$ among the imputations – or possibly preimputations, since we might have to give up demands for individual rationality in some games when considering Shapley values.

**1.1. The bargaining set** $\mathcal{M}_1^{(i)}$. In some situations, for example in cases where two disjoint coalitions $S$ and $T$ satisfy $v(S) + v(T) > v(N)$, the demand for Pareto optimality of the solution cannot easily be made compatible with the choice of an imputation as solution. Also with regard to applications such a demand would not be reasonable, since it seems to rely on the assumption that if some coalition can give its members a better result, it will be formed. This is however subject to some debate, since other considerations may play a role, and such other conditions (not easily recognized as belonging to game theory proper) are quite important in empirical or experimental game situations. Among such considerations one could mention that of "security": A given payoff can emerge as the result of the game even when some players could obtain more by alternative coalition formation, namely when the players believe that such a coalition would be upset by other players inside or outside it.

This can be illustrated by an example: Let $(N, v)$ be given by $N = \{1, 2, 3\}$ and

$$v(\{1\}) = v(\{2\}) = v(\{3\}) = v(\{1, 2, 3\}) = 0,$$
$$v(\{1, 2\}) = v(\{1, 3\}) = 100, \ v(\{2, 3\}) = 50.$$

Clearly, in this game the imputations are not particularly interesting as potential solutions (there is only one imputation, namely $(0, 0, 0)$ ). The game must result in the formation of coalitions that can achieve something different from 0.

Following the tradition in this field, we introduce some terminology: a *payoff configuration*, shorthand a PC, is a pair $(z, \mathcal{P})$, where $z$ is a payoff vector and $\mathcal{P}$ a partition of $N$, also called a *coalition structure*. Consider now the PC $((80, 20, 0), \{1, 2\}, \{3\})$. If this is proposed as the outcome of the game, then player 2 might argue that the PC $((0, 21, 29), \{1\}, \{2, 3\})$ is a possibility as well. We say that this PC is an *objection* (for player 2) against the first PC.

This objection is obviously directed against player 1 (who is to receive less than in the original one, while the other two players get more), and moreover, player 1 cannot come up with any PC which would suggest that player 2 cannot demand a larger payoff than the original 20. Therefore, the first PC is *unstable*.

However, if instead we consider the PC $(75, 25, 0), \{1, 2\}, \{3\})$, we get that any objection of player 2, such as for example $((0, 26, 24), \{1\}, \{2, 3\})$, can be answered by player 1 by a *counterobjection* (an objection against the objection), in the present example this might be the PC $((75, 0, 25), \{1, 3\}, \{2\})$, just as every objection of player 1 (such as $((76, 0, 24), \{1, 3\}, \{2\})$ ) can be answered by player 2 with a counterobjection (such as $((0, 25, 25), \{1\}, \{2, 3\})$ ). The PC considered here is *stable*.

The set of stable PCs can be considered as possible outcomes of a bargaining procedure, namely as the outcomes which remain after an initial phase of objections and counterobjections. We formalize these notions below.

Let $(N, v)$ be a game. An *individually rational payoff configuration* (shorthand an IRPC) is a pair $(z, \mathcal{P})$, where $z$ is an individually rational payoff vector and $\mathcal{P}$ is a partition of $N$, that is

$$\mathcal{P} = \{C_1, \ldots, C_r\}, \ C_i \cap C_j = \emptyset, \ i \ne j, \ \cup_{k=1}^r C_k = N,$$

such that $\sum_{i \in C_k} z_i = v(C_k)$ for $k = 1, \ldots, r$.

Let $(z, \mathcal{P})$ be an IRPC and let $k$ and $l$ be different players in a coalition $C_j$ belonging to $\mathcal{P}$. An *objection* for $k$ against $l$ in $(z, \mathcal{P})$ is a pair $(y, C)$, where $C$ is a coalition and $y = (y_i)_{i \in C}$ a distribution of $v(C)$ among the members of $C$, such that

$$k \in C, \ l \notin C, \ k, l \in C_j,$$
$$\sum_{i \in C} y_i = v(C),$$
$$y_k > z_k, \ y_i \ge z_i, \ \text{all} \ \in C.$$

As before, let $(z, \mathcal{P})$ be an IRPC and let $(y, C)$ be an objection for $k$ against $l$ in $(z, \mathcal{P})$. A pair $(w, D)$, where $D$ is a coalition and $w = (w_i)_{i \in D}$ a distribution of $v(D)$ on its members, is a *counterobjection* against the objection $(y, C)$ if

$$l \in D, \ k \notin D,$$
$$\sum_{i \in D} w_i = v(D),$$
$$w_i \ge y_i, \ \text{all} \ i \in D, \ w_i > y_i, \ \text{all} \ i \in C \cap D.$$

DEFINITION 1 *An IRPC $(z, \mathcal{P})$ is stable if for every objection to $(z, \mathcal{P})$ there is a counterobjection. The set of stable IRPCs is called the bargaining set and is denoted by $\mathcal{M}_1^{(i)}$.*

The somewhat peculiar notation has historical roots: The superscript $(i)$ indicates individual rationality, and the subscript 1 tells that this bargaining set is only one of several possible, even though it is the only one which has survived over the years.

The bargaining set was introduced in Aumann and Maschler [1964]. In most games the bargaining set $\mathcal{M}_1^{(i)}$ will be a rather large set. It is easily seen that if $\text{Core}(N, v) \ne \emptyset$, then this set is contained in $\mathcal{M}_1^{(i)}$ (or, to be more precise, if

$z \in \text{Core}(N, v)$, then $(z, \{N\}) \in \mathcal{M}_1^{(i)}$ – note that the partition here is the standard one consisting only of the grand coalition). Indeed, for $z \in \text{Core}(N, v)$ we have that $\sum_{i \in C} z_i \geq v(C)$ for every coalition $C$, and therefore there are no individuals $k$ and $l$ such that $k$ has an objection against $l$ in $(z, \{N\})$. Since there are no objections, it is trivially true that every objection has a counterobjection.

Actually, $\mathcal{M}_1^{(i)}$ is large enough to be nonempty. The following result is due to Peleg.

THEOREM 1 *For every coalition structure $\mathcal{P}$ there is a payoff vector $z$ such that $(z, \mathcal{P}) \in \mathcal{M}_1^{(i)}$.*

We shall not prove this theorem. The slightly weaker statement $\mathcal{M}_1^{(i)} \neq \emptyset$ will follow from our later results about the nucleolus.

**1.2. The kernel.** In spite of the intuitive appeal behind the bargaining set $\mathcal{M}_1^{(i)}$, it is not very well suited for practical purposes. Since it is so relatively large, it may then be worthwhile to look for selections which may be easier to work with, and several such selections have been singled out in the literature. We consider here the *kernel* introduced in Davis and Maschler [1965].

As our point of departure for the following discussion, we take the excess function $e : C \times A \to \mathbb{R}$ introduced previously as

$$e(S, z) = v(S) - \sum_{i \in S} z_i$$

for $S \in C, z \in A$. For two different players $i$ and $j$, let

$$s_{ij}(z) = \max\{e(S, z) \mid S \in C, i \in S, j \notin S\}.$$

Let $(z; \mathcal{P})$ be an IRPC. We say that player $i$ *outweighs* player $j$ if they belong to the same coalition $C_k$ in $\mathcal{P}$ and

$$z_j > v(\{j\}), \ s_{ij}(z) > s_{ji}(z).$$

DEFINITION 2 *For any game $(N, v)$ and any partition $\mathcal{P}$ of $N$, the kernel of $\mathcal{P}$, denoted $K(v, \mathcal{P})$ or just $K(\mathcal{P})$, is the set of all payoff vectors $z$ such that $(z; \mathcal{P})$ is an IRPC and for every pair $(i, j)$, $i$ does not outweigh $j$.*

We check that the kernel $K(\mathcal{P})$ is indeed a subset of $\mathcal{M}_1^{(i)}$. This is a consequence of the following lemma.

LEMMA 1 *Let $(z; \mathcal{P})$ be an IRPC such that $z \in K(\mathcal{P})$. Then $(z; \mathcal{P}) \in \mathcal{M}_1^{(i)}$.*

PROOF: Suppose that there is an objection of $k$ against $l$ in $(z; \mathcal{P})$, that is a pair $(y; C)$ such that

$$k \in C, \ l \notin C, \ k, l \in C_j,$$
$$\sum_{i \in C} y_i = v(C),$$
$$y_k > z_k, \ y_i \geq z_i, \ \text{all } i \in C.$$

Then $v(C) = \sum_{i \in C} y_i > \sum_{i \in C} z_i$, so that $s_{kl}(z) \geq e(C, z) > 0$. Since $k$ does not outweigh $l$, we must have either $z_l = v(\{l\})$ or $s_{lk}(z) \geq s_{kl}(z)$.

If $z_l = v(\{l\})$, then $(z_l; \{l\})$ is a counterobjection. If $z_l > v(\{l\})$, we get from $s_{kl}(z) \geq e(C, z)$ that there is a coalition $D$ with $l \in D$, $k \notin D$, such that

$$v(D) - \sum_{i \in D} z_i \geq v(C) - \sum_{i \in C} z_i.$$

Let $\beta = y_k - z_k > 0$, in the objection $(y; C)$ the other members of $C$ get in total

$$v(C) - \sum_{i \in C} z_i - \beta < v(D) - \sum_{i \in D} z_i.$$

Consequently, there is $w = (w_i)_{i \in D}$ with $w_i = z_i$ for $i \notin C \cap D$ and

$$\sum_{i \in C \cap D} w_i = \sum_{i \in C \cap D} z_i + (v(D) - \sum_{i \in D} z_i),$$

so that $w_i > z_i$ for $i \in C \cap D$. We conclude that $(w; D)$ is a counterobjection. $\square$

It may be noticed that the definition of $K(\mathcal{P})$ can be given in a somewhat shorter form: Let $I(\mathcal{P})$ denote the set of individually rational payoff vectors associated with the partition $\mathcal{P}$,

$$I(\mathcal{P}) = \left\{ z \in \mathbb{R}^N \;\middle|\; z_i \geq v(\{i\}), \; i \in N, \; \sum_{i \in C_k} z_i = v(C_k), \; k = 1, \ldots, r \right\},$$

where $\mathcal{P} = \{C_1, \ldots, C_r\}$. We then have that

$$K(\mathcal{P}) = \{z \in I(\mathcal{P}) \mid s_{ij}(z) - s_{ji}(z))(z_i - v(\{i\})) \leq 0, \; i, j \in C_k, \; C_k \in \mathcal{P}\}.$$

In a similar way, we can define a *prekernel* $K^0(\mathcal{P})$ as

$$K^0(\mathcal{P}) = \{z \in I(\mathcal{P}) \mid s_{ij}(z) - s_{ji}(z) = 0, \; i, j \in C_k, \; C_k \in \mathcal{P}\}.$$

It is seen that $K^0(\mathcal{P}) \subseteq K(\mathcal{P})$.

This formulation is convenient if we want to show that kernel and prekernel are invariant under strategic equivalence: Suppose that the given game $(N, v)$ is replaced by the game $av + \mu$, where $a > 0$ is a real number and $\mu$ is a trivial game (that is an additive set function). Replacing payoff vectors $z$ in $v$ by $av + m$, with $m = (\mu(\{1\}), \ldots, \mu(\{n\}))$, then

$$z \in K(v, \mathcal{P}) \Leftrightarrow az + m \in K(av + \mu, \mathcal{P}),$$
$$z \in K^0(v, \mathcal{P}) \Leftrightarrow az + m \in K^0(av + \mu, \mathcal{P}).$$

The proof is straightforward and left to the reader. As a consequence of this result, we may assume $v(\{i\}) = 0$ for all $i$ in the following.

The kernel $K(v, \{N\})$, that is the kernel belonging to the particular coalitional structure for which the partition consists only of the grand coalition, is of special interest; it will be denoted $K(v)$ in the following (and similarly, the prekernel $K^0(v, \{N\})$ with the grand coalition as only coalition is denoted $K^0(v)$ ).

It can be shown that if the game $(N, v)$ is monotonic in the sense that for all $S, T \in C, S \subseteq T \Rightarrow v(S) \leq v(T)$, then $K^0(v) = K(v)$. Also the class of convex games stands ahead by its nice properties: For convex games, $K(v)$ consists of a single element (which will be characterized further in the next section), and this element belongs to Core$(N, v)$, which by the way defines the only stable IRPC belonging to the partition $\{N\}$. The latter results are however not quite easy to prove, and we refer the reader to the literature for proofs.

## 2 The nucleolus

Consider a game $(N, v)$, and as usual, let $A$ be the set of imputations (so that we are using an implicit assumption that the coalition structure is $\{N\}$. We assume that $A \neq \emptyset$.

Let $\theta : A \to \mathbb{R}^{2^N - 1}$ be the mapping from imputations to vectors with as many coordinates as there are coalitions, defined by

$$\theta(z) = (e(S_1, z), \ldots, e(S_k, z)),$$

where $k = |C|(= 2^N - 1)$ and the numbers $e(S, z)$ are arranged in nonincreasing order, so that

$$i < j \Rightarrow e(S_i, z) \geq e(S_j, z)$$

for $i, j = 1, \ldots, k$. Thus, for another imputation $z'$ the coalitions having their excess at 1st, 2nd, $n$th place in $\theta(z')$ will normally be different from those having excess at 1st, 2nd, $n$th place in $\theta(z)$.

We now introduce the lexicographic ordering $<_L$ on the set $\{\theta(z) \mid z \in A\}$, so that we have

$$\theta(z) <_L \theta(y)$$

if there is some $j \in \{1, \ldots, k\}$ such that for $i < j$ the $i$th coordinate of $\theta(z)$ equals that of $\theta(y)$, and the $j$th coordinate of $\theta(z)$ is smaller that the $j$th coordinate of $\theta(y)$.

DEFINITION 3 *The nucleolus of* $(N, v)$, *denoted* $\mathcal{N}(v)$, *is the set of imputations* $z$ *such that* $\theta(z)$ *is minimal for* $<_L$ *on* $\theta(A)$.

The nucleolus was introduced in Schmeidler [1969]. In contrast to the previously considered solution concepts, there is no (or at least no convincing) "story" behind the definition of the nucleolus. On the other hand it is interesting due to its properties, and it may be used also to give insight in properties of several other solution concepts.

Before considering its properties, we establish an existence result.

THEOREM 2 *For every game* $(N, v)$, *if* $A \neq \emptyset$, *then* $\mathcal{N}(v) \neq \emptyset$.

PROOF: Let $\theta^i(z)$ be the $i$th coordinate of $\theta(z)$. This coordinate can also be written as

$$\theta^i(z) = \max_{C' \subseteq C, |C'|=i} \min_{S \in C'} e(S, z),$$

since the $i$th largest excess is obtained by taking for each subfamily consisting of $i$ coalitions the one coalition in the subfamily having the smallest excess, and then maximizing over all such subfamilies.

From this expression it is seen that $\theta^i$ is a continuous function, since taking maximum and minimum over a finite set of continuous functions preserves continuity. We define subsets $A_1 \supseteq A_2 \supseteq \cdots \supseteq A_k$ of $A$ inductively as follows:

$$A_1 = \{z \in A \mid \theta^1(z) \le \theta^1(y), \text{ all } y \in A\},$$

(so that $A_1$ consists of all imputations minimizing $\theta^1$), and for $i = 2, \ldots, k$,

$$A_i = \{z \in A_{i-1} \mid \theta^i(z) \le \theta^i(y), \text{ all } y \in A_{i-1}\}.$$

Since $A$ is nonempty (by our assumption) and compact, $\theta^1$ attains a minimum on $A$, so that $A^1 \ne \emptyset$. As $A^1$ is a closed subset of $A$, it is compact, and then $\theta^2$ has a minimum on $A^1$, meaning that $A^2 \ne \emptyset$. Proceeding in this way, we finally get that $\theta^k$ has a minimum on $A^{k-1}$, so that $A^k \ne \emptyset$. Since $A^k = \mathcal{N}(v)$ by our above argument, we have shown that $\mathcal{N}(v) \ne \emptyset$. $\qquad\square$

The nucleolus turns out to be a singleton, so that it assigns a unique payoff vector as solution to any given game.

THEOREM 3 *For each game $(N, v)$ the nucleolus $\mathcal{N}(v)$ consists of at most one element.*

PROOF: It is easily seen that if two different imputations $z$ and $y$ both have the property that $\theta(z)$ and $\theta(y)$ are minimal for $<_L$ on $\theta(A)$, then we must have $\theta(z) = \theta(y)$ ($<_L$ is a complete order).

Consider now the imputation $w = \frac{1}{2}z + \frac{1}{2}y$. Let $\eta$ be the mapping which sends arbitrary $k$-vectors (where $k = |C| = 2^n - 1$) to $k$-vectors with the same coordinate values but in descending coordinate order. We have then

$$\theta(z) = \eta((e(S, z)_{S \in C}).$$

We claim that for all $k$-vectors $a$ and $b$,

either $\eta(a + b) <_L \eta(a) + \eta(b)$ or $\eta^t(a + b) = \eta^t(a) + \eta^t(b)$, all $t \in \{1, \ldots, k\}$.    (1)

To show that (1) holds, let

$$\eta^t(a) = a^{i_t}, \quad \eta^t(b) = b^{j_t}, \quad \eta^t(a + b) = a^{k_t} + b^{k_t}.$$

Since $a^{k_1} \le a^{i_1}$ and $b^{k_1} \le b^{j_1}$, we must have $a^{k_1} + b^{k_1} \le a^{i_1} + b^{j_1}$; if the inequality sign is sharp, we have $\eta(a + b) <_L \eta(a) + \eta(b)$. Otherwise we have equality sign, and in that case $a^{k_1} = a^{i_1}$, $b^{k_1} = b^{j_1}$. After this, we proceed in the same way with the remaining coordinates, proving our claim.

In the case we are treating, we have $a = (e(S, z))_{S \in C}$, $b = (e(S, y))_{S \in C}$, $a + b = (2v(S) - \sum_{i \in S} z_i - \sum_{i \in S} y_i)_{S \in C}$. We have that $\eta(a) = \theta(z)$, $\eta(b) = \theta(y)$, and $\eta(a+b) = 2\theta(w)$.

We now apply (1): If $2\theta(w) <_L \theta(z) + \theta(y)$, then we get from $\theta(z) = \theta(y)$ that $2\theta(w) <_L 2\theta(y)$ or $\theta(w) <_L \theta(y)$ contradicting that $z \in \mathcal{N}(v)$.

Consequently, the other alternative must hold, and for arbitrary $t \in \{1, \ldots, k\}$ we have

$$2\theta^t(w) = v(S_{k_t}) - \sum_{i \in S_{k_t}} z_i + v(S_{k_t}) - \sum_{i \in S_{k_t}} y_i,$$

$$\theta^t(z) = v(S_{i_t}) - \sum_{i \in S_{i_t}} z_i, \quad \theta^t(y) = v(S_{i_t}) - \sum_{i \in S_{i_t}} y_i.$$

By (1),

$$v(S_{k_t}) - \sum_{i \in S_{k_t}} z_i = v(S_{i_t}) - \sum_{i \in S_{i_t}} z_i, \quad v(S_{k_t}) - \sum_{i \in S_{k_t}} y_i = v(S_{j_t}) - \sum_{i \in S_{j_t}} y_i,$$

and from $\theta(z) = \theta(y)$ we get that

$$v(S_{k_t}) - \sum_{i \in S_{k_t}} z_i = v(S_{k_t}) - \sum_{i \in S_{k_t}} y_i$$

or $\sum_{i \in S_{k_t}} z_i = \sum_{i \in S_{k_t}} y_i$. As $t$, and thereby the coalition $S_{k_t}$, was arbitrary, we have that $\sum_{i \in S} z_i = \sum_{i \in S} y_i$ for all $S$, and we conclude that $z = y$. □

From now on we identify the set $\mathcal{N}(v)$ with its unique element. We shall have a closer look at the relationship between the nucleolus and other solution concepts. For this we assume throughout that the nucleolus exists (which according to Theorem 3 amounts to the assertion that the game admits a nonempty set of imputations).

THEOREM 4 *$\mathcal{N}(v)$ is contained in any nonempty $\varepsilon$-core.*

PROOF: Let $y = \mathcal{N}(v)$. Clearly, an imputation $z$ belongs to the $\varepsilon$-core for some $\varepsilon > 0$ if and only if $e(S, z) \leq \varepsilon$ for all $S$, or alternatively if and only if $\theta(z) \leq (\varepsilon, \ldots, \varepsilon)$. Since $\mathcal{N}(v)$ is a lexicographical minimum, the largest of all the $e(S, y)$ (for $S \in \mathcal{S}$) is smaller than $\varepsilon$, and consequently $y$ is in the $\varepsilon$-core. □

The following theorem shows that the nucleolus always belongs to the kernel (which again is a subset of the bargaining set).

THEOREM 5 *$\mathcal{N}(v) \in K(v)$.*

PROOF: Suppose that $y = \mathcal{N}(v) \notin K(v)$. Then there must be $i$ and $j$ such that $i$ outweighs $j$, that is $y_j > 0$ and $s_{ij}(y) > s_{ji}(y)$. Let

$$\delta = \frac{1}{2} \min\{y_j, s_{ij}(y) - s_{ji}(y)\}.$$

We show that the imputation $y^\delta$ defined by

$$y_i^\delta = y_i + \delta, \quad y_j^\delta = y_j - \delta, \quad y_k^\delta = y_k, \quad k \neq i, j,$$

(so that $y^\delta$ is obtained from $y$ by transferring $\delta$ from $j$ to $i$) satisfies $\theta(y^\delta) <_L \theta(y)$:
    Let $\mathcal{L} = \{S \mid e(S, y) \geq s_{ij}(y) \text{ and } j \in S \text{ if } i \in S\}$, and let $l = |L|$. Then
$$\theta^l(y) \geq s_{ij}(y), \text{ but } \theta^{l+1}(y) = s_{ij}(y),$$
since, by the definition of $s_{ij}(y)$, there is at least one coalition $R$ such that $i \in R$, $j \notin R$, and $e(R, y) = s_{ij}(y)$. For this coalition $R$ we have $e(R, y^\delta) = s_{ij}(y) - \delta$. By transferring $\delta$ we have increased the excess for coalitions with $j$ but without $i$ to at most $s_{ji}(y) + \delta$, which is $< s_{ij}(y)$. Thus, $\theta^{l+1}(y^\delta) < s_{ij}(y)$. The excess for coalitions in $\mathcal{L}$ is not reduced, and consequently $\theta(y^\delta) <_L \theta(y)$, contradicting that $N(v) = y$.  □

From Theorem 5 we get first of all that the nucleolus belongs to the bargaining set $M_1^{(i)}$, and secondly, when combined with Theorem 4, that the core (if it is nonempty) and the kernel has nonempty intersection, since the nucleolus belongs to both.

## 3   The reduced game property

In this section, we consider a system of axioms for abstract solutions to TU games which characterizes the nucleolus, much in the same way as we earlier considered axioms for the Shapley value. An important role is played by the so-called *reduced game property*, which we proceed to define.

As in the previous case, the point of departure is an abstract solution $\phi$, a mapping which to every TU game $(N, v)$ assigns a particular payoff vector $\phi(N, v)$. For any feasible payoff vector $x$ in $(N, v)$, that is a payoff vector with $\sum_{i \in N} x_i = v(N)$, and for any player $i$, we define the *i-reduced game at $x$* as the game $(N \setminus \{i\}, v_x^i)$, where
$$v_x^i(S) = \max\{v(S), v(S \cup \{i\}) - x_i\}$$
for all coalitions $S \subseteq N \setminus \{i\}$. Intuitively, the reduced game arises if the players agree to letting $i$ leave the conflict upon paying him what he would have got according to the payoff vector $x$. Coalitions can then use the power of this player (including her in the coalition) by paying her the amount $x_i$.

A solution is said to have the *reduced game property* if for each game $(N, v)$, each feasible payoff vector $x$ in $(N, v)$ and each player $i$, we have
$$\phi_j(N \setminus \{i\}, v_{\phi(v)}^i) = \phi_j(v)$$
for all $j \in N \setminus \{i\}$. Thus, the reduced game property says that if the solution is used, then it makes no difference whether the players get their payoff according to the solution right away or reduce the game and use the solution on the smaller game.

The nucleolus has the reduced game property; this follows from the result below:

THEOREM 6 *Let $(N, v)$ be a game, and let $N(v)$ be its nucleolus, and assume $N_i(v) > 0$ for all $i$. Then*
$$N_j(v_{N(v)}^i) = N_j(v)$$
*for all $i, j$ with $i \neq j$.*

PROOF: Let $a$ be the vector of excesses in the game $v^i_{N(v)}$,

$$a_k = \max \left\{ v(S), v(S \setminus \{i\}) - N_i(v) - \sum_{j \in S} N_j(v) \right\}$$

written in descending order. Assume that there is a payoff vector $z'$ in $v^i_{N(v)}$ such that the associated excess vector satisfies

$$b_1 < a_1.$$

We have for each $S \in C$ that the excess of the original game satisfies

$$e(S, N(v)) = v(S) - \sum_{j \in S} N_j(v) = v(S) - N_i(v) - \sum_{j \in S \setminus \{i\}} N_j(v)$$

$$\leq v^i_{N(v)}(S \setminus \{i\}) - \sum_{j \in S \setminus \{i\}} N_j(v) \leq a_1,$$

since $a_1$ is the largest excess in $v^i_{N(v)}$, and correspondingly we have for the payoff vector $(z', N_i(v))$ in the game $(N, v)$, that for all $S \subset N$

$$e(S; (z', N_i(v))) \leq b_1.$$

Summing up, we have that the largest coordinate in the excess vector associated with $(z', N_i(v))$ is smaller than the largest coordinate in the excess vector associated with $N(v)$, a contradiction.

To show the general case we prove that if the game $v^i_{N(v)}$ has a payoff vector $z$ such that the associated excess vector satisfies

$$b_1 = a_1, \ldots, b_{r-1} = a_{r-1}, b_r < a_r$$

for some $r > 1$, then there is also a payoff vector $z'$ with associated excess vector $b'$ such that

$$b_1 = a_1, \ldots, b_{p-1} = a_{p-1}, b_p < a_p$$

for $p < r$. For this, let $S$ be a coalition with $e(S, z) = b_r$, and let $\hat{S} = \{i \in S \mid z_i > 0\}$. If there is $i \in \hat{S}$ such that $i \notin T$ for all $T$ with $e(T, z) = a_1$, then we can choose $j \in T$ for some such $T$ and $z'$ with

$$z'_i = z_i - \varepsilon, \ z'_j = z_j + \varepsilon, \ z'_h = z_h, h \neq i, j,$$

where $\varepsilon > 0$ is small enough so that $e(S, z') < a_1$.

Assume therefore that $\hat{S} \subseteq T$ for every $T$ with $e(T, z) = a_1$, and let $T_1$ be such a coalition. If $z_i > 0$ for some $i \in N \setminus T_1$, we consider the payoff vector $z'$ with

$$z'_i = z_i - \varepsilon, \ z'_j = z_j + \frac{\varepsilon}{|\hat{S}|} \text{ for } j \in \hat{S}, \ z'_h = z_h \text{ otherwise.}$$

For $\varepsilon > 0$ small enough we have $\max_S e(S, z') < a_1$. If $z_i = 0$ for all $i \in N \setminus T_1$, we must have $a_1 = e(T_1, z) = v(T_1) - v(N)$, and if $\sum_{i \in T_1} N_i(v) < v(N)$, we get that $e(T_1, N(v)) > a_1$, so that $N_i(v) = 0$ for $i \in N \setminus T_1$, a contradiction which concludes our proof. $\qquad \square$

The converse of Theorem 6 is perhaps more interesting, and indeed it is the main result of this section. It was obtained in Sobolev [1975], see also Sobolev and Pechorsky [1983].

THEOREM 7 *Let $\phi$ be a solution for side-payment games, such that $\phi$ has the reduced game property, and assume that $\phi(N, v)$ for all $(N, v)$ with $|N| = 2$ is given by the rule*

$$x_i^0 = v(N) - v(\{j\}) + \frac{v(\{1\}) + v(\{2\}) - v(N)}{2} \tag{2}$$

*for $i, j = 1, 2, i \neq j$. Then $\phi(N, v) = \mathcal{N}(v)$, that is $\phi$ is the nucleolus.*

PROOF: By induction in $n$, the number of players in $(N, v)$. For $n = 2$ the result follows immediately from the formula in (2).

Assume that the theorem has been shown for all games $(N, v)$ with $|N| = n \geq 2$, and let $(N, v)$ be an arbitrary game with $|N| = n + 1$. Let $(x_1, \ldots, x_{n+1}) = \phi(N, v)$, and let $a$ be the vector of excesses $e(S, x)$, $S \subseteq N$, ordered by decreasing coordinates.

Let $x' = (x'_1, \ldots, x'_{n+1})$ be a payoff vector with $b$ the associated vector of ordered excesses, and assume that $b$ is lexicographically smaller than $a$; as in the proof of Theorem 6 we may assume that $b_1 < a_1$.

If $x'_i = x_i$ for some $i \in N$, consider the $i$-reduced game $(N\setminus\{i\}, v_x^i)$; we know by the reduced game property and our induction hypothesis that $\phi(N\setminus\{i\}, v_x^i) = (x_j)_{j \neq i} = \mathcal{N}(v_x^i)$. Let the excess vector in the game $v_x^i$ w.r.t. $(x_j)_{j \neq i}$ and $(x'_j)_{j \neq i}$, be $\hat{a}$ and $\hat{b}$, respectively. We show that $\hat{b}_1 < \hat{a}_1$, and this will give us a contradiction: We have $a_1 = \max_{S \subseteq N} e(S, x) = v(S) - \sum_{i \in S} x_i$ for some $S$ and $x$. If $i \in S$, then $a_1 = v(S) - x_i - \sum_{j \in S\setminus\{i\}} x_j = v_x^i(S\setminus\{i\}) - \sum_{j \in S\setminus\{i\}} x_j$, since $v(S\setminus\{i\}) > v(S) - x_i$ would mean that $S\setminus\{i\}$ yields a greater excess. We therefore have $\max_k \hat{a}_k \geq a_1$; on the other hand $v_x^i(S'\setminus\{i\}) - \sum_{j \in S'\setminus\{i\}} x_j > a_1$ would mean that either $S'\setminus\{i\}$ or $S' \cup \{i\}$ has greater excess in the game $v$ than $S$, a contradiction, so that $\max_k \hat{a}_k = a_1$. Correspondingly, we have that $\max_k \hat{b}_k = b_1$, and we have that $b$ is lexicographically smaller than $a$.

Suppose now that $x'_i \neq x_i$ for all $i$. We show that there is another payoff vector $x''$ such that the corresponding excess vector is lexicographically smaller than $a$. Let $S$ be such that $e(S, x') = b_1$. If $x'_i < x_i$ for some $i \in S$, then $x''$ can be found by redistributing inside $S$, so that $x''_i = x_i$ and $\sum_{i \in S} x''_i = \sum_{i \in S} x_i$. Assume therefore that $x'_i \geq x_i$, all $i \in S$. If there is more than one $i \in S$ with $x'_i > x_i$, choose one of these, $i_0$, and define $x''$ by

$$x''_i = x'_i + \frac{x_i - a_i}{n}, \ i \neq i_0, \ x''_{i_0} = x_{i_0}.$$

If a coalition $T$ has an excess $\geq a_i$ at this payoff vector, then $T$ must have an excess which is $> a_1 - (a_1 - b_1) = b_1$ at the payoff vector $x'$, a contradiction. Assume finally that there is only one $i \in S$ with $x'_i > x_i$, choose $j \notin S$ and consider $x''$ defined by $x''_i = x'_i, x''_j = x_j + (x'_i - x_i)$, and $x'' \leq e(T, x) + (a_1 - b_1)$, so that we again have that $x''$ is lexicographically smaller than $a$.                                                                □

The reduced game property, which has played an important role in our characterization of the nucleolus, has later been reformulated and given slightly different forms so as to be useful also for axiomatic characterizations of other solutions, see e.g. Peleg and Südholter [2007].

# 4 The τ-value

**4.1. Introduction.** The solution concepts which have been discussed in the previous chapters are all products of the game theoretic research activity of the period up to 1960, with a few later extensions. This may give the impression that the field has been fully developed, so that one should not expect new solution concepts but rather further work (and extension to NTU games) on the old, already existing solution concepts.

This is however not the case. In the present chapter we look closer at a solution concept of later date, namely the τ-value introduced in Tijs [1981]. The τ-value shows by its very nature that the cooperative game theory is still quite open, and new developments can occur. The considerations leading to the τ-value is quite different from those leading to core, Shapley value and nucleolus, so that one may speak of an independent theory. We shall consider this as well as an axiomatic characterization of the τ-value, which may be of independent interest and presents another approach to this solution concept.

**4.2. The upper vector and the gap function.** We begin with some definitions which can also be used in other connections. For a cooperative TU game $(N, v)$ we define the *upper vector* $b^v \in \mathbb{R}^N$ as

$$b_i^v = v(N) - v(N \setminus \{i\}).$$

The upper vector gives an intuitive upper limit to what a player may expect to obtain from participating in the game; if $x_i > b_i^v$ for an imputation $x$ and a player $i$, then $\sum_{j \neq i} x_j < v(N \setminus \{i\})$, so that the other players would be better off without $i$. In other words $b_i^v$ is player $i$'s marginal contribution to the grand coalition of all players.

Next step is the introduction of the *gap function* $g^v : \mathcal{S} \to \mathbb{R}$ given by

$$g^v(S) = \sum_{j \in S} b_j^v - v(S)$$

for $S \in \mathcal{S}$. The value $g^v(S)$ of this function at the coalition $S$ is called the gap of the coalition $S$ in the game $v$. What is measured by $g^v(S)$ (except that it gives the difference betweeen the "ideal" payoff vector as expressed by the upper vector and the strength of the coalition – in a certain sense how much the coalition falls short of achieving the maximal payoff), is indicated by the following trivial result:

LEMMA 2 *Let $(N, v)$ be a TU game, where $g^v(S) < 0$ for a coalition $S$. Then $\mathrm{Core}(N, v)$ is empty.*

PROOF: We show that $g^v(S) \geq 0$ if $\mathrm{Core}(N, v) \neq \emptyset$. Let $x$ belong to the core and let $S$ be an arbitrary coalition. We then have

$$v(S) \leq \sum_{j \in S} x_j \leq \sum_{j \in S} b_j^v$$

or $g^v(S) \geq 0$.　　　　　　　　　　　　　　　　　　　　　　　□

If the core is to be nonempty, then the upper vector must overvalue the possibilities of the coalitions; or alternatively, if a coalition is so powerful that it can achieve more than the upper vector, then one cannot hope for finding nondominated imputations.

The gap function and the upper vector are well behaved under addition and multiplication by scalars of the underlying characteristic functions, so that

$$b^{av+cw} = ab^v + cb^w, \quad g^{av+cw} = ag^v + cg^w,$$

where $a, c \in \mathbb{R}$ and $v, w$ are the characteristic functions for TU games with fixed player set $N$.

If $v$ is superadditive, we have that

$$b_i^v = v(N) - v(N\setminus\{i\}) \geq v(\{i\}),$$

so that $b^v$ is individually rational; however, in many cases we have $\sum_{i \in N} b_i^v > v(N)$ (or equivalently $g^v(N) > 0$), so that $b^v$ cannot be obtained and therefore is not an imputation; in this case we can see it as a utopia payoff. The grand coalition $N$ will have to perform a *reduction* of magnitude $g^v(N) = \sum_{i \in N} b_i^v - v(N)$ in order to get a payoff vector which is actually an imputation. The next question is then how to distribute this reduction among the players.

In the following we assume that the share of each player in the reduction is determined by the smallest gaps for coalitions containing this player. The argumentation for this assumption is as follows: In order to convince the other players in the coalition, $i$ must promise each coalition member $j$ her utopia-payoff $b_j^v$, provided that they go together with $i$ in the coalition $S$. The remainder $v(S) - \sum_{j \in S, j \neq i} b_j^v$, is kept by player $i$. But we have

$$v(S) - \sum_{j \in S, j \neq i} b_j^v = b_i^v - g^v(S);$$

the maximal amount which a player $i$ can get by collecting followers in this way, is therefore $= b_i^v - \min_{S:i \in S} g^v(S)$. In other words, the concession made by $i$ (with respect to $b_i^v$), is

$$\lambda_i^v = \min_{S:i \in S} g^v(S).$$

The vector $\lambda^v = (\lambda_1^v, \ldots, \lambda_n^v)$ is called the concession vector. If the gap function is nonnegative, its components make sense as a maximal concession, since player $i$ can realize the payoff $b_i^v - \lambda_i^v$. Thus $\lambda_i^v$ expresses how far player $i$ will go in accepting her part of the total reduction $g^v(N)$.

In the sequel we assume that the game is such that individual concessions match the total reduction (if not, new arguments are necessary). Formally, this is usually achieved by introducing the class of *quasi-balanced* games $(N, v)$ by the condition

$$g^v(S) \geq 0 \text{ for all } S, \quad \sum_{i \in N} \lambda_i^v \geq g^v(N).$$

The terminology is meaningful, as it can be seen from the following:

LEMMA 3 *Let $(N, v)$ be balanced. Then $(N, v)$ is quasi-balanced.*

PROOF: Since $(N, v)$ is balanced, we have Core$(N, v) \neq \emptyset$, and from Lemma 2 we get that $g^v(S) \geq 0$ for all $S$. Further we notice that if $x \in$ Core$(N, v)$, then $b_i^v - \lambda_i^v \leq x_i$ for all $i$: By the definition of $\lambda^v$ there is $S$ and $i$ such that $\lambda_i^v = g^v(S)$, and we then have

$$\lambda_i^v = g^v(S) = \sum_{j \in S} b_j^v - v(S) \geq \sum_{j \in S} (b_j^v - x_j) \geq b_i - x_i,$$

where the last inequality follows from $b_j^v - x_j \geq 0$ and $i \in S$.

When $b_i^v - \lambda_i^v \leq x_i$ for all $i$ and arbitrary $x \in$ Core$(N, v)$, we must have

$$\sum_{j \in N} (b_j^v - \lambda_j^v) \leq v(N),$$

or

$$\sum_{j \in N} \lambda_j^v \geq \sum_{j \in N} b_j^v - v(N) = g^v(N).$$

But this shows that $(N, v)$ is quasi-balanced. $\qquad\square$

**4.3. The $\tau$-value.** At this moment we have done all the preliminary work for the definition of the $\tau$-value. Let $(N, v)$ be a quasi-balanced game. We define the $\tau$-value $\tau(v)$ of the game $(N, v)$ by

$$\tau(v) = \begin{cases} b^v & g^v(N) = 0, \\ b^v - g^v(N)[\sum_{i \in N} \lambda_i^v]^{-1}\lambda^v & g^v(S) > 0. \end{cases}$$

Intuitively the $\tau$-value is the imputation which emerges when the reduction $g^v(N)$ is distributed among players in the proportion given by their concession vector.

Alternatively (and in accordance with the way the $\tau$-value was first defined) one can find the $\tau$-value on the line segment between the utopian payoff $b^v$ and the total reduction $b^v - \lambda^v$ (which typically sums to less than $v(N)$) as the unique payoff which is efficient (sums to $v(N)$) and therefore an imputation. This approach bears some similarity with Kalai-Smorodinsky bargaining solution, a similarity which turns out not to be wholly superficial.

The existence problem for the $\tau$-value was taken care of by the very definition – it is only defined for quasi-balanced games, and for these games it is well-defined and unique. Finally the $\tau$-value is stable under strategic equivalence: If we multiply $v$ by a non-negative constant $\alpha$ and add an additive game $u$, we get $b^{av+u} = ab^v + u$ and $g^{av+u} = ag^v$, where we have identified the additive game $u$ with the payoff vector $(u(\{1\}), \ldots, u(\{n\}))$. From this we get that $\lambda^{av+u} = a\lambda^v + \lambda^u$; it is easily checked that $av + u$ is quasi-balanced, and that the $\tau$-value of $av + u$ is $a\tau(v) + u$.

Having thus checked the fundamental conditions for a solution concept to be worthwhile considering, we may start analyzing its relations to other solutions. Here in particular the relationship to the Shapley value seems important.

It is rather obvious from the definition that the $\tau$-value is symmetric (cf. the discussion in relation to the Shapley value), so that if $\sigma$ is a permutation of the players in $N$, then the $\tau$-value of the game $(\sigma v)$ is $\sigma(\tau(v))$. Further it may be shown that the $\tau$-value has the dummy property: Let $i \in N$ be a dummy for $v$, so that $v(S) - v(S\backslash\{i\}) = v(\{i\})$ for all $S$ containing $i$; then we have in particular that $b_i^v = v(N) - v(N\backslash\{i\}) = v(\{i\})$, from which we obtain that $\tau_i(v) = v(\{i\})$.

We have therefore that the $\tau$-value satisfies the axioms for the Shapley value except the additivity axiom. This shows firstly that it may be of interest to consider non-additive versions of the Shapley-value, and next it hints that there must be some condition which can replace additivity in an axiomatic characterization of the $\tau$-value. We shall however not pursue this line of investigation further for the moment.

**4.4. The $\tau$-value and the core.** As usually it is interesting to compare the new solution concept to other solutions, and in our case in particular to the core.

Consider the 4-person game $v$ given by

$$v(\{1,2\}) = v(\{1,3\}) = v(\{2,3\}) = v(\{1,2,3\}) = 1, \ v(\{1,2,3,4\}) = \frac{3}{2}$$

and $v(S) = 0$ for all other coalitions. The core of this game is the set consisting of the single element

$$\left(\frac{1}{2}, \frac{1}{2}, \frac{1}{2}, 0\right),$$

and it coincides with the nucleolus. Further it may be found that the Shapley value of $v$ is

$$\phi(v) = \left(\frac{13}{24}, \frac{13}{24}, \frac{13}{24}, -\frac{1}{8}\right).$$

Unfortunately it is not an imputation, since the game is not superadditive.

Let us find the $\tau$ value: First of all we determine the upper vector

$$b^v = \left(\frac{3}{2}, \frac{3}{2}, \frac{3}{2}, \frac{1}{2}\right),$$

and then the gap function $g^v$, where we obtain

$$g^v(\{1\}) = g^v(\{2\}) = g^v(\{3\}) = \frac{3}{2}, g^v(\{4\}) = \frac{1}{2}$$

for the singleton coalitions, and $g^v(S)$ equal to $2, 7/2$ og $7/2$, depending on whether $S$ has 2, 3 or 4 members.

Based on this we find that

$$\lambda^v = \left(\frac{3}{2}, \frac{3}{2}, \frac{3}{2}, \frac{1}{2}\right),$$

and now we get

$$\tau(v) = \frac{3}{10}b^v = \left(\frac{9}{20}, \frac{9}{20}, \frac{9}{20}, \frac{3}{20}\right).$$

We have thus shown that $\tau(v)$ does not belong to $\mathrm{Core}(N, v)$.

It is possible to elaborate on the example: If we make the game $v$ superadditive in the obvious way, namely as

$$\hat{v}(S) = \max\left\{\sum_{j=1}^{k}(S_j)|\{S_1,\ldots,S_k\} \text{ a partition of } S\right\},$$

we obtain a new (and obviously superadditive) game with

$$\hat{v}(\{1,2,4\}) = \hat{v}(\{1,3,4\}) = \hat{v}(\{2,3,4\}) = 1$$

and $\hat{v}(S) = v(S)$ otherwise.

It is seen immediately that $\text{Core}(N,\hat{v}) = \text{Core}(N,v)$, but we get a new Shapley value,

$$\phi(\hat{v}) = \left(\frac{11}{24}, \frac{11}{24}, \frac{11}{24}, \frac{1}{8}\right).$$

To find $\tau(\hat{v})$ we repeat the computations from before to get

$$b^{\hat{v}} = \left(\frac{1}{2}, \frac{1}{2}, \frac{1}{2}, \frac{1}{2}\right),$$

$$g^{\hat{v}}(\{1,2\}) = g^{\hat{v}}(\{1,3\}) = g^{\hat{v}}(\{2,3\}) = 0,$$

$$g^{\hat{v}}(\{1,4\}) = g^{\hat{v}}(\{2,4\}) = g^{\hat{v}}(\{3,4\}) = 1,$$

and $g^{\hat{v}}(S) = 1/2$ if $S$ has 1,3 or 4 members. From this we obtain

$$\lambda^{\hat{v}} = \left(0,0,0,\frac{1}{2}\right),$$

and get

$$\tau(\hat{v}) = \left(\frac{1}{2}, \frac{1}{2}, \frac{1}{2}, 0\right),$$

which is the single element in the core.

## 5 Problems

**1.** Consider the game $(N,v)$ with $N = \{1,2,3\}$, $v(\{i\}) = 0$ for all $i$, $v(\{1,2\}) = 20$, $v(\{1,3\}) = 30$, $v(\{2,3\}) = 40$, $v(N) = 42$. Find $M_1^i$ for the coalition structures $\{\{1\}, \{2\}, \{3\}\}$, $\{\{1,2\}, \{3\}\}$, $\{\{1,3\}, \{2\}\}$, $\{\{1\}, \{2,3\}\}$, $\{\{1,2,3\}\}$.

**2.** A *bankruptcy problem* is defined as a triple $(N, E, c)$, where $N = \{1,\ldots,n\}$ is a set of creditors or claimants, $E \in \mathbb{R}_+$ an estate to be divided among the claimants, and $in\mathbb{R}_+^N$ a vector of claims $c_i$ for $i \in N$. A *division rule* is a map $\delta$ which to each bankruptcy problem $(N, E, c)$ assigns a vector $x \in \mathbb{R}_+^N$ with $\sum_{i \in N} x_i = E$. Examples are:

(i) The proportional rule $P$: For each $(N, E, c)$, $P(c, E) = \lambda c$, where $\lambda = E/\sum_{i \in N} c_i$.

(ii) The adjusted proportional rule $A$: Define for each $i$ the *minimal right of claimant* $i$ by $m_i(N, E, c) = \max\{E - \sum_{j \neq i} c_j, 0\}$ and let

$$A(N, E, c) = m(c, E) + P(N, E - \sum_{j \in N} m_j(N, E, c), \hat{m}),$$

where $\hat{m}_i = \min\left\{c_i - m_i(N, E, c), E - \sum_{j \in N} m_j(N, E, c)\right\}$.

A bankruptcy problem gives rise to a TU game $(N, v_{(N,E,c)})$ with $v_{(N,E,c)}(S) = \max\{E - \sum_{i \in N \setminus S} c_i, 0\}$. Show that the adjusted proportional rule for $(N, E, c)$ corresponds to the $\tau$-value of $v_{(N,E,c)}$.

**3.** (The Talmud problem). An example of a (discrete) rule for solving bankruptcy problems (cf. Problem 2) is the rule given in the Talmud for only a finite number of bankruptcy problems, as shown in the table below, where the alternative values of $E$ are shown in the first column, and where the claims of three claimants are in the first row:

|     | 100 | 200 | 300 |
| --- | --- | --- | --- |
| 100 | $33\frac{1}{3}$ | $33\frac{1}{3}$ | $33\frac{1}{3}$ |
| 200 | 50 | 50 | 100 |
| 300 | 50 | 100 | 150 |

The table may be taken as coming from the Talmud rule $T$, defined as follows

(i) if $\sum_{i \in N} \frac{c_i}{2} \geq E$, then $T_i(N, E, c) = \min\left\{\frac{c_i}{2}, \lambda\right\}$, where $\lambda$ is such that

$$\sum_{i \in N} \min\left\{\frac{c_i}{2}, \lambda\right\} = E,$$

(ii) if $\sum_{i \in N} \frac{c_i}{2} \leq E$, then $T_i(N, E, c) = c_i - \min\left\{\frac{c_i}{2}, \lambda\right\}$, where $\lambda$ is such that

$$\sum_{i \in N}\left[c_i - \min\left\{\frac{c_i}{2}, \lambda\right\}\right] = E.$$

Show that the Talmud rule for $(N, E, c)$ corresponds to the pre-nucleolus (which is not necessarily an imputation) of $v_{(N,E,c)}$.

**4.** Let $(N, v)$ be the 7-person game with $v(S) = 1$ for all coalitions containing any of the sets $\{1, 2, 4\}, \{2, 3, 5\}, \{3, 4, 6\}, \{4, 5, 7\}, \{5, 6, 1\}, \{6, 7, 2\}, \{7, 1, 3\}$, and $v(S) = 0$ otherwise (cf. Maschler [1992], von Neumann and Morgenstern [1944]).

Show that the kernel of $(N, v)$ consists of line segments from the payoff $\left(\frac{1}{7}, \cdots \frac{1}{7}\right)$ to payoffs where the members of each of the above coalitions share its value equally.

**5.** Let $(N, v)$ be defined by $N = \{1, 2, 3\}$, $v(\{i\}) = 0$, all $i$, $v(\{1, 2\}) = 4$, $v(\{1, 3\}) = 3$, $v(\{2, 3\}) = 2$, and $v(N) = 6$. Show that the kernel consists of the point $(3, 2, 1)$. Find the reduced games after removing each of the players, and check the consistency.

**6.** A weighted graph game is a TU game $(N, v)$, where

$$v(S) = \begin{cases} 0 & |S| < 2, \\ w_S & |S| = 2, \\ \sum_{T \subset S} w_T & |S| > 2. \end{cases}$$

(One may illustrate the game with a graph, where the nodes are players, and where edges between nodes are coalitions consisting of the relevant two players.

Show that the $\tau$-value of the game is the imputation $x = (x_1, \ldots, x_n)$ with $x_i = \frac{1}{2} \sum_j w_{\{i,j\}}$.

# Chapter 14

# Solutions of NTU Games: The Core

## 1  Introduction

We now return to general cooperative games as described in Chapter 10, where we introduced the characteristic function in its general form for games without side payment. Recall that a cooperative NTU game, or a cooperative game without side payment, is defined as a pair $(N, V)$, where $N = \{1, \dots, n\}$ is a finite set of players, and where $V$, the characteristic function of the game, assigns to each coalition $S \subset N$, $S \neq \emptyset$, a set $V(S) \subset \mathbb{R}^S$ of payoff vectors. We interpret elements of $x \in V(S)$ as utility payoffs, which can be obtained by the coalition $S$ for its members if they act in a coordinated way, which however is not specified.

We assume throughout in the following that the characteristic function has the property

$$[x \in V(S), \ y_i \leq x_i, \ \text{all } i \in S] \Rightarrow y \in V(S),$$

for all $S$, in the literature known as *comprehensiveness*, and furthermore it is usually assumed that $V(N)$ is closed and upper bounded.

As in the previous chapter we may define *payoff vectors* for the game $(N, V)$ as vectors $x \in \mathbb{R}^N$ (a utility for each player), as the outcome of the game. A payoff vector $x$ is *individually rational* if $x_i \geq \max V(\{i\})$ for every $i$ (so that player $i$ will not be better off by discontinuing the cooperation and acting on her own) and *Pareto optimal* if $x \in V(N)$ and there is no $y \in V(N)$, so that $y \geq x$, $y \neq x$. An *imputation* is (as previously) a payoff vector which is both Pareto optimal and individually rational.

Continuing our transfer of previously introduced concepts, we can extend the concept of dominance to NTU games: A payoff vector $x$ dominates another payoff vector $y$ if there is a coalition $S$ so that $x_S = (x_i)_{i \in S}$ belongs to $V(S)$ and $x_i > y_i$ for all $i \in S$. Using this definition, we can extend all solution concepts pertaining to TU games and involving only the notion of domination, to the more general class of NTU games. For example, *a vNM solution to $(N, V)$ is a subset $V$ of imputations*, which are internally and externally stable, meaning that they satisfy the conditions

$$\forall x \in V, \ \nexists \ y \in V : x \operatorname{Dom} y,$$

$$\forall y \notin V, \exists x \in V : x \text{ Dom } y.$$

As for TU games, the vNM solution of an NTU game is not unique, and the existence problem has not become simpler in the context of NTU games. There are few positive results about vNM solutions, so we turn instead to the other solution concept based on domination, the core.

## 2   The core, convex games

The core of an NTU game can as the vNM solution be defined immediately from the dominance relation, namely as the set of nondominated imputations,

$$\text{Core}(N, V) = \{x \in V(N) \mid x \text{ is Pareto optimal, } \nexists\, S, y : y \in V(S), y_i > x_i, i \in S\}.$$

Thus the core consists of the Pareto optimal payoff vectors which cannot be dominated by any coalition.

Our first task concerning the core of an NTU game is to investigate whether a given cooperative game has a nonempty core. For this it is convenient to introduce a slightly changed version of the characteristic function, which for every coalition takes values which are subsets of $\mathbb{R}^N$ (rather than $\mathbb{R}^S$). We obtain that simply by defining the *revised characteristic function* $\widetilde{V}$ by

$$\widetilde{V}(S) = V(S) \times \mathbb{R}^{N\backslash S},$$

for each coalition $S$, so that we assign arbitrary utility payoffs to the players which are anyway not members of $S$. From $\widetilde{V}$ the original characteristic function $V$ can be recovered by projecting $\widetilde{V}(S)$ on the coordinate subspace corresponding to the players in $S$, so the two representations of a cooperative game are equivalent, and $\widetilde{V}$ is the more useful one in the present context.

A cooperative NTU game $(N, V)$ is (ordinally) convex if

$$\widetilde{V}(S) \cap \widetilde{V}(T) \subseteq \widetilde{V}(S \cup T) \cup \widetilde{V}(S \cap T).$$

This condition says that if a payoff vector is such that its $S$-part can be realized by $S$ and its $T$-part by $T$, then either the intersection of the two can get this payoff vector (or rather the part corresponding to its coordinates) or the union can make it.

A remarkable property of the convex games is obtained by constructing a reduction of the game, which reminds of the construction of reduced games seen in our treatment of TU games. In the present situation the construction is even more convincing, since we reduce the game by removing a player without reference to a particular payoff vector.

Formally the *i-reduced game* of $(N, V)$ is defined as the game $(N\backslash\{i\}, V^{)i()}$, where

$$V^{)i(}(N) = V(N\backslash\{i\}) \cap \text{cl}\left\{x \in \mathbb{R}^{N\backslash\{i\}} \mid (x, x_i) \in V(N), x_i \geq \sup V(\{i\})\right\}$$
$$V^{)i(}(S) = V(S) \cup \text{cl}\left\{x \in \mathbb{R}^S \mid (x, x_i) \in V(S \cup \{i\}), x_i > \sup V(\{i\})\right\}$$

for $S \subset N\backslash\{i\}$, $S \neq \emptyset$. The reduced game has a straightforward interpretation: Suppose that the players consider bribing a player to retire from the contest. In order for the player to agree, she should receive at least $a_i = \sup V(\{i\})$. The remaining players may then proceed bargaining about all results which are such that player $i$ gets exactly the amount $a_i$, which is needed for keeping her out. This will give us the $i$-reduced game.

The following holds for the $i$-reduction of a convex game:

LEMMA 1 *Let $(N, V)$ be a convex NTU game, let $i \in N$, and let $(N\backslash\{i\}, V^{)i(})$ be the i-reduction of $(N, V)$. Then $(N\backslash\{i\}, V^{)i(})$ is convex.*

PROOF: Let $S$ and $T$ be arbitrary coalitions in $N\backslash\{i\}$ and assume that $x \in \widetilde{V}^{)i(}(S) \cap \widetilde{V}^{)i(}(T)$. We may assume that $S$ and $T$ both are different from $N\backslash\{i\}$ (since in the opposite case the condition for convexity is satisfied trivially). There are four possibilities, since we have

either (a) $(x, a_i) \in \widetilde{V}(S)$
or (b) $(x, x'_i) \in V(S \cup \{i\})$ for some $x'_i > a_i = \sup V(\{i\})$

and similarly

either (c) $(x, a_i) \in \widetilde{V}(T)$
or (d) $(x, x''_i) \in V(T \cup \{i\})$ for some $x''_i > a_i = \sup V(\{i\})$.

We consider the case where (b) and (d) are satisfied; the other combinations are checked in the same way.

If $S \cap T = \emptyset$, we have $(S \cup \{i\}) \cap (T \cup \{i\}) = \{i\}$. But no $(x, x_i)$ with $x_i > a_i$ is in $\widetilde{V}(\{i\})$, so the convexity of $V$ gives us that $(x, \tilde{x}_i) \in \widetilde{V}(S \cup T \cup \{i\})$ for $\tilde{x}_i \leq \min\{x'_i, x''_i\}$. Since $\tilde{x}_i > a_i$, we have $x \in V^{)i(}(S \cup T)$ according to the definition of $V^{)i(}$.

If $S \cup T \neq \emptyset$ we have

$$(x, x_i) \in \widetilde{V}(S \cap T) \cup \widetilde{V}(S \cup T \cup \{i\}),$$

for $\tilde{x}_i \leq \min\{x'_i, x''_i\}$, from which it follows that $x \in \widetilde{V}^{)i(}(S \cap T) \cup \widetilde{V}^{)i(}(S \cup T)$. □

This stability of the class of convex games under the operation of reduction has an important consequence [Greenberg, 1985], which is the reason that we investigate this concept.

THEOREM 1 *Let $(N, V)$ be a convex NTU game. Then the core of $(N, V)$ is nonempty.*

PROOF: We use induction on $n$, the number of players. For $n = 1$ and $n = 2$ the result is straightforward. Assume that it has been shown for all games with at most $n - 1$ players, $n \geq 3$, and consider an arbitrary convex game $(N, V)$ with $|N| = n$.

Choose $i \in N$ arbitrarily, and let $a_i = \sup V(\{i\})$. By the induction hypothesis we have that $\text{Core}(N\backslash\{i\}, V^{)i(})$ is nonempty. Choose $x \in \text{Core}(N\backslash\{i\}, V^{)i(})$; it is now easy to check that $(x_i, a_i) \in \text{Core}(N, V)$. □

It may be noticed that the proof implicitly gives a method for finding elements of the core, namely by successive elimination of players in a given order, whereby one ends up with a non-dominated payoff. From this payoff one can then move to another one, also non-dominated, which is Pareto optimal. This procedure parallels the approach that we used for convex TU games.

## 3 Balanced games

As it was to be expected from our results about TU games, the result about nonemptiness of the core for convec games can be extended to a larger class of games. It seems logical to attempt a generalization of the balancedness condition which could be used for NTU games.

As before there are several possibilities which all appear in the literature. The most direct generalization is the so-called cardinal balancedness. Recall (from Section 3.3 of Chapter 12) that a family $C$ of coalitions is balanced if there are weights $(\lambda_S)_{S \in C}$ such that $\sum_{S \in C, i \in S} \lambda_S = 1$ for all $i$, or alternatively, if

$$e_N \in \text{conv}\left(\{e_S \mid S \in C\}\right),$$

where for each $S \subseteq N$, $S \neq \emptyset$, the vector $e_S \in \mathbb{R}^N$ has coordinates $(e_S)_i = 1$ for $i \in S$ and $(e_S)_i = 0$ for $i \notin S$. An NTU game $(N, V)$ is cardinally balanced if for every balanced family $C$ with weights $(\lambda_S)_{S \in C}$,

$$\sum_{S \in C} \lambda_S \widetilde{V}(S) \subseteq V(N).$$

This class appears as the most immediate extension of balancedness from TU to NTU games; however it is not the one which yields the most interesting results, so we proceed to another one.

The alternative approach to balanced games (also called ordinally balanced games) is as follows: The game $(N, V)$ is balanced if

$$\bigcap_{S \in C} \widetilde{V}(S) \subseteq \widetilde{V}(N)$$

for every balanced family $C$.

The basic result about balanced NTU games is due to Scarf (1967):

THEOREM 2 *Let $(N, V)$ be a balanced NTU game. Then $\text{Core}(N, V)$ is nonempty.*

The proof of this theorem is obtained by a slight detour, since the crucial part is the following much less intuitive statement, referring to an arbitrary game, balanced or not.

LEMMA 2 *Let $(N, V)$ be an NTU game with $\sup V(\{i\}) > 0$ for all $i$. If $\text{Core}(N, V)$ is empty, then there is a balanced family $C$ of coalitions and a Pareto optimal payoff vector $x \in V(N)$ such that $x$ is dominated by all $S \in C$.*

PROOF: For every coalition $S$ we define the set

$$G_S = \{z \in \Delta_N \mid \forall x \in V(N) : [z = x/\|x\|] \Rightarrow \exists y \in V(S), y_i > x_i, i \in S\},$$

consisting of the payoffs, normalized so that their sum is 1, which are dominated by something in $V(S)$. Every $G_S$ is an open set, and since the core of $(N, V)$ is empty, every payoff vector must be dominated, so that every vector $z$ from $\Delta_N$ must belong to some $G_S$. The family $(G_S)_{S \subset N, S \neq N}$ is therefore an open covering of $\Delta_N$. Since $\sup V(\{i\}) > 0$, every payoff vector $x$ with $x_i = 0$ will be dominated by $\{i\}$ and therefore belong to $G_{\{i\}}$.

The result now follows directly from Lemma 3 below. $\qquad \square$

The following lemma generalizes a wellknown result (the "KKM lemma") by the three mathematicians Knaster, Kuratowski and Mazurkiewicz [Knaster et al., 1929].

LEMMA 3 (Generalized KKM-lemma) *Let $(G_S)_{S \subset N, S \neq N}$ be an open covering of $\Delta_N$ such that $\Delta_{N \setminus \{i\}}$ is covered by $G_{\{i\}}$, all $i \in N$. Then there is a balanced family $C$ of coalitions, so that*

$$\bigcap_{S \in C} G_S \neq \emptyset.$$

PROOF: Define a new covering $(\widetilde{G}_S)_{S \subset N, S \neq N}$ by

$$\widetilde{G}_{\{i\}} = G_{\{i\}}, \quad \widetilde{G}_S = G_S \cap \{z \in \Delta_N \mid \forall i \in N : x_i > \varepsilon\}, \quad |S| > 1,$$

where $\varepsilon > 0$ is chosen such that all $x \in \Delta_N$ with $x_i \leq \varepsilon$ are contained in $G_{\{i\}}$, $i = 1, \ldots, n$. For every $S$ we may choose a continuous function $\psi_S : \Delta_N \to [0, 1]$ such that $\psi_S(z) > 0$ for $z \in \widetilde{G}_S$ and $\psi_S(z) = 0$ for $z \notin \widetilde{G}_S$. We may assume that the $\psi_S$ are normalized such that

$$\sum_{S \subset N, S \neq N} \psi_S(z) = 1$$

for all $z \in \Delta_N$. Consider now the mapping $g : \Delta_N \to \Delta_N$ given by

$$g(z) = \sum_{S \subset N, S \neq N} \psi_S(z) e_S.$$

We claim that there exists an $z^0$ with $g(z^0) = e_N$; if this is true, then the family $\{S | \psi_S(z) > 0\}$ is balanced and dominates exactly the Pareto optimal payoff corresponding to the vector $z^0$.

To prove the claim we use the map $h : \Delta_N \to \mathbb{R}^N$ given by

$$h(z) = z + \mu[g(z) - e_N], \quad \mu > 0$$

which moves the point $z$ in the direction given by the difference between $g(z)$ and $e_N$. To obtain a mapping which sends $\Delta_N$ into itself, we compose it with the function $k : \{z \in \mathbb{R} \mid \sum_{i \in N} z_i = 1\} \to \Delta_N$ defined by

$$k(z) = \begin{cases} z & z \in \Delta_N, \\ e_N + \lambda(z)(z - e_N) & \text{otherwise} \end{cases}$$

where $\lambda(z) = \min\{\lambda \mid e_N + \lambda(z - e_N) \in \triangle_N\}$ (so that the vector $z$ is pushed against $e_N$ untill it reaches $\triangle_N$. It is easily seen that $k$ is continuous; the composed map $f = k \circ h$ is therefore also continuous. We also see that $f(z) = h(z)$ for all $z$ in the interior of $\triangle_N$.

We then use Brouwer's fixed point theorem on $f : \triangle_N \rightarrow \triangle_N$, whereby we obtain a point $z^0$ with $f(z^0) = z^0$. Assume first that $z^0$ is on the boundary of $\triangle_N$. From our construction we have that only the functions $\psi_{\{i\}}$ for $i \in N$ can be different from 0. Let $I = \{i \mid \psi_{\{i\}}(z^0) > 0\}$; if $I = N$ we have a balanced family, namely $(\{i\})_{i\in N}$, which dominates $z^0$, and we are done. Otherwise there is at least one $j \in N$, which is not in $I$, so that the vectors $(e_{\{i\}})_{i\in I}$ belong to a proper subsimplex of $\triangle_N$. From our construction we know that $z_i^0 = 0$ implies $i \in I$, so $g(z^0)$ cannot belong to the same $(n-1)$-dimensional subsimplex as $z^0$, and consequently $h(z^0)$ cannot be situated on the line through $e_N$ and $z^0$. We conclude that $z^0$ is not a boundary point.

If $z^0$ belongs to the interior of $\triangle_N$, we have $z^0 = h(z^0)$, from which we get that $g(z^0) = e_N$, which as earlier noticed gives us the desired balanced family of coalitions with nonempty intersection.                                    $\square$

Now we can easily get Scarf's theorem:

PROOF OF THEOREM 2: We may assume that $\sup V(\{i\}) = a_i > 0$ for all 0 (otherwise we add a sufficiently large positive vector $a$ to all $\overline{V}(S)$, which does not change the question of existence of a nonempty core).

Consider a sequence of games $(N, V^t)_{t=1}^\infty$ such that $V^t(S) = V(S)$ for all $t$ and

$$V^t(N) = \left(1 + \frac{1}{t}\right) V(N).$$

For every $t$ we have that $\text{Core}(N, V^t) \neq \emptyset$. If namely the core was empty, according to Lemma 3 there would be a balanced family $C$ and a Pareto optimal payoff vector $x$ dominated by all $S \in C$. Choose a vector $y$ with $y_i > x_i$, all $i$, such that $y \in \overline{V}^t(S)$ for all $S \in S$. Since $(N, V)$ is balanced, we have

$$y \in \bigcap_{S\in C} \widetilde{V}^t(S) = \bigcap_{S\in C} \widetilde{V}(S) \subseteq V(N),$$

which contradicts that $y$ is bigger in all coordinates than $x$, which was Pareto optimal in $\left(1 + \frac{1}{t}\right) V(N)$.

We conclude that for each $t$ there is $x^t \in \text{Core}(N, V^t)$. Since the sequence $(x^t)_{t=1}^\infty$ can be chosen from a compact subset of $\mathbb{R}^N$, it will have a subsequence converging to some $x^0 \in V(N)$. It is seen immediately that $x^0 \in \text{Core}(N, V)$.                                    $\square$

For later use we notice that the generalized KKM-lemma holds also for coverings $\{G_S^i \mid i \in I_S, S \subset N, S \neq N\}$, where $I_S$ is a finite index set (so that there can be several open sets for each $S$). The proof is as above and gives a balanced family of coalitions $C$ and a $z^0$ contained in one of the sets $G_S^i$, for each $S \in C$.

## 4   Extensions of Scarf's theorem

As it can be seen from the formulation of Theorem 2, the balancedness condition is a sufficient but not necessary condition for nonemptyness of the core, so that the situation for NTU games is different from that of TU games. This is not just a question of proving the converse; it is easily seen that many games that are not balanced (among which some convex games) have nonempty core.

The logical next step is therefore to consider possible improvements of the result of Theorem 2. The first type of improvement pertains to the notion of balanced families of coalitions. Let $\pi = (\pi_S)_{S \subset N}$ be a family of coalitions from $\Delta_N$ such that $\pi_{\{i\}} = e_{\{i\}}$ for all $i$. A family of coalitions $C$ is $\pi$-balanced if

$$\pi_N \in \operatorname{conv}(\{\pi_S | S \in C\}).$$

An NTU game $(N, V)$ is $\pi$-balanced if

$$\bigcap_{S \in C} \widetilde{V}(S) \subseteq V(N)$$

for every $\pi$-balanced family $C$.

The method of the previous section can be applied with only notational changes to obtain the following generalization, which is due to Billera [1970]:

THEOREM 3 *Let $(N, V)$ be $\pi$-balanced. Then $\operatorname{Core}(N, V) \neq \emptyset$.*

Still this does not give us conditions which are both necessary and sufficient: There are examples of games which are not $\pi$-balanced for any $\pi$ but which have nonempty core. In order to get both necessary and sufficient conditions one has to generalize the balancedness condition slightly more, and in addition take into account that there are some rather simple operations on games under which the property of having a nonempty core is respected, namely the following:

LEMMA 4 *(i) Assume that $\operatorname{Core}(N, V) \neq \emptyset$ and $(N, V')$ are such that $V(S) \subseteq V'(S)$ for all $S$, $V(N) = V'(N)$. Then $\operatorname{Core}(N, V') \neq \emptyset$.*

*(ii) Let $(N, V^t)_{t=1}^{\infty}$ be a sequence of NTU games with $\operatorname{Core}(N, V^t) \neq \emptyset$, all $t$, and let $(N, V)$ be a game such that $V^t(N) = V(N)$ all $t$, and $\operatorname{cl} V^t(S) \to \operatorname{cl} V(S)$ for $t \to \infty$, all $S$. Then $\operatorname{Core}(N, V) \neq \emptyset$.*

We shall not go into details about convergence of the closed sets $\operatorname{cl} V^t(S)$ (one may restrict to suitable compact subsets and then use the Hausdorff metric to define convergence), since the message is quite intuitive. Both of the statements of the lemma are straightforward and their proofs are left to the reader.

We need a slight elaboration of the balancedness condition, proposed by Keiding and Thorlund-Petersen [1987]: Let $>$ be a relation defined on all coalitions, such that $>$ is acyclic and the coalitions $\{i\}$, $i \in N$ are all minimal for $>$. A game $(N, V)$ is said to be $(\pi, >)$-balanced if for every $\pi$-balanced family $C$ we have that

$$x \in \bigcap_{S \in C} \operatorname{cl} \widetilde{V}(S),$$

and if $x \notin \text{int } \widetilde{V}(S)$ for every coalition $S$ such that $S > T$ for a $T \in C$, then $x \in V(N)$.

As it can be seen, $(\pi, >)$-balancedness is a weaker condition than $\pi$-balancedness, since a payoff which is dominated by all coalitions in a $\pi$-balanced family, cannot necessarily be sustained by $N$. This is the case only when it is dominated by a $\pi$-balanced family, all coalitions of which are $>$-minimal among the dominating coalitions.

It may be shown that $(\pi, >)$-balanced games have nonempty core. The approach is the by now standard one, used in the previous section, whereby some care must be taken, since the coverings used in the proof do not necessarily consist of open sets. We shall not consider these details, since anyway the concept of $(\pi, >)$-balancedness is lacking in intuitive content.

It is more interesting to notice that by the introduction of the constructions from Lemma 4 one gets all the games with nonempty core. To get a short formulation of this statement, we introduce the notion of a weakly $(\pi, >)$-balanced game for an NTU game $(N, V)$, which is the limit (in the sense of Lemma 4) of a sequence $(N, V^t)_{t=1}^{\infty}$ of $(\pi, >)$-balanced games. The following holds:

THEOREM 4 *Let $(N, V)$ be an NTU game. Then $Core(N, V) \neq \emptyset$ if and only if there is $\pi$, $>$, and a weakly $(\pi, >)$-balanced game $(N, V')$ such that $V(S) \subseteq V'(S)$, all $S$, and $V(N) = V'(N)$.*

For another approach to the problem of finding necessary *and* sufficient conditions for nonemptiness of the core, see Predtetchinski and Herings [2004].

## 5   The partnered core

In this section, we consider payoffs in the core which have some further properties. More specifically, we shall be concerned with the *partnered core*, cf. Reny and Wooders [1996].

Let $(N, V)$ be an NTU game; we assume in this section that $V(S)$ is closed for every coalition $S$. A payoff vector $x \in \mathbb{R}^N$ is sustained by the coalition $S$, if $x \in \widetilde{V}(S)$. We shall have a closer look at core elements sustained by particular families of coalitions.

Let $C$ be a family of coalitions from $N$. For each $i \in N$ we may consider the set $P_C(i)$ of $i$'s partners in $C$, defined by

$$P_C(i) = \{j \in N | \forall S \in C : i \in S \Rightarrow j \in S\}.$$

In other words, $j$ is $i$'s partner if $j$ is in each coalition in which $i$ participates.

If the family $C$ satisfies the condition

$$\forall i, j \in N : j \in P_C(i) \Rightarrow i \in P_C(j),$$

so that every player is a partner of her partners, then the family $C$ is said to be partnered. A possible interpretation of this condition goes as follows: If a given payoff vector $x$ is sustained by a partnered family (which means that $\{S \mid x \in \widetilde{V}(S)\}$ is partnered), then there is a stable agreement about $x$, since nobody can claim a larger share than any other player with the argument that she is more important for achieving the payoff vector $x$.

We can now define the partnered core $\mathrm{Core}_P(N, V)$ as the set of payoff vectors in $\mathrm{Core}(N, V)$, which are sustained by a partnered family. In the remaining part of this section we shall consider the question of whether the partnered core is nonempty. This is not a consequence of the previous results, since we have added an additional condition on the family of coalitions.

We would like to have a result saying that balanced games have a nonempty partnered core. Without further conditions, this is however not true. Consider the game $(\{1, 2, 3\}, V)$, where

$$V(\{1, 2, 3\}) = \{x \in \mathbb{R}^3 \mid x_i \leq 1, i = 1, 2, 3\}$$
$$V(\{2, 3\}) = \{(x_2, x_3) \in \mathbb{R}^2 \mid x_i \leq 1, i = 2, 3\}$$
$$V(\{1, 3\}) = \{(x_1, x_3) \in \mathbb{R}^2 \mid x_i \leq 1, i = 1, 3\}$$
$$V(\{1\}) = \{x_1 \in \mathbb{R} \mid x_1 \leq 1\},$$

and $\widetilde{V}(S) = \mathbb{R}^3_-$ for all other coalitions. The only element of the core is $(1, 1, 1)$, which is sustained by $\{\{1, 2, 3\}, \{1\}, \{1, 3\}, \{2, 3\}\}$. This family is not partnered. If 2 is in a coalition from the family (that is in the coalitions $\{1, 2, 3\}$ and $\{2, 3\}$), then so is 3. But 3 participates in a coalition without 2, namely $\{1, 3\}$. We conclude that the partnered core is empty.

However, this game is balanced: The only balanced family which can be made of coalitions for which $V(S)$ is nontrivial, is $\{\{1\}, \{2, 3\}\}$. If $x \in V(S)$ for both $S$ in this family, we have $x_1 \leq 1$ og $x_2, x_3 \leq 1$, so that $x \in V(\{1, 2, 3\})$.

To get a result about existence of a partnered core in balanced games we need further assumptions. It turns out that we can do with assumptions which exclude that $V(N)$ contains core elements which Pareto dominates other payoff vectors on the boundary of $V(N)$, which in their turn are not dominated by any coalition. This is a rather weak assumption about the shape of $V(N)$; we shall see later that assumptions about $V(N)$ play an important role in several connections.

A family $C$ of coalitions is said to be *strictly balanced* if it is balanced with strictly positive balancing weights, or alternatively, if $e_N$ belongs to the relative interior of the convex hull of the vectors $e_S$ for $S \in C$,

$$e_N \in \mathrm{ri} \, \mathrm{conv}(\{e_S \mid S \in C\}).$$

LEMMA 5 *Let $(N, V)$ be an NTU game.*

(i) *If a family $C$ of coalitions in $(N, V)$ is strictly balanced, then it is partnered.*

(ii) *If $C_1$ and $C_2$ are strictly balanced, then $C_1 \cup C_2$ is strictly balanced.*

PROOF: (i) Let $(\lambda_S)_{S \in C}$ be balancing weights of $C$. Assume that $j \in P_C(i)$, $j \neq i$, so that for every $S \in C$ that $i \in S \Rightarrow j \in S$. Then

$$\sum_{S \in C : j \in S} \lambda_S \geq \sum_{S \in C : i \in S} \lambda_S = 1. \tag{1}$$

If there is $T \in C$ such that $j \in T$ but $i \notin T$, then by (1), $\lambda_T = 0$, contradicting the assumptions. We conclude that $i \in P_C(j)$.

(ii) Choose systems of positive weights $(\lambda_S^1)_{S \in C_1}$ and $(\lambda_S^2)_{S \in C_2}$, and define $\lambda_S$ for $S \in C_1 \cup C_2$ by

$$\lambda_S = \begin{cases} \frac{\lambda_S^1}{2} & S \in C_1 \backslash C_2 \\ \frac{\lambda_S^2}{2} & S \in C_2 \backslash C_1 \\ \frac{\lambda_S^1}{2} + \frac{\lambda_S^2}{2} & S \in C_1 \cap C_2 \end{cases}$$

We then have

$$\sum_{S \in C_1 \cup C_2 : i \in S} \lambda_S = \sum_{S \in C_1 \backslash C_2 : i \in S} \frac{\lambda_S^1}{2} + \sum_{S \in C_2 \backslash C_1 : i \in S} \frac{\lambda_S^2}{2} + \sum_{S \in C_1 \cap C_2 : i \in S} \left[ \frac{\lambda_S^1}{2} + \frac{\lambda_S^2}{2} \right]$$

$$= \frac{1}{2} \sum_{S \in C_1 : i \in S} \lambda_S^1 + \frac{1}{2} \sum_{S \in C_2 : i \in S} \lambda_S^2 = 1$$

for every $i \in N$.                                                                                    □

The lemma can be put to the following use: Let $x \in V(N)$ be a payoff vector. Then there is a maximal strictly balanced family $C_x$ of coalitions which sustains $x$ (so that this family cannot be extended with one or more coalitions). This follows immediately from the fact that the union of all strictly balanced families sustaining $x$ is itself a strictly balanced family according to the lemma.

THEOREM 5 *Let* $(N, V)$ *be a balanced NTU game. Assume that for all* $x \in \text{Core} V(N)$ *there are no non-dominated payoff vectors* $y \neq x$ *in* $\text{bd} V(N)$ *such that* $y_i \leq x_i$*, all i. Then* $\text{Core}_P(N, V) \neq \emptyset$.

PROOF: Let $\mathfrak{C}$ be the set of all strictly balanced families $C$ of coalitions in $(N, V)$ and let

$$V^* = \bigcup_{C \in \mathfrak{C}} \bigcap_{S \in C} \widetilde{V}(S).$$

Define the game $(N, V^*)$ by

$$V^*(N) = V^*, \quad V^*(S) = V(S) \text{ for } S \neq N.$$

For each $x \in \text{Core}(N, V^*)$ we have that if $x \notin \text{Core}(N, V)$, then there is an imputation $x' \in \text{Core}(N, V)$ which Pareto-dominates $x$, and by our assumption we must have $x_i' > x_i$ for all $i$. It then follows that $\{N\}$ is the only family of coalitions sustaining $x'$ (otherwise $x$ would be dominated), so that $x'$ belongs to $\text{Core}_P(N, V)$. As a consequence, it is enough to show that $\text{Core}_P(N, V*) \neq \emptyset$. For notational convenience, we assume in the following that $V = V^*$.

Suppose that $\text{Core}_P(N, V) = \emptyset$. Since $(N, V)$ is balanced, the core of $(N, V)$ is nonempty, so that each $x \in \text{Core}(N, V)$ must belong to $\tilde{V}(S)$ for some coalition $S$ not in $C_x$. Let $X$ be the boundary of $V(N)$ relative to $\mathbb{R}_+^N$ and construct a covering of $X$ as follows: For each $S$ we let $G_S$ be all the payoff vectors in $X$, which are dominated by $S$. Further, for each $S$ let $F_S$ be given by

$$F_S = \{x \in \text{Core}(N, V) \mid x \in V(S), S \notin C_x\}.$$

Then $X$ is covered by the sets $G_S$ and $F_S$, for $S \subset N, S \neq N$.

For each $\varepsilon > 0$, let

$$B_S^\varepsilon = \{x \in X \mid \exists y \in F_S : \|x - y\| < \varepsilon\},$$

and let $D_S^\varepsilon = G_S \cup F_S$. The sets $D_S$ constitute an open covering of $X$, and by the generalized KKM-lemma there are $x^\varepsilon \in X$ and a balanced family $C^\varepsilon$ such that $x^0 \in \cap_{S \in C^\varepsilon} D_S^\varepsilon$. Replacing by a subfamily if necessary we may assume that $C^\varepsilon$ is strictly balanced. We cannot have $x^\varepsilon \in G_S$ for all $S \in C^\varepsilon$, since $(N, V)$ is balanced, so there is $S^\varepsilon \in C^\varepsilon$ such that $x^\varepsilon \in B_{S^\varepsilon}^\varepsilon$.

Let $\varepsilon$ go to 0. Taking a subsequence if necessary, we get that $(x^\varepsilon)$ converges to some $x^0$ and that $C^\varepsilon$ becomes constant, with $C'\varepsilon = C^0$ and $S^\varepsilon = S^0$, from a certain step, we get in the limit that $x^0 \in F_{S^0}$ is sustained by a strictly balanced family $C^0$ containing $S^0$, a contradiction. We conclude that $\text{Core}_P(N, V) \neq \emptyset$. $\qquad\square$

It may be noticed that the proof exploited balancedness several times, in particular it was used at the beginning that strictly balanced families of coalitions are also partnered. Therefore, the result does not generalize in any straightforward way to $\pi$-balanced games.

## 6   Axiomatic characterization of the core

At several occasions during our discussion of solutions for TU games, we have considered axiomatic characterization of solutions. An axiomatic foundation for a solution is an advantage in cases where the solution itself is not intuitively clear or does not differ in an obvious way from other solutions, since the axioms specify the properties which should be satisfied by the solution in question.

From this point of view it may perhaps not be self-evident that we should look for an axiomatic characterization of the core, which stands out from the other solutions by having a very intuitive background. But one may view the axiomatization as a general research program which aims at constructing a system of solutions to cooperative games, whereby the individual solutions emerge when one combines axioms from suitable – and presumably small – collection of axioms. So far the program is far from being carried out fully. Historically the interest in axiomatic characterizations of NTU games is relatively recent, beginning with the characterization by Aumann [1985] of the Shapley NTU value, to which we return later.

As earlier we understand by a solution a mapping $\Phi$, which to every game $(N, V)$ assigns a subset $\Phi(N, V)$ of $V(N)$. As it can be seen, we allow for the solution of a given game to be a set rather than a single payoff vector, as indeed we have to do if it shall apply to the core, which is seen above usually consists of more than one payoff vector. The class $\mathcal{G}$ of games for which $\Phi$ is defined must be sufficiently big to allow for all the constructions that we need in setting up the axioms. In particular, $\mathcal{G}$ contains games with arbitrary numbers of players.

We now turn to a description of the axioms for $\Phi$:

A game $(N, V)$ is trivial if

$$V(S) = \{x \in \mathbb{R}^S \mid x_i \leq 0, i \in S\}$$

for all $S \neq N$.

AXIOM 1. (Triviality) *If $(N, V)$ is trivial and $V(N) \cap \mathbb{R}^N_+ = \{0\}$, then $\Phi(N, V) = \{0\}$.*

This axiom seems uncontroversial. There are few other reasonable solutions to a trivial game.

AXIOM 2. (Strategic equivalence) *For each game $(N, V)$ and each vector $a \in \mathbb{R}^N$,*

$$\Phi(N, V + \{a\}) = \Phi(N, V) + \{a\}.$$

Here $(N, V + \{a\})$ is the game given by $(V + \{a\})(S) = V(S) + \{a_S\}$ for all $S$. S This is again an axiom, which can hardly give rise to problems, since in most cases it is considered as desirable that a solution is stable under this type of transformations.

AXIOM 3. (Antimonotonicity) *If $(N, V)$ and $(N, U)$ are games with $V(S) \subseteq U(N)$ for all $S$, and $V(N) = U(S)$, then $\Phi(N, U) \subseteq \Phi(N, V)$.*

With this axiom we have introduced one of the more essential demands to the solution; intuitively the axiom says that games with powerful coalitions have few solutions.

AXIOM 4. (Continuity) *If $(N, V), (N, U)$ is a game with $\mathrm{cl} V(S) = \mathrm{cl} U(S)$, all $S$, and $V(N) = U(N)$, then $\Phi(N, V) = \Phi(N, U)$.*

The axiom says that if we in a given game take the closure of all sets $V(S)$, then the solution is unaffected. This again seems acceptable, in particular since details about closedness of sets $V(S)$ have a technical nature, and they are not easy to decide upon in practice. However, the axiom turns out to be rather central for the argumentation in the sequel, so it may be appropriate to be aware of its presence. The last axiom is probably the least self-evident one, and on the other side the one which most decidedly characterizes the core. We need some additional notions.

For a given abstract solution $\Phi$ we can associate with each game $(N, V)$ its $\Phi$-derived game $(N, D_\Phi V)$ in the following way: For coalitions $S$ with $|S| = 1$, such that $S = \{i\}$ for some $i \in N$, we put $(D_\Phi V)(\{i\}) = V(\{i\})$. Assume that $(D_\Phi V)(S')$ has been defined for all coalitions $S' \subseteq S$ with $S' \neq S$. We then have a game $(S, V')$ defined by

$$V'(S') = \begin{cases} \mathrm{cl}V(S) & S' = S, \\ (D_\Phi V)(S') & \text{otherwise.} \end{cases} \qquad (2)$$

The set $(D_\Phi V)(S)$ can now be defined as

$$(D_\Phi V)(S) = \mathrm{cl}\{x' \in \mathbb{R}^S | \exists x \in \Phi(S, V') : x_i' \leq x_i, \ i \in S\}.$$

If $(D_\Phi V)(S')$ is not defined for some $S' \subset S$, then $(D_\Phi V)(S)$ is not defined.

The $\Phi$-derived game $(N, D_\Phi V)$ can be seen as a result of cleaning the sets $V(S)$ of elements, which are irrelevant in the sense that the coalition could not attain them without getting problems with its own subcoalitions, at least if $\Phi$ must be used as a means of achieving a compromise, both when negotiating with $N$ and in the internal argumentations of the coalition. Otherwise put, if the arguments of coalition $S$, pertaining to its ability to achieve a certain payoff vector $y$, is to carry any weight with the grand coalition, then the attempts by $S$ to achieve $y$ should not be met with opposition by some of $S$'s own subcoalitions, using the same arguments against $S$ as those of $S$ against $N$.

This interpretation of the $\Phi$-derived game gives the background for the last of our axioms, namely the following:

AXIOM 5. (Independence of $\Phi$-irrelevant alternatives) *If the game $(N, V)$ is such that $(N, D_\Phi V)$ is defined, then $\Phi(N, D_\Phi V) = \Phi(N, V)$.*

To characterize solutions satisfying axioms $1 - 5$, we begin by considering a particular type of games with little practical relevance, which however turns out to be useful here. A game $(N, V)$ is a *corner game* if there is $a \in \mathbb{R}^N$ so that

$$V(S) = \{x \in \mathbb{R}^S \mid \exists i \in S : x_i < a_i\}$$

for all $S \neq N$. The vector $a$ is then the *corner* of $(N, V)$ (the name comes from the geometric appearance of the game, cf. Fig. 1).

LEMMA 6 *Suppose that $\Phi$ satisfies the axioms 1,2,4,and 5. Let $(N, V)$ be a corner game with corner $a$, such that $V(N) \cap \{x \in \mathbb{R}^N \mid x_i \geq a_i\} = \{a\}$. Then $\Phi(N, V) = \{a\}$.*

PROOF: By Axiom 2 we may assume that $a = 0$. We may also assume that $V(S) \backslash \mathbb{R}_+^S = \{0\}$ for all $S$ (since by Axiom 4 we can remove part of the boundary of each $V(S)$ for $S \neq N$ without changing $\Phi$'s values).

We begin by looking at the trivial game $(N, U)$ with $U(N) = V(N)$. By Axiom 1, $\Phi(N, U) = \{0\}$. We now claim that $(N, D_\Phi V) = (N, U)$. For coalitions $S'$ with $|S'| = 1$ we clearly have that

$$(D_\Phi V)(S') = U(S') = \{x \in \mathbb{R}^{S'} \mid x_i \leq 0, \ i \in S'\},$$

so that our claim is true. Suppose that it is true for all proper subcoalitions of $S$, where $S \neq N$. Then the game $(S, V')$ in (1) is trivial, and $V'(S) \cap \mathbb{R}_+^S = \{0\}$, so by

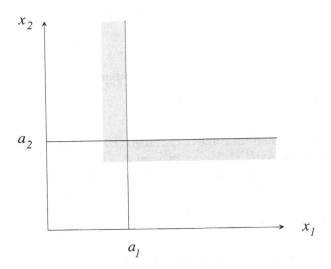

**Fig. 1.** A corner game: The set $V(\{1,2\})$ consists of all vectors $(x_1, x_2)$ in $\mathbb{R}^2$ except those for which $x_1 \geq a_1$ and $x_2 \geq a_2$.

Axiom 1 $\Phi(S, V') = \{0\}$, from which $(D_\Phi V)(S) = \{x \in \mathbb{R}^S | x_i \leq 0,\ i \in S\}$, proving our claim.

Altogether we therefore have that $\Phi(N, U) = \{0\} = \Phi(N, D_\Phi V) = \Phi(N, V)$, where the last equality sign comes from applying Axiom 5. $\qquad\square$

It is now easy to obtain the following result.

THEOREM 6 *Let $\Phi$ be a solution which satisfies Axioms 1 – 5, and assume that $\Phi$ is minimal among all such solutions. Then $\Phi = \mathrm{Core}$.*

PROOF: It is easily checked that Core satisfies Axioms 1 – 5. Conversely, let $\Phi$ be a solution satisfying the axioms. We show that $\Phi(N, V) \subset \mathrm{Core}(N, V)$ for all games $(N, V)$.

If $(N, V)$ is a game with empty core, the inclusion is trivial, so we may assume that the core is nonempty, For each $a \in \mathrm{Core}(N, V)$ we consider the corner game $(N, V^a)$ with corner $a$ and $V^a(N) = V(N)$. By Lemma 6 we have that $\Phi(N, V^a) = \{a\}$. Since $a$ belongs to $\mathrm{Core}(N, V)$, no coalition $S$ dominates $a$, and therefore

$$V(S) = \{x \in \mathbb{R}^S \mid \exists i \in S : x_i \leq a_i\}$$

for all $S \neq N$. From Axiom 3 we then obtain that $\Phi(N, V^a) \subset \Phi(N, V)$, i.e. $a \in \Phi(N, V)$. As $a$ was chosen arbitrarily in $\mathrm{Core}(N, V)$, it follows that $\mathrm{Core}(N, V) \subset \Phi(N, V)$. $\qquad\square$

The axiomatization of the core shown here has been supplemented by several others, for more details the reader is referred to Peleg and Südholter [2007].

## 7 Problems

**1.** A game $(N, V)$ is exact if for each coalition $S$, there is $x \in \text{Core}(N, V)$ such that $x$ belongs to the boundary (relative to $\mathbb{R}_+^S$) of $V(S)$.

   Show that if $(N, V)$ is convex, then it is exact.

**2.** Give examples of NTU games which are

   (1) convex and not balanced,
   (2) balanced and not convex,
   (3) neither convex nor $\pi$-balanced for any $\pi$ but have nonempty core.

**3.** Let $(N, V)$ be an NTU game and let $q \in \mathbb{R}_+^N, q \neq 0$. Define a TU game $v_q$ by

$$v_q(S) = \sup_{x \in V(S)} q_S \cdot x,$$

where $q_S = (q_i)_{i \in S}$, and show that if $x \in \text{Core}(N, v_q) \cap V(N)$, then $x \in \text{Core}(N, V)$. Find conditions on the sets $V(S)$ such that $\text{Core}(N, V) = \text{Core}(N, v_q) \cap V(N)$.

**4.** Let $(N, V)$ be an NTU game. Define the *extended core* of $(N, V)$, denoted $\text{Core}_e(N, V)$ as follows: Let $\Lambda(N, V)$ be the set of numbers $\lambda \geq 1$ such that the game $(N, V_\lambda)$ with $V_\lambda(S) = V(S)$ for $S \neq N$ and $V_\lambda(N) = \lambda V(N)$ has nonempty core, and let

$$\text{Core}_e(N, V) = \bigcap_{\lambda \in \Lambda(N,V)} \lambda^{-1} \text{Core}(N, V_\lambda).$$

Find conditions on $(N, V)$ such that it has a nonempty extended core.

**5.** (The core of an exchange economy) Consider a society of $n$ consumers having endowments $\omega_i \in \mathbb{R}^\ell$ of $\ell$ different commodities and utility functions $u_i$ defined on bundles $x_i \in \mathbb{R}_+^\ell$, all $i$. An allocation is an array $(x_1, \ldots, x_n)$ of bundles with $\sum_{i=1}^n x_i = \sum_{i=1}^n \omega_i$. A coalition $S$ can *improve* an allocation $(x_1, \ldots, x_n)$ if it can reallocate its endowments so as to make every coalition member better off, and the *core* of the economy consists of the allocations for which no coalition has an improvement. Write down the formal definition of the core.

   An allocation $(x_1, \ldots, x_n)$ is an *equilibrium* at the price system $p \in \mathbb{R}_+^\ell, p \neq 0$, if $\sum_{i=1}^n x_i = \sum_{i=1}^n \omega_i$ and for each $i$, $p \cdot x_i = p \cdot \omega_i$ and $u_i(x_i') > u_i(x_i)$ implies that $p \cdot x_i' > p \cdot \omega_i$ (no better bundle can be obtained with the given budget). Show that an equilibrium belongs to the core. Give an example (with $n = \ell = 2$) of a core allocation which is not an equilibrium.

   An $r$-replica economy is an exchange economy with $r$ consumers having the same endowment and utility function as consumer $i$ in the original economy. An allocation $(x_1, \ldots, x_n)$ in the original economy gives rise to an ("equal-treatment") allocation in the $r$ replica economy, whereby each of the $r$ consumers identical to consumer $i$ gets $x_i$, for $i = 1, \ldots, n$.

   Show that if an allocation $(x_1, \ldots, x_n)$ is such that the corresponding equal-treatment allocation belongs to the core of all the replica economies, then there is a price system $p$ such that $(x_1, \ldots, x_n)$ together with $p$ constitutes an equilibrium (Hint: use separation of suitable convex sets, whereby convexity is obtained using the core property).

6. (Market games) A market is an array $\mathcal{E} = (X_i, Y_i, \omega_i, u_i)_{i \in N}$, where for each $i \in N$

- $X_i \subseteq \mathbb{R}_+^\ell$ is nonempty, closed and convex,
- $Y_i \subseteq \mathbb{R}^\ell$ is nonempty, closed and convex, and $Y_i \cap \mathbb{R}_+^\ell = \{0\}$,
- $\omega_i \in X_i - Y_i$ (meaning that there is $y_i \in Y_i$ such that $y_i + \omega_i \in X_i$),
- $u_i : X_i \to \mathbb{R}$ is continuous and concave

(in the interpretation, $X_i$ is the consumption set and $Y_i$ the production set of agent $i$, the third condition shows that survival is possible without trade).

An NTU *market game* is an NTU game $(N, V)$ such that for some market $\mathcal{E} = (X_i, Y_i, \omega_i, u_i)_{i \in N}$,

$$V(S) = \left\{ (z_i)_{i \in S} \,\middle|\, \exists x_i \in X_i, y_i \in Y_i, z_i \le u_i(x_i), i \in S, \sum_{i \in S}(x_i - y_i) = \sum_{i \in N} \omega_i \right\}.$$

Show that market games are cardinally balanced and even *totally (cardinally) balanced*, in the sense that for each $S$, the game $(N, V|_S)$, where $V|_S$ is the restriction of the characteristic function to subsets of $S$, is cardinally balanced.

Show that market games are *compactly convexly generated*: For each $S$, the set $V(S)$ can be written as

$$V(S) = C_S - \mathbb{R}_+^S,$$

where $C_S$ is a convex and compact subset of $\mathbb{R}^S$.

Show that if an NTU game is compactly convexly generated and totally balanced, then it is a market game for some market $\mathcal{E}$.

# Chapter 15

# Values of NTU Games

## 1 The Shapley NTU value

Following the approach to TU games laid out in previous chapters, the next solution concept to be treated must be one which corresponds to the Shapley value. But in contrast to what we saw when dealing with the vNM solution and the core, a generalization to NTU games is not straightforward. Indeed, it is not obvious that there exists a meaningful generalization, and extensions of the Shapley value from TU to NTU games was proposed only after some years. In the present section, we discuss the proposal for an extension put forward by Shapley himself [Shapley, 1969]. Another extension is considered later in this chapter.

Clearly, we cannot hope for the possibility of defining a Shapley value directly by a formula, since this would demand a simple expression of the values of the characteristic function in term of numbers. The idea which eventually proved fruitful is to approximate the given game locally (in a sense to be made precise) by TU games, and then find the Shapley values of these local TU games. If this Shapley value defines an imputation in the original game, it is an NTU Shapley value. There can, as it might be seen already from this intuitive description, be more than one value for a given game.

We now make the above notions more precise. Our point of departure is a given NTU game $(N, V)$. Apart from the usual assumptions (every $V(S)$ is comprehensive, $V(N)$ is upper bounded and closed) we shall also assume that

(a) $V(N)$ is convex, and for each $x \in \text{bd}V(N)$ the set $\{x\} - [\mathbb{R}_+^N \setminus \{0\}]$ is contained in the interior of $V(N)$,

(b) $V(S)$ is closed and upper bounded for each $S \subset N$.

(c) $(N, V)$ is superadditive, so that $V(S) \times V(T) \subset V(S \cup T)$ for disjoint coalitions $S, T$.

When $V(N)$ is convex, then every point $x \in \text{bd}_{\mathbb{R}_+^N} V(N)$ (the boundary of $V(N)$ relative to $\mathbb{R}_+^N$) has a supporting hyperplane, i.e. the set

$$P(x) = \{\lambda \in \Delta_N \mid \lambda \cdot (x - x') \geq 0 \text{ for all } x' \in V(N)\}. \tag{1}$$

(we may choose the coefficients of the supporting hyperplane in $\Delta_N$ since $V(N)$ is comprehensive).

The correspondence $P : \mathrm{bd}_{\mathbb{R}_+^N} V(N) \rightrightarrows \Delta_N$ has closed graph (this is easily checked) and convex values. Since its values are subsets of the compact set $\Delta_N$, it is upper hemicontinuous.

Our next step is as follows: For each $\lambda \in \Delta_N$ we define the TU game $(N, v_\lambda)$ by

$$v_\lambda(S) = \max \left\{ \sum_{i \in S} \lambda_i x_i \,\Big|\, x \in V(S) \right\}. \tag{2}$$

Due to the superadditivity of $(N, V)$ (which we assumed under (c)), every TU game $v_\lambda$ is superadditive. By our earlier results, $\phi(v_\lambda)$, the Shapley value of $v_\lambda$, gives an imputation in $(N, v_\lambda)$, i.e.

$$\sum_{i \in N} \phi_i(v_\lambda) = v_\lambda(N) = \max\{\lambda \cdot x \mid x \in V(N)\}.$$

This is however not equivalent to $\phi(v_\lambda) \in V(N)$. Only if the latter holds, we have defined a payoff vector in $(N, V)$, which may be considered as an NTU version of the Shapley value.

Thus we define the *set of Shapley NTU values* in $(N, V)$ as

$$\Lambda(V) = \{x \in V(N) \mid \exists \lambda \in P(x) : x = \phi(v_\lambda)\}.$$

The construction above can be interpreted in the following way: Assume that we introduce exchange rates between the utility scales of the individual players, such that one unit of $i$-utility corresponds to $\lambda_i/\lambda_j$ units of $j$-utility. Given such exchange rates the coalitions can determine their strength in this new synthetical utility unit, and in this way we have obtained a TU game from the given game. One may now find a solution of the TU game, and *if this solution does not demand actual transfers of utility*, it may reasonably be considered as the NTU version of the solution concept used on the TU games. In our derivation above the latter solution concept was the Shapley value, but it is seen that the same reasoning may be used with other types of solutions.

One problem clearly remains: Are there Shapley NTU values in the game $(N, V)$? Theorem 1 shows that there are.

THEOREM 1 *Let $(N, V)$ be an NTU game, which satisfies the conditions (a)-(c) above. Then the set of Shapley NTU values for $(N, V)$ is nonempty.*

PROOF: We begin by constructing a map $h : \Delta_N \to \mathrm{bd}_{\mathbb{R}_+^N} V(N)$, defined as

$$h(z) = \max\{\mu z \mid \mu > 0, \mu z \in V(N)\}. \tag{3}$$

This map is well-defined and continuous, and it takes an arbitrary $z$ to a Pareto optimal payoff vector in $(N, V)$. Composed with $P(\cdot)$ defined in (1) we get a correspondence $P \circ h : \Delta_N \to \Delta_N$, which is upper hemicontinuous (if $U$ is an open set containing $P(h(z))$, then there is an open neighborhood $W$ of $z$ such that $P(h(z')) \subset U$ for all $z' \in W$) with convex values.

Next, we consider the map $\lambda \mapsto v_\lambda$, where $v_\lambda$ is treated as a point in $\mathbb{R}^S$, $S$ being the family of coalitions in $(N, V)$. By our construction, this map is continuous. We then compose with $\phi$, getting a map

$$f : \triangle_N \rightarrow \mathbb{R}^N_+ \backslash \{0\}. \tag{4}$$

Also the second component of this map is continuous, since the Shapley value is given by a formula which contains only addition and multiplication. Finally we take the image by this map to $\triangle_N$ dividing by the coordinate sum (the inverse map of $h$), so that we finally get the continuous map $f^* : \triangle_N \rightarrow \triangle_N$,

$$f^*(\lambda) = \left( \sum_{i \in N} \phi_i(v_\lambda) \right)^{-1} \phi(v_\lambda). \tag{5}$$

We now have a correspondence $\Phi : \triangle_N \times \triangle_N$ given by

$$\Phi(z, \lambda) = (f^*(\lambda), P(z)), \tag{6}$$

which is upper hemicontinuous with convex and compact values. By Kakutani's fixed point theorem (see e.g. Aliprantis and Border [1999] Corollary 16.51) there is $(z^0, \lambda^0)$ such that $(z^0, \lambda^0) \in \Phi(z^0, \lambda^0)$. This means that $\lambda^0$ supports in the point $h(z^0)$, and that $z^0$ is the Shapley value of the TU game $(N, v_{\lambda^0})$. We thus have that $h(z^0)$ is a Shapley NTU value. $\qquad \square$

## 2 Axiomatic characterization of the Shapley NTU value

One of the factors which have contributed towards making the Shapley value one of the most widely applied solutions for TU games is that it has a simple and useful axiomatic characterization. It was therefore natural that an extension of this axiomatization to NTU games was searched for; it is actually rather surprising that such an extension did not appear before the work by Aumann [1985], which will be considered here.

As earlier we let $\Phi$ be an abstract solution which to every game $(N, V)$ assigns a subset $\Phi(N, V)$ of $V(N)$. In what follows the set of players $N$ is kept fixed, so we may skip explicit reference to $N$.

Our first axiom is rather uncontroversial, stating that $\Phi$ is defined on all games in the class of games considered (the one introduced in the previous section).

Axiom 1. $\Phi(V) \neq \emptyset$.

Axiom 2 is a counterpart of the axiom of Pareto optimality in the TU case: Solutions should belong to the boundary of $V(N)$ (relative to $\mathbb{R}^N_+$), which by our general assumptions is identical to the set of Pareto optimal payoff vectors.

Axiom 2. $\Phi(V) \subset \mathrm{bd}_{\mathbb{R}^N_+} V(N)$.

The following axiom generalizes the additivity property of the TU Shapley value. In its basic idea this generalization is straightforward, only care must be taken since the sum of two solutions might not be a subset of the boundary of the sets feasible for $N$ in the sum game. The axiom is often called *conditional additivity*.

AXIOM 3. *Let $U = V + W$ (dvs. $U(S) = V(S) + W(S)$ for all $S$). Then*

$$[\Phi(V) + \Phi(W)] \cap \text{bd}_{\mathbb{R}_+^N} U(N) \subset \Phi(U).$$

We also need an NTU-version of the dummy axiom. As earlier we let $e_T$ be the unanimity game corresponding to the coalition $T$; considered as an NTU game we write it as $U_T$, where

$$U_T(S) = \begin{cases} \{x \in \mathbb{R}^S \mid \sum_{i \in S} x_i \le 1\} & T \subseteq S, \\ \{x \in \mathbb{R}^S \mid \sum_{i \in S} x_i \le 0\} & \text{otherwise.} \end{cases}$$

The vector $1_T \in \mathbb{R}^N$ is defined by

$$(1_T)_i = \begin{cases} 1 & i \in T, \\ 0 & i \notin T. \end{cases}$$

AXIOM 4. $\Phi(U_T) = \dfrac{1}{|T|} 1_T.$

The last axioms have no counterpart in the TU case, and they are needed to make the necessary connection between TU and NTU games.

AXIOM 5. *If $U = clV$ (i.e. $U(S) = clV(S)$ for all $S$), then $\Phi(U) = \Phi(V)$.*

This continuity assumption can hardly give rise to problems – it seems intuitively reasonable that it should not change the outcome whether one adds the closure of the sets $V(S)$. But the last of the axioms is less immediately acceptable, since here we are moving very close to the way in which the Shapley NTU value is related to the Shapley value for TU games.

AXIOM 6. *Let $\lambda = (\lambda_1, \ldots, \lambda_n) \in \mathbb{R}_+^N$ be a vector of positive numbers, and let the game $\lambda V$ be given by*

$$(\lambda V)(S) = \{(\lambda_i x_i)_{i \in S} \mid (x_i)_{i \in S} \in V(S)\}.$$

*Then $\Phi(\lambda V) = \lambda \Phi(V)$.*

We conclude with a version of the much used assumption of *independence of irrelevant alternatives* (IIA).

AXIOM 7. *If $V(N) \subset W(N)$ and $V(S) = W(S)$ for all $S \ne N$, then $\Phi(W) \cap V(N) \subset \Phi(V)$.*

The axiom states that if we reduce $W(N)$ without changing $W(S)$ for all proper subcoalitions $S$, and if a payoff vector in the solution for $W$ still belongs to $V(N)$, then it is also a solution for $V$. Although IIA assumptions have considerable intuitive appeal, it is not always clear that they are or should be satisfied.

We have now presented Aumann's system of axioms. Our first task is to check that $\Lambda$, the Shapley NTU value, satisfies the axioms.

LEMMA 1 $\Lambda$ *satisfies Axioms 1 – 7.*

PROOF: Axiom 1 follows from Theorem 1. Axiom 2 is an immediate consequence of the definition, and Axiom 5 likewise follows from the way in which $\Lambda$ was defined.

Suppose that $U = V + W$, and $x^1 \in \Phi(U)$, $x^2 \in \Phi(V)$ are such that $x^1 + x^2 \in \mathrm{bd}_{\mathbb{R}^N_+} U(N)$. Since we have assumed that $U(N)$ has unique support in each point of $\mathrm{bd}_{\mathbb{R}^N_+} U(N)$ (and this goes also for $V$ and $W$), the support $\lambda$ must be the same in $x^1$, $x^2$ and $x^1 + x^2$. But this means that $\lambda x^1 = (\lambda_1 x^1_1, \dots, \lambda_n x^1_n)$ is the Shapley value of the $\lambda$-scaled TU game $v_\lambda$, and $\lambda x^2$ is the Shapley value of the $\lambda$-scaled TU game $w_\lambda$. By the additivity of the Shapley value $x^1 + x^2$ is the Shapley value of $v_\lambda + w_\lambda$, which is the $\lambda$-scaled TU game for $U = V + W$. But this tells us that $x^1 + x^2$ is a Shapley NTU value for $U$.

Axiom 4 follows immediately from the dummy axiom for the Shapley value applied to the $(1,\dots,1)$-scaled game $U_T$ (which is identical to $U_T$, also a TU game). Axiom 6 is obvious from the definition of the Shapley NTU value. Finally, for Axiom 7 we have that if $x \in \Lambda(W)$, then $\lambda x$ is the Shapley value of $w_\lambda$ for $\lambda$ the unique support in $x$. If $x \in V(N)$, then $\lambda$ must also be a support of $\mathrm{bd}_{\mathbb{R}^N_+} V(N)$ in $x$, so that also $x$ is a Shapley NTU value for $V$. $\square$

Clearly the other implication has our main interest. We begin with a partial result:

LEMMA 2 *Let $\Phi$ be a solution which satisfies Axioms 1–6. Then $\Phi(V) \subseteq \Lambda(V)$ for all $V$.*

PROOF: The proof has two parts. First of all we show that $\Phi$ agrees with the Shapley value at all games which corresponds to TU games. Then we get the desired inclusion using our knowledge of TU games.

(1) Let $V$ be a game, which corresponds to a TU game $v$ (so that $V(S) = \{x \in \mathbb{R}^S \mid \sum_{i \in S} x_i \le v(S)\}$). For each real number $\alpha$ we let $V^\alpha$ be the game corresponding to $\alpha v$. In particular this means, that $V^0$ is the game corresponding to the game whose characteristic function is identically 0; this game plays a key role in the sequel.

We have from Axiom 4 that $\Phi(U_N) = \{n^{-1} 1_N\}$, so

$$\Phi(V^0) + \{n^{-1} 1_N\} = \Phi(V^0) + \Phi(U_N) \subset \Phi(V^0 + U_N),$$

where we also used Axiom 3. Since $V^0 + U_N = U_N$ ($V^0$ can be added to every TU game without changing anything), we have that

$$\Phi(V^0) + \{n^{-1} 1_N\} = \{n^{-1} 1_N\}.$$

Adding the set $\Phi(V^0)$ (which is not $\emptyset$ by Axiom 1) to a point different from 0 and getting the point itself means that this set must contain only the point 0, that is

$$\Phi(V^0) = \{0\},$$

which gives us most of part (1) of the proof: We have that $V + V^{-1} = V^0$ (easily seen from the TU representations), so additivity gives

$$\Phi(V) + \Phi(V^{-1}) = \{0\},$$

and since each of these sets consists of a single point, we have

$$\Phi(V^{-1}) = -\Phi(V).$$

Using the scaling axiom (Axiom 6) with $\lambda = (\alpha, \ldots, \alpha)$, we get $\Phi(V^\alpha) = \alpha\Phi(V)$. Thus, $\Phi(V^\alpha) = \alpha\Phi(V)$ for all $\alpha$, positive or negative.

We now look at the scaled unanimity games $U_T^\alpha$; from Axiom 4 and linearity we get $\Phi(U_T^\alpha) = (\alpha/|T|)1_T = \phi(\alpha e_T)$, where $\phi$ is the Shapley value for TU games. As is well-known, every TU game can be expressed as a linear combination of unanimity games, $v = \sum_T \alpha_T e_T$, and for the associated NTU games we have that $V = \sum_T U_T^{\alpha_T}$. In total we get that

$$\Phi(V) = \Phi\left(\sum_T U_T^{\alpha_T}\right) = \sum_T \Phi\left(U_T^{\alpha_T}\right) = \sum_T \{\phi(\alpha_T e_T)\} = \{\phi(v)\},$$

from which we conclude that $\Phi$ agrees with the Shapley value on all TU games.

(2) Now let $x \in \Phi(V)$; it follows from Axiom 4 that $x \in \mathrm{bd}_{\mathbb{R}_+^N} V(N)$, and by uniqueness of support there is a vector $\lambda$ of scaling factors with positive coordinates at $x$. We may assume that $\lambda = (1, \ldots, 1)$ (otherwise we look at $\lambda V$ instead).

Consider the game $V + V^0$. Taking the closure $\mathrm{cl}(V + V^0)$, we get a TU game (the other important property of $V^0$), and it is straightforward that this game is $v_\lambda$. We therefore have

$$\lambda x = x \in \Phi(V) \cap \mathrm{bd}_{\mathbb{R}_+^N}(V + V^0)(N) = (\Phi(V) + \Phi(V^0)) \cap \mathrm{bd}_{\mathbb{R}_+^N}(V + V^0)(N)$$
$$= \Phi(\mathrm{cl}(V + V^0)) = \{\phi(v_\lambda)\}.$$

We get that $x = \phi(v_\lambda)$ for $\lambda$ a support in $x \in \mathrm{bd}_{\mathbb{R}_+^N} V(N)$, or in other words, $x \in \Lambda(V)$, which is what should be proved.  □

We now have a characterization of the Shapley NTU value as the largest (for pointwise inclusion) solution correspondence satisfying Axioms 1–6. If we want the Shapley NTU value to emerge in exact form, we need also Axiom 7.

THEOREM 2 *Let $\Phi$ be a solution which satisfies Axioms 1-7. Then $\Phi$ is identical to $\Lambda$, the Shapley NTU value.*

PROOF: We know from Lemma 2 that $\Phi(V) \subseteq \Lambda(V)$. Let $y \in \Lambda(V)$ be arbitrary with associate vector of scaling factors $\lambda$, such that $\lambda y = \phi(v_\lambda)$.

Let $V_\lambda$ be the NTU game corresponding to $v_\lambda$, and define $W$ by

$$W(S) = \begin{cases} V_\lambda(N) & S = N, \\ \lambda \mathrm{cl}V(S) & S \neq N. \end{cases}$$

We see that $\lambda y$ is a Shapley NTU value for $W$.

We have $V_\lambda = \mathrm{cl}(W + V^0)$ and $\mathrm{bd}_{\mathbb{R}_+^N}(W + V^0)(N) = \mathrm{bd}_{\mathbb{R}_+^N} W(N)$, so that

$$\{\phi(v_\lambda)\} = \Phi(V_\lambda) = \Phi(\mathrm{cl}(W + V^0)) = \Phi(W + V^0).$$

From additivity we further get that $\Phi(W + V^0)$ contains $[\Phi(W) + \Phi(V^0)] \cap \mathrm{bd}_{\mathbb{R}^N_+}(W + V^0)(N)$, and since the latter is equal to $\Phi(W)$, we have that

$$\Phi(W) = \{\phi(v_\lambda)\} = \{\lambda y\}.$$

We now use Axiom 7 with $\lambda \mathrm{cl} V$ instead of $V$ and get that $\lambda y \in \Phi(\lambda \mathrm{cl} V) = \lambda \Phi(V)$ (where we also used Axioms 5 og 6). This gives us $y \in \Phi(V)$, and since $y \in \Lambda(V)$ was chosen arbitrarily, we get the desired conclusion. $\qquad\square$

## 3  The Harsanyi NTU value

The extension of the Shapley value to NTU games, which we have treated above, is not the only possible one. There are several ways in which one can proceed from TU games to the more general class of NTU games while keeping the intuitive content of the Shapley value. In the following we look at an approach which is fundamentally different from the previous one, even if the formalism looks much the same.

This is the *generalized bargaining solution* proposed by Harsanyi [1963]. At the outset it was conceived more as an extension of the Nash bargaining solution to NTU games with nontrivial subcoalitions, which then turned out to give the Shapley value for TU games. In the years following it became more common to see it as an extension of the Shapley value to NTU games.

**3.1. Intuitive background.** As before, we are dealing with the class $G$ of NTU games, such that for each point of $\mathrm{bd}_{\mathbb{R}^N_+} V(N)$ there is a well-defined and unique (except for normalization) support. Adopting the convention $\max_i \|\lambda_i\| = 1$ for these supports, we use the notation $\lambda(V, x)$ for the unique support in the point $x \in \mathrm{bd}_{\mathbb{R}^N_+} V(N)$.

We begin with a closer look at the Nash bargaining solution, something which may seem rather far from the problem at hand, but which as mentioned above is nevertheless closely connected with it. Thus, let $(U, t)$ be a two-person bargaining problem, where $U \subset \mathbb{R}^2_+$ is the set of feasible utility payoffs, and $t \in U$ is the status quo or disagreement payoff. We assume that the bargaining problem is solved in accordance with the Nash bargaining solution, so that the result $(u_1^*, u_2^*)$ maximizes the Nash product $(u_1 - t_1)(u_2 - t_2)$ over $U$.

We now extend the bargaining solution with an initial phase, where the two players can choose their disagreement payoffs. We assume that this is done indirectly, in the sense that each player chooses a strategy $\theta_i \in \Theta_i$, whereafter a disagreement payoff

$$(t_1, t_2) = F(\theta_1, \theta_2)$$

will be determined. Here $F : \Theta_1 \times \Theta_2 \to U$ is a given function.

Assume that $U$ has unique support in each boundary point. If the Nash bargaining solution is $u = (u_1, u_2)$, then for $\lambda = \lambda(U, u)$ (the support of $U$ at $u$) we have trivially that

$$\lambda \cdot u \geq \lambda \cdot u' \text{ for all } u' \in U$$

(this is the support property), and furthermore, we get that

$$\lambda_1(u_1 - t_1) = \lambda_2(u_2 - t_2); \tag{7}$$

this is a consequence of the way in which we find the Nash bargaining solution: transform the problem by scaling according to $\lambda$ and move the origin to $t$, then the solution is on the diagonal, which is what is expressed in (7)). Alternatively, it can be written as

$$\lambda_i t_i - \lambda_j t_j = \lambda_i u_i - \lambda_j u_j, \ i = 1, 2, \ j \neq i. \tag{8}$$

Assume that player $i$ chooses $\theta_i$ such that she obtains the best possible result in the subsequent Nash bargaining. We claim that this corresponds to attaining the optimal value of

$$\lambda_i t_i - \lambda_j t_j,$$

$i = 1, 2, \ j \neq i$. Indeed, assume first that player $i$ has a strategy $\theta_i$, which given the choice $\theta_j^0$ by the other player gives $(t_i', t_j') = F(\theta_i, \theta_j^0)$ with

$$\lambda_i t_i' - \lambda_j t_j' > \lambda_i t_i - \lambda_j t_j = \lambda_i u_i - \lambda_j u_j.$$

In this case we would have

$$\lambda_i u_i' - \lambda_j u_j' = \lambda_i t_i' - \lambda_j t_j' > \lambda_i u_i - \lambda_j u_j$$

for the associated Nash solution $(u_i', u_j')$ (according to (8)), and therefore

$$\lambda_i(u_i' - u_i) > \lambda_j(u_j' - u_j).$$

But from $\lambda_i u_i' + \lambda_j u_j' < \lambda_i u_i + \lambda_j u_j$ we get $\lambda_i(u_i' - u_i) < \lambda_j(u_j - u_j')$ or

$$-\lambda_i(u_i' - u_i) > -\lambda_j(u_j - u_j') = \lambda_j(u_j' - u_j),$$

from which $\lambda_j(u_j' - u_j) < 0$. But if $u_j' < u_j$, then $u_i' > u_i$, so that player $i$ by choosing $\theta_i$ can get a larger final payoff, contradicting the Nash equilibrium property. We conclude that $\lambda_i t_i - \lambda_j t_j$ has attained its maximal value.

Notice that the reformulation, by which the game can be seen as one in which player $i$ should obtain maximal value of $\lambda_i t_i - \lambda_j t_j$, gives us a 2 person zero-sum game (the payoff of player $j$ is minus the payoff of player $i$). The advantage of looking at the conflict as a 2-person zero-sum game is that the latter has a *value* $\lambda_i t_i^0 - \lambda_j t_j^0$, obtained as the payoff to the first player in any Nash equilibrium. This value can be considered as the rational result of the conflict at hand.

The original NTU game seems to have been lost during this development, but it enters now: For each coalition $S$ we consider the conflict between $S$ and its complement $N \backslash S$. The set of possible outcomes is in principle all payoffs in $V(N)$, but in accordance with what we did when defining the Shapley NTU value, we will look at the situation locally at a given payoff $x \in \mathrm{bd}_{\mathbb{R}_+^N} V(N)$, so that we can make the utility payoff vectors comparable by $\lambda x$ for $\lambda = \lambda(V(N), x)$. The bargaining will then result in a certain amount to $S$ and an amount to $N\backslash S$. What the coalitions get, depend on the threat or disagreement points.

If we consider a TU game with characteristic function $v$, we look at the Shapley value $\phi(v)$ as the result of a long series of negotiations and bargaining between the coalition $S$ and its complements: We have that

$$\phi_i(v) = \sum_{S:i\in S} \frac{(s-1)!(n-s)!}{n!}[v(S) - v(S\backslash\{i\}]$$

$$= \sum_{S:i\in S} \frac{(s-1)!(n-s)!}{n!}v(S) - \sum_{S:i\in S} \frac{(s-1)!(n-s)!}{n!}v(S\backslash\{i\}$$

$$= \sum_{S:i\in S} \frac{(s-1)!(n-s)!}{n!}v(S) - \sum_{T:i\in S} \frac{(n-s)!(s-1)!}{n!}v(N\backslash T)$$

$$= \sum_{S:i\in S} \frac{(s-1)!(n-s)!}{n!}[v(S) - v(N\backslash S],$$

where we have written $S\backslash\{i\}$ as $N\backslash T$ and performed summation over $T$ with $n-s+1$ elements; this yields a summation over exactly the same terms.

If we consider each term $[v(S) - v(N\backslash S)]$ as a possible result of negotiation between $S$ and its complement (which gets the opposite of what $S$ gets), then the interpretation of $v$ acquires a central importance. The strategy for our development of a solution for NTU games is to let $v$ be determined by the underlying threat structure.

To this purpose we introduce a third aspect (apart from bargaining and relation to Shapley value) of the Harsanyi solution, namely the idea that players receive their final result as a system of *dividends* (positive or negative) from the coalitions in which they participate,

$$u_i = \sum_{S:i\in S} w_i^S, \ i = 1, \ldots, n.$$

The main property here is that dividends are distributed equally (or rather, in proportions determined by the vector $\lambda$). The intuitive explanation is as follows: If everything is a result of bargaining, then also the mutual relationship between $i$ and $j$ (i.e. the final payoffs $u_i$ and $u_j$) are outcomes of bargaining. As we have seen, the result of bargaining must satisfy

$$\lambda_i(u_i - t_i) = \lambda_j(u_j - t_j),$$

where $t_i$ and $t_j$ are the threats which result from the best possible threat strategies of the players. Making more precise what a threat means, we have that $t_i$ must be a signal for what $i$ can obtain without $j$ (and similarly for $j$), that is

$$t_i = \sum_{S:i\in S, j\notin S} w_i^S, \ t_j = \sum_{S:j\in S, i\notin S} w_j^S,$$

Using that $u_i$ and $u_j$ has the dividend structure mentioned above, we get that

$$\lambda_i \sum_{S:i,j\in S} w_i^S = \lambda_j \sum_{S:i,j\in S} w_j^S.$$

In other words, when taking into account the scale parameters in $\lambda$, the dividend payment to the players of the coalitions to which they belong is the same.

**3.2. Formal definition of the Harsanyi NTU value.** We have now outlined the principles which are used in formulating the Harsanyi NTU value: The final payoff must consist of dividend payments for the participation of players in coalitions, whereby the players are remunerated proportionally for their participation in each coalition. The distribution of what is available in each coalition is a result of a bargaining between this coalition and its complement.

In order to obtain a shorter definition of the solution concept outlined in the previous section, we take another route and pay more attention to the properties which the proposed solution should have.

First of all we change slightly our usual point of view in the discussion of solutions, since now we are not only searching for a payoff vector $x \in \mathbb{R}^N$ with suitable properties, but instead a suitable *payoff configuration*, by which we understand a family $x = (x^S)_{S\subseteq N}$ with $x^S \in \mathbb{R}^S$ for each $S$. In the final event it will still be the part of the payoff configuration concerning the grand coalition $N$, that is the component $x^N \in \mathbb{R}^N$, which matters to us, but on our way to it, discussing the properties, it is convenient that also the other components $x^S$ are available.

In our new terminology we have that a payoff configuration $x = (x^S)_{S\subseteq N}$ is a Harsanyi NTU value of $(N, V)$ if there exists a vector $\lambda = (\lambda_1, \ldots, \lambda_n)$ with positive coordinates and a family $(w^S)_{S\subseteq N}$ of real numbers, such that

    (i) $x^S \in \mathrm{bd}_{\mathbb{R}_+^N} V(S)$ for every $S \subseteq N$,
    (ii) $\lambda = \lambda(V, x)$ (the support of $V(N)$ at $x^N$),
    (iii) $\lambda_i x_i^S = \sum_{T\subseteq S:i\in T} w^T$.

The set of Harsanyi NTU values associated with $(N, V)$ (i.e. all payoff configurations satisfying (i)-(iii) above) is denoted $H(V)$.

Apart from some notational coincidence there seems to be very little relation between the background story given in the previous section and the precise definition of the solution concept. This corresponds to some degree to the development of the literature from Harsanyi's own definition to subsequent applications, where it turns out to be convenient to abstract from some of the circumstances leading up to the definition and instead to concentrate on the strictly necessary.

The three conditions (i)-(iii) are usually referred to as *efficiency, utilitarianism and fairness*, respectively. For (i) and (iii), this is straightforward: When all $x^S$ and in particular $x^N$ must be on the boundary of $V(S)$, then we get an allocation which is Pareto optimal. Fairness described the property that all members of a coalition gets the same dividend from this coalition (which is (iii)) when the utility scales are made comparable using $\lambda$. The term 'utilitarianism' comes from the fact that the support property of $\lambda$ can be reformulated as the property that $x^N$ maximizes $\sum_{i \in N} \lambda_i x_i$ on $V(N)$. The expression to be maximized is the sum of the (rescaled) utilities, and one speaks of utilitarian decision making when choosing the alternative which maximizes the sum of the utilities of the relevant individuals.

One might wonder what became of the Shapley value, which is not mentioned in the definition. It is there, however.

LEMMA 3 *Let* $x = (x^S)_{S \subseteq N}$ *be a Harsanyi NTU value for* $(N, V)$ *with support* $\lambda$ *and corresponding numbers* $(w^S)_{S \subseteq N}$. *Let the TU game* $(N, v)$ *be defined by*

$$v(T) = \sum_{i \in T} \lambda_i w_i^T, \ T \subseteq N;$$

*then*

$$\lambda_i x_i^S = \phi_i(S, v_{|S}),$$

*where* $(S, v_{|S})$ *is the restriction of* $v$ *to* $S$ *and its subcoalitions. Conversely, any payoff configuration satisfying this property will be a Harsanyi value.*

PROOF: For each $T$ we have

$$v(T) = \sum_{R \subseteq T} |R| w^R,$$

due to (iii) above, since each coalition $R$ contained in $T$ pays the same dividend $w^R$ to all its members. In particular,

$$v(T) - v(T \setminus \{i\}) = \sum_{R \subseteq T: i \in R} |R| w^R.$$

Computing $\phi_i(S, v_{|S})$, we get

$$\phi_i(S, v_{|S}) = \sum_{T: i \in T} \frac{(t-1)!(s-t)!}{s!} \left[ \sum_{R \subseteq T: i \in R} |R| w^R \right].$$

Here summation is performed first over all $T \subseteq S$ containing $i$ and then for each $T$ over all $R \subseteq T$ containing $i$. For given $R$ we get a sum over all $T$ between $R$ and $S$ of $w^R$ multiplied by suitable coefficients, in total

$$\sum_{R \subseteq T \subseteq S} r \frac{(t-1)!(s-t)!}{s!},$$

where $|R| = r$. There are $\binom{s-r}{s-t}$ coalitions $T$ with $t$ members between $R$ and $S$, so the expression reduces to

$$\sum_{t=r}^{s} \frac{r(t-1)!(s-r)!}{(t-r)!s!} = \frac{(s-r)!r}{s!} \sum_{t=r}^{s} \frac{(t-1)!}{(t-r)!} = \frac{(s-r)!r(r-1)!}{s!} \sum_{t=r}^{s} \frac{(t-1)!}{(t-r)!(r-1)!}$$

$$= \binom{s}{r}^{-1} \sum_{t=r}^{s} \binom{t-1}{r-1} = \binom{s}{r}^{-1} \sum_{t'=r-1}^{s-1} \binom{t'}{r} = \binom{s}{r}^{-1} \binom{s}{r} = 1,$$

where the final equality is obtained using a standard formula for summation of binomial coefficients. We thus have that

$$\lambda_i x_i^S = \sum_{T \subseteq S : i \in T} w^T = \phi_i(S, v_{|S}).$$

For the converse implication we define the family $(w^S)_{S \subseteq N}$ as follows: For coalitions consisting of one player only we have trivially that

$$w^{\{i\}} = \phi_i(\{i\}, v_{|\{i\}}) = v(\{i\}) = \lambda_i x_i;$$

assume that $w^T$ is defined for all proper subcoalitions of $S$ and define $w^S$ by

$$|S| w^S = v(S) - \sum_{T \subset S} w^T;$$

by the above computation this is consistent with

$$\phi_i(S, v_{|S}) = \sum_{T \subseteq S : i \in T} w^T,$$

from which it is seen that the family $(w^S)_{S \subseteq N}$ has the property (iii). $\qquad \square$

For later use we notice that the definition of $H(V)$ (apart from the efficiency condition) does not involve the sets $V(S)$ for $S \neq N$. Otherwise put, we have the following lemma.

LEMMA 4 *Let $(N, V)$ and $(N, W)$ be NTU games, and suppose that the payoff configuration $x = (x^S)_{S \subseteq N}$ belongs to $H(V)$. If $x \in bd_{\mathbb{R}_+^N} W$ (i.e. $x^S \in bd_{\mathbb{R}_+^S} V(S)$ for all $S$) and $\lambda(V, x) = \lambda(W, x)$, then $x \in H(W)$.*

PROOF: Simple check of the conditions (i) – (iii). $\qquad \square$

An alternative approach to the Harsanyi NTU value, and one which is closer to what we did when deriving the Shapley NTU value, is the following: For a given scaling vector $\lambda \in \mathbb{R}_{++}^N$ we may construct a system of dividends $(w_S^\lambda)_{S \subseteq N}$ as follows: For coalitions with only one member, i.e. $S = \{i\}$ for some $i \in N$, we first define $z[\lambda]^{\{i\}}(t) = t / \lambda_i$, and we let

$$w_{\{i\}}^\lambda = \max\{t \mid z[\lambda]^{\{i\}}(t) \in V(\{i\})\},$$

and in similar way we define $z[\lambda]^S(t)$ for arbitrary $S$, assuming that $w_T^\lambda$ is defined for all proper subcoalitions $T$ of $S$, by

$$z[\lambda]_i^S(t) = \left( \sum_{T \subset S : i \in T} w_T^\lambda + t \right) \Big/ \lambda_i.$$

Next we set

$$w_S^\lambda = \max \left\{ t \mid z[\lambda]^S(t) \in V(S) \right\}.$$

The payoff configuration $x[\lambda] = (x[\lambda]^S)_{S \subseteq N}$ defined by

$$x[\lambda]^S = z[\lambda]^S(w_S^\lambda), \quad S \subset N,$$

is a candidate for being a Harsanyi NTU value for $(N, V)$. Indeed it satisfies the conditions (i) and (iii) above by construction, so the only condition to be verified is (ii). If $\lambda$ is a support of $V(N)$ in $x[\lambda]^N$, or, in our notation,

$$\lambda = \lambda(V, x[\lambda]),$$

then all three conditions are satisfied, and $x[\lambda]$ is a Harsanyi NTU value. The converse statement clearly holds as well.

We can use this for two purposes: First of all it gives us a method of proof of existence of Harsanyi NTU values for a sufficiently large family of NTU games (for this the construction above must be supplemented by a fixed point argument). We shall not go further into such considerations, instead we notice another consequence of the derivation.

LEMMA 5 *Let $(N, V)$ be such that $0 \in bd_{\mathbb{R}^N_+} V(S)$ for all $S \subset N$. Then $H(V) = \{0\}$ (the payoff configuration with $0 \in \mathbb{R}^S$ as payoff to every coalition is the unique Harsanyi NTU value).*

PROOF: For each $\lambda$ in $\mathbb{R}^N_{++}$, the above construction gives that $z[\lambda]^S = 0$ og $w^\lambda_S = 0$ for all $S$. In particular this holds for the uniquely determined $\lambda$, which supports $V(N)$ in the point $0$, so we have that $0$ is a Harsanyi NTU value. It is clearly the only one, since the construction gives only this payoff configuration. □

Our last results in this section have to do with games where there is only one possible support (independent of the point in $V(N)$):

LEMMA 6 *Let $(N, V)$ be a game for which*

$$V(N) = \left\{ x \in \mathbb{R}^N \mid \sum_{i \in N} \mu_i x_i \leq c \right\}$$

*for $\mu \in \mathbb{R}^N_{++}$ and $c \in \mathbb{R}$ (i.e. $V(N)$ is a halfspace). Then $(N, V)$ has a unique Harsanyi NTU value.*

PROOF: There is only one possible support, namely $\mu$ (suitably normalized). The above procedure gives a uniquely determined payoff configuration $x[\mu]$, so the Harsanyi NTU value is uniquely determined. □

This result has a consequence which was to be expected but which still needs a proof.

LEMMA 7 *Let $(N, V)$ be a game corresponding to a TU game $(N, v)$. Then $(N, V)$ has a unique Harsanyi NTU value $x = (x^S)_{S \subset N}$ given by*

$$x^S = \phi_i(S, v_{|S}), \ S \subset N.$$

PROOF: It follows from Lemma 6 that $x = (x^S)_{S \subset N}$ defined above is a Harsanyi NTU value. Uniqueness now follows immediately from Lemma 6. □

## 4   Axiomatic characterization of the Harsanyi NTU value

As was the case for the Shapley NTU value, there is an axiomatization of the Harsanyi NTU value, due to Hart [1985]. In its structure it is rather close to Aumann's axioms, but there are important differences (and indeed Axioms 5 – 6 below are *not* satisfied by the Shapley NTU value).

Our point of departure is – as always – an abstract solution correspondence. It is natural in view of our previous considerations to define this correspondence in such a way that its values are payoff *configurations* rather than as in earlier axiomatizations payoff vectors.

We thus have a correspondence $\Psi$, which to every game $(N, V)$ in $\mathcal{G}$ gives a set $\Psi(V)$ of payoff configurations in $(N, V)$; as on earlier occasions we shall omit the explicit reference to $N$, which plays no significant role here.

The first axiom is an efficiency property; it corresponds to condition (i) in the definition of the Harsanyi NTU value:

AXIOM 1.   (Efficiency) $\Psi(V) \subseteq \mathrm{bd}_{\mathbb{R}_+^N} V$, *i.e. for* $x = (x^S)_{S \subseteq N} \in \Psi(V)$ *we have that* $x^S \in \mathrm{bd}_{\mathbb{R}_+^N} V(S)$, *all* $S \subseteq N$.

Also the next axiom is an old acquaintance, which is necessary in any characterization of the Harsanyi (or Shapley) NTU value:

AXIOM 2.   (Scaling) *For every* $\lambda \in \mathbb{R}_{++}^N$, $\Psi(\lambda V) = \lambda \Psi(V)$.

Here we have (as earlier) used the notation $\lambda A$ (where $A$ is a subset of $\mathbb{R}^N$) for the set $\{(\lambda_1 x_1, \ldots, \lambda_n x_n) \mid x = (x_1, \ldots, x_n) \in A\}$, and the game $\lambda V$ is given by $(\lambda V)(S) = \lambda V(S)$ for all $S$).

Likewise it is clear that some version of an additivity axiom must be used.

AXIOM 3.   (Conditional additivity) *Let* $U = V + W$. *Then*

$$[\Psi(V) + \Psi(W)] \cap \mathrm{bd}_{\mathbb{R}_+^N} U \subseteq \Psi(U).$$

The formulation of the axiom is the same as in the earlier version, but here it refers not to payoff vectors but to the whole payoff configuration.

The last of the common axioms is independence of irrelevant alternatives, which now looks as follows:

AXIOM 4.   *Let* $V \subseteq W$ *(i.e.* $V(S) \subseteq W(S)$ *for all* $S$*). Then*

$$\Psi(W) \cap V(N) \subseteq \Psi(V).$$

As before the axiom states that if we have something belonging to the solution of the "big" game, which for each coalition selects something that is possible in the smaller game, then it also belongs to the solution of this smaller game.

The following axioms are specific for our present problem, and they are not very surprising given this problem. The first of them refers to the unanimity games which we have used earlier. If we use the notation $U_{T,c}$ for the NTU game corresponding to the TU game $(N, ce_T)$ (for which the value of the characteristic function at a coalition is $c$ if the coalition contains $T$ and $0$ otherwise), we can write the axiom as shown below.

AXIOM 5. (Unanimity) $\Psi(U_{T,c}) = \{z\}$, where $z^S = (c/|T|)1_T$ if $T \subseteq S$, $0$ otherwise.

The axiom seems reasonable, at least for the part concerned with the payoff vector (which is also known from the earlier axiomatization). Here it is demanded that the same principle holds for all coalitions.

The last axiom specifies a dummy property, saying that if nobody can do anything, then nobody gets anything.

AXIOM 6 (0-inessential games) If $0 \in \mathrm{bd}_{\mathbb{R}^N_+} V$, then $0 \in \Psi(V)$.

The relevance of this system of actions in the case considered follows from the following result.

THEOREM 3 *The Harsanyi NTU value H satisfies Axioms 1 – 6.*

PROOF: As said above, Axiom 1 is a repetition of condition (i) in the definition of the Harsanyi NTU value. Axiom 2 follows directly from the conditions (ii) and (iii).

For Axiom 3 we consider $u = v + w$, where $v$ is in $H(V)$ and $w \in H(W)$. If $u \in \mathrm{bd}_{\mathbb{R}^N_+} U$, then $u^N$ must maximize $\lambda \cdot u$ on $U(N)$, where $\lambda = \lambda(U(N), x^N)$, and then

$$v^N \text{ maximizes } \lambda \cdot v \text{ on } V(N),$$
$$w^N \text{ maximizes } \lambda \cdot w \text{ on } W(N),$$

so that $\lambda = \lambda(V, v) = \lambda(W, w)$. Then it remains only to show $u \in H(U)$ checking the linearity of condition (iii) for fixed $\lambda$, and this is straightforward.

Axiom 4 follows since if $x \in H(W) \cap \mathrm{bd}_{\mathbb{R}^N_+} V$, then we have in particular that $\lambda(V(N), x^N) = \lambda(W, x)$, and then we get the statement from Lemma 6.

Axiom 5 is a direct consequence of Lemma 7. Finally we get Axiom 6 from Lemma 5. $\qquad\square$

The interesting part of the axiomatization is the converse of Theorem 3. First we show that $\Psi$ induces the obvious solution, namely the Shapley value, in all subgames which are derived from at TU game.

LEMMA 8 *Let V be a game which corresponds to a TU game v. If $\Psi$ satisfies the Axioms 1, 3 and 5, then*

$$\Psi(V) = \{(\phi(S, v_{|S}))_{S \subseteq N})\}.$$

PROOF: If $V$ is derived from a unanimity game, the result follows immediately from Axiom 5.

For arbitrary $V, W$ corresponding to TU games $v, w$ we have that $U = V + W$ corresponds to $v + w$. For these games we further have that $\mathrm{bd}_{\mathbb{R}^N_+} U = \mathrm{bd}_{\mathbb{R}^N_+} V + \mathrm{bd}_{\mathbb{R}^N_+} W$, so using Axioms 1 and 3 we find that $\Psi(V) + \Psi(W) \subseteq \Psi(U)$.

We now use that every TU game can be written as a sum of unanimity games, so that an arbitrary game $V$ is a sum of games of the form $U_{T,c}$. From the additivity we get

$$\{(\Phi(S, v_{|S}))_{S \subseteq N}\} = \sum_{T,c} \Psi(U_{T,c}) \subseteq \Psi(V)$$

(the first equality comes from those of TU games). This gives us one of the inclusions and we need only show the opposite inclusion.

Let $V'$ be the game which corresponds to the TU game $-v$ (where $v$ is the TU game corresponding to $V$). Then $V' + V$ is the game corresponding to the trivial unanimity game $U_0$ (the constant $c$ is 0, and clearly the reference coalition is unimportant). Using Axiom 3 we get

$$\Psi(V) + \Psi(V') \subseteq \Psi(U_0) = \{0\},$$

where the last equality is a consequence of Axiom 5. We now use the first part of the proof, giving us that $\{(\phi(S, v_{|S}))_{S \subseteq N}\} \subseteq \Psi(V)$, and similarly $\{(-\phi(S, v_{|S}))_{S \subseteq N}\} \subseteq \Psi(V')$. But two nonempty sets, the sum of which is a singleton, must both be singletons, and this gives us the uniqueness.                                                                □

Notice that the technique of proofs differs somewhat from what was used in Aumann's axiomatization of the Shapley NTU value, even though it actually is the same that happens. The difference is that we need not check additivity also for negative signs; this is already a direct consequence of Axiom 5, which thus is seen to be rather powerful.

We are now ready for the main result of this section.

THEOREM 4 *Let $\Psi$ be an abstract solution correspondence satisfying Axioms 1 – 6. Then $\Psi = H$, the Harsanyi NTU value.*

PROOF: Initially, it is shown that $\Psi(V) \subseteq H(V)$. Begin with $x \in \Psi(V)$, let $\lambda = \lambda(V(N), x)$, and define $\mu \in \mathbb{R}_{++}^N$ by $\mu_i = 1/\lambda_i$ for all $i \in N$. Now consider the games $V_1$ defined by

$$V_1(S) = \left\{ y \in \mathbb{R}^S \mid y \leq x^S \right\}, \; S \neq N,$$
$$V_1(N) = V(N),$$

and $V_2$ defined by

$$V_2(S) = \left\{ y \in \mathbb{R}^S \mid \sum_{i \in S} \lambda_i y_i \leq \lambda_i x_i^S \right\}$$

for all $S \subseteq N$. Using Axiom 4 and noting that $V_1 \subseteq V$ and $x \in V_1$, we get that $x \in \Psi(V_1)$. From the definitions we further have that $V_2 = V_1 + \mu U_0$, and since $U_0$ comes from a unanimity game, we have $\Psi(U_0) = \{0\}$, from which by Axiom 2 $\Psi(\mu U_0) = \{0\}$. The additivity axiom now gives that

$$x \in [\Psi(V_1) + \Psi(\mu U_0)] \cap \text{bd}_{\mathbb{R}_+^N} V_2 \subseteq \Psi(V_2).$$

Looking next at the game $\lambda V_2$, we see that it has the form

$$\lambda V_2(S) = \left\{ y \in \mathbb{R}^S \,\Big|\, \sum_{i \in S} \lambda_i(\mu_i y_i) \leq \sum_{i \in S} \lambda_i x_i^S \right\},$$

so that it corresponds to the TU game given by the characteristic function $v$ with $v(S) = \sum_{i \in S} \lambda_i x_i^S$ for every $S$.

We now have from Lemma 8 that $\Psi(\lambda V_2) = \{\lambda x\} = H(\lambda V_2)$ or $\Psi(V_2) = H(V_2)$. Since $H(V)$ only depends on the support to $V(N)$ in $x^N$ (which is $\lambda$), and we have $x \in \mathrm{bd}_{\mathbb{R}_+^N} V$, it follows that $x \in H(V)$.

To show the opposite inclusion we begin with $x \in H(V)$, and we let $\lambda = \lambda(V, x)$. Then we define $W_1$ by

$$W_1(S) = \left\{ y \in \mathbb{R}^S \,\Big|\, \sum_{i \in S} \lambda_i y_i \leq \sum_{i \in S} \lambda_i x_i^S \right\}, \text{ all } S,$$

$W_2$ by

$$W_2(S) = \{ y \in \mathbb{R}^S | y \leq x^S \}, \ S \neq N,$$

$W_2(N) = W_1(N)$, and $W_3$ by $W_3(S) = V(S)$, $S \neq N$, $W_3(N) = W_1(N)$. Finally $W$ is defined by

$$W(S) = V(S) - \{x^S\}$$

for all $S$. We have $0 \in \mathrm{bd}_{\mathbb{R}_+^N} W$, so by Axiom 6, $0 \in \Psi(W)$.

From $x \in H(V)$ it follows that $x \in H(W_1)$ (since $x \in \mathrm{bd}_{\mathbb{R}_+^N} W_1$ and $x^N$ have the same support in $W_1(N)$ as in $V(N)$. Turning to $\lambda W_1$, we get (as above)

$$(\lambda W_1)(S) = \left\{ y \in \mathbb{R}^S \,\Big|\, \sum_{i \in S} \lambda_i(\lambda_i^{-1}) y_i \leq \sum_{i \in S} \lambda_i x_i^S \right\}$$

for all $S$, so $\lambda W_1$ corresponds to a TU game with characteristic function $v(S) = \sum_{i \in S} \lambda_i x_i^S$, and therefore $\lambda x$ (which is the Harsanyi NTU value of the game $\lambda W_1$ and as such obtained by taking the Shapley value of all $v_{|S}$) is also contained in $\Psi(\lambda W_1)$, i.e. $x \in \Psi(W_1)$.

Since $W_2 \subseteq W_1$ and $x \in W_2$, we have $x \in \Psi(W_2)$. Further we have $W_3 = W_2 + W$, so from additivity (and $0 \in \mathrm{bd}_{\mathbb{R}_+^N} W$) we obtain

$$x = x + 0 \in [\Psi(W_2) + \Psi(W)] \cap \mathrm{bd}_{\mathbb{R}_+^N} W_3 \subseteq \Psi(W_3).$$

Now we need only to notice that $V \subseteq W_3$; since $x \in V$ we have $x \in \Psi(V)$, which gives the desired inclusion. $\qquad\square$

## 5   Problems

**1.** The procedure used in Section 1, constructing of a TU game $v_\lambda$ as in (2) and applying a solution concept for TU games, followed by selection of a specific $\lambda$ such that the solution to $v_\lambda$ belongs to $V(N)$, known as the *Shapley transfer principle,* can be used for other solution concepts than the core.

Show that the core of $(N, V)$, defined in the standard way of Chapter 14, may not coincide with the NTU core found using the Shapley transfer principle.

**2.** [Roth, 1980] Consider the game $(N, V)$ with $N = \{1, 2, 3\}$, for which

$$V(\{i\}) = \{z_i \mid z_i \leq 0\}, i = 1, 2, 3,$$
$$V(\{1, 2\}) = \left\{(z_1, z_2) \mid z_i \leq \tfrac{1}{2}, i = 1, 2\right\},$$
$$V(\{1, 3\}) = \left\{(z_1, z_3) \mid z_1 \leq \tfrac{1}{4}, z_3 \leq \tfrac{3}{4}\right\},$$
$$V(\{2, 3\}) = \left\{(z_2, z_3) \mid z_2 \leq \tfrac{1}{4}, z_3 \leq \tfrac{3}{4}\right\},$$
$$V(N) = \mathrm{conv}\left(\left\{\left(\tfrac{1}{2}, \tfrac{1}{2}, 0\right), \left(\tfrac{1}{4}, 0, \tfrac{3}{4}\right), \left(0, \tfrac{1}{4}, \tfrac{3}{4}\right)\right\}\right) - \mathbb{R}^3_+.$$

Show that the payoff vector $\left(\tfrac{1}{2}, \tfrac{1}{2}, 0\right)$ is the only reasonable outcome if players are payoff maximizers.

Show that the payoff vector $\left(\tfrac{1}{3}, \tfrac{1}{3}, \tfrac{1}{3}\right)$ is a Shapley NTU value. Comment on the results.

**3.** (Value allocations in an exchange economy, cf. Problem 5 in Chapter 14) Suppose that a society $N$ consisting of $n$ consumers have endowments $\omega_i \in \mathbb{R}^\ell$ of $\ell$ different commodities, and utility functions $u_i$ defined on bundles $x_i \in \mathbb{R}^\ell_+$, all $i$. An allocation is an array $(x_1, \ldots, x_n)$ of bundles with $\sum_{i=1}^n x_i = \sum_{i=1}^n \omega_i$. An allocation $(x_1, \ldots, x_n)$ is an *equilibrium* at the price system $p \in \mathbb{R}^\ell_+, p \neq \emptyset$, if $\sum_{i=1}^n x_i = \sum_{i=1}^n \omega_i$ and for each $i$, $p \cdot x_i = p \cdot \omega_i$ and $u_i(x_i') > u_i(x_i)$ implies that $p \cdot x_i' > p \cdot \omega_i$ (no better bundle can be obtained with the given budget).

Assume in the following that each $u_i$ is quasi-concave, i.e. $\{x' \mid u_i(x') \geq u_i(x)\}$ is convex for each $x \in \mathbb{R}^\ell_+$. For $p \in \mathrm{int}\mathbb{R}\ell_+$ a price system, $u$ a quasi-concave utility function and $x \in \mathbb{R}^\ell_+$ a bundle, let $M(u, p, x) = \min_{x'}\{p \cdot x' \mid u(x') \geq u(x)\}$ be the minimum cost of buying a bundle as good as $x$ at prices given by $p$, and let

$$v(p, S) = \max\left\{\sum_{i \in S} M(u_i, p, x_i) \,\Big|\, \sum_{i \in S} x_i = \sum_{i \in S} \omega_i\right\}.$$

Show that for each $p \in \mathrm{int}\mathbb{R}^\ell_+$, $(N, v(p, \cdot)$ is a TU game.

Let $\phi(v(p, \cdot))$ be the Shapley value of $(N, v(p, \cdot))$. An allocation $(x_1, \ldots, x_n)$ is a *value allocation* if there is $p \in \mathrm{int}\mathbb{R}^\ell_+$ such that $M(u_i, p, x_i) = \phi_i(v(p, \cdot))$ for $i = 1, \ldots, n$.

Show that if $(x_1, \ldots, x_n)$ is an equilibrium at some price system $p$, then it is a value allocation. Construct an example showing that not all value allocations are equilibria at some price system.

**4.** [Nakayama, 1983] Let $(N, V)$ be an NTU game such that for each coalition $S$, the set $V(S)$ is compactly generated, i.e. $V(S) = K_S - \mathbb{R}_+^S$ for some compact set $K_S$. Assume also that $V(S)$ contains a point $x$ with $x_i \geq \max V(\{i\})$ for all $i \in S$, and that $\max V(\{i\}) > 0$ for all $i$.

Let $w \in \Delta_N$ with $w_i > 0$, all $i$, be a given ("weight") vector, and for each coalition $S$, let

$$h(w, S) = \max\{h \mid hw_S \in V(S)\},$$

where $w_S = (w_i)_{i \in S}$ is the restriction of $w$ to $S$, and let $W^*$ be the subset of weight vectors $w$ for which $h(w, N)w_i \geq \max V(\{i\})$ for all $i$. Define the *excess* of the coalition $S$ at $w$ as $e(w, S) = \sum_{i \in S}(h(w, S) - h(w, N))w_i$. Give an interpretation of this excess.

Let $\theta(w)$ be the vector of all the excesses $e(w, S)$ in non-increasing order. If $w^*$ minimizes the vector $\theta(w)$ in the lexicographic order, then it is a *nucleoulus weight vector*, and the payoff $h(w^*, N)w^*$ is a *nucleolus* for $(N, V)$.

Show that the nucleolus defined in this way is nonempty and singlevalued. Show that if the core of $(N, V)$ is nonempty, then it contains the nucleolus.

# Chapter 16

# The Theory of Game Forms

## 1 Introduction

In this chapter, we return to the concept of a *game form* introduced at the very beginning of our treatment of game theory. It may be argued, that game forms are the most basic concept, describing as it does the rules of the game without involving the preferences over outcomes of the players. Therefore, game forms matter in *setting up rules* for a particular behavior, which leads to the *implementation problems* touched upon in the context of auction and mechanism theory. Also, game forms show up in other contexts, as will be indicated in the course of the present chapter.

We begin with a game form $G = (N, (S_i)_{i \in N}, \pi)$ (cf. Definition 1 of Chapter 1), where $\pi : S_1 \cdots S_n \to A$ is the outcome function, taking strategy arrays $s = (s_1, \ldots, s_n)$ to elements $\pi(s)$ of $A$. When the outcome space $A$ matters for the discussion, we shall make it eksplicit using the notation $G = (N, (S_i)_{i \in N}, \pi; A)$. For the moment, we shall assume that $A$ is a finite set.

Let $Q(A)$ be the set of *quasi-orders* (complete, transitive and reflexive partial orders) on the set $A$. An array $Q_N = (Q_1, \ldots, Q_n)$ in $Q(A)^N$ (known in the literature as a *profile* of preferences) gives rise to a game $G(Q_N) = (N, (S_i)_{i \in N}, (u_i)_{i \in N})$, where for each $i \in N$, $u_i$ is a utility representation of $Q_i$. We thus have a map

$$G(\cdot) : Q(A)^N \to \Gamma,$$

where $\Gamma$ is the set of games. It is convenient for the discussion below to extend the description of a game so as to include an outcome map, so that the game $G(Q_N)$ is written as $(N, (S_i)_{i \in N}, (u_i)_{i \in N}, \pi)$. We call this game an *extended game* and by a slight abuse of notation, $\Gamma$ will denote the set of all such extended games.

Suppose that $\phi$ is a (single-valued) solution concept for normal form games, so that $\phi$ assigns a strategy array $\phi(G) = (s_1, \ldots, s_n) \in S_1 \times \cdot \times S_n$ to each game $\Gamma = (N, (S_i)_{i \in N}, (u_i)_{i \in N})$ . Composing $\phi$ with the outcome map $\pi$ of $G$, we get an outcome $\pi(\phi)(G)$ in $A$. Thus, every array $Q_N \in Q(A)^N \to A$ induces (via the solution $\phi$) a specific choice of alternative from $A$. We denote by $h$ the function defined by the commutative diagram

$$Q(A)^N$$
$$G(\cdot) \downarrow \qquad \overset{h}{\cdots\cdots\cdots}$$
$$\Gamma \xrightarrow[\pi \circ \phi]{} A$$

(1)

A map from preference profiles to alternatives is called a *social choice function*. It may represent a formalization of the collective ethics of a society: given that the individuals assess the alternatives in $A$ as described by $Q_N$, the society should then select $h(Q_N)$. In our present context, the social choice function was derived from a game-theoretical solution concept, so it represents the outcome of choosing under the rules specified by $G$.

Changing our way of looking at the diagram in (1), we may consider the social choice function $h$ as the primitive concept, and the problem then turns to one of designing a game form $G$ with associated map $G(\cdot)$ such that the diagram

$$Q(A)^N \xrightarrow{h} A$$
$$G(\cdot) \vdots \qquad \nearrow \pi \circ \phi$$
$$\Gamma$$

(2)

is commutative. This is the classical implementation problem: Find rules such that the actions of the individuals within the framework of these rules perform with some initially stated ethics of social choice.

Now, formulating the problem as we did in (1) and (2) is too ambiguous for several reasons. We know already that solutions to games are rarely single-valued, so that $\pi \circ \phi$ is a correspondence or multimap, assigning to each game $G$ a subset of $A$. Keeping the social choice function $h$, we are faced with a problem of *partial implementation*, finding a game form such that for each profile, there is a solution to the associated game which coincides with society's choice, or

$$Q(A)^N \xrightarrow{h} A$$
$$G(\cdot) \vdots \qquad \nearrow \pi \circ \phi$$
$$\Gamma$$

(3)

where the difference from (2) is indicated by the double heads of the arrow indicating that the mapping is multivalued.

But there is an additional reason why use of social decision functions is not appropriate. If we are given a social choice function $h : Q(A)^N \to A$ which we want to implement in the sense of (2) or (3), then a first candidate for an implementing game form would be a *direct* one, i.e. a game form of the type $G^h = (N, Q(A)^N, \pi)$, where the strategy set of each player is the set of quasi-orders on $A$. Notice that the outcome map of this mechanism is itself a social choice function. Clearly, the reasoning behind the revelation principle carries over to this situation and tells us that if $h$ can be implemented (in $\phi$-solutions) by some game form, then it can be implemented by a direct mechanism and in such a way that the true preferences are $\phi$-solution strategies.

If as solution $\phi$ we choose Nash equilibrium, then the idea of a direct mechanism where truth-telling is an equilibrium strategy for all preference profiles will pose problems, since it contradicts a result about manipulability of social choice to be outlined below. A social choice function $h : Q(A)^N \to A$ is *manipulable* if there is a preference profile $Q_N$, an individual $i$ and a preference $R_N \in Q(A)$ such that

$$h(R_N, Q_{N\setminus S}) \; Q_i \; h(Q_N),$$

so that truth-telling is not a Nash equilibrium for $G^h$.

The following result is known as the Gibbard-Satterthwaite theorem (Gibbard [1973], Satterthwaite [1975]).

THEOREM 1 *Let $h : Q^N \to A$ be a social choice function, which is not manipulable. Assume that $h$ is onto and that $|A| \geq 3$. Then $h$ is dictatorial, i.e. there is $i \in N$ such that for all $Q_N = (Q_1, \dots, Q_n) \in Q(A)^N$,*

$$h(Q_N) \; Q_i \; a, \text{ all } a \in A.$$

We give a proof of Theorem 1 only for the special case of linear orders (so that indifference classes are singletons) and leave the extension to the general case to the exercises.

PROOF OF THEOREM 1: We prove the theorem in several steps:

*Step 1.* First of all we show that for $B \subset A$ a proper subset of $A$, if $Q_N \in \mathcal{L}(A)$ is such that $B \; Q_i \; A \setminus B$ for all $i$ (so that all elements in $B$ are ranked above all other elements), then $h(Q_N) \in B$. Indeed, suppose that $h(Q_N) \notin B$, and let $R_N$ be such that $h(R_N) \in B$. Define the sequence $Q^{(0)} Q_N^{(1)}, \dots, Q_N^{(n)}$ with $Q_N^0 = Q_N$ and

$$Q_i^{(j)} = \begin{cases} Q_i^{(j-1)} & i \neq j, \\ R_i & i = j \end{cases}$$

(so that one preference at a time is shifted from $Q_i$ to $R_i$). Let $k \in \{1, \dots, n\}$ be the smallest index such that $h(Q^{(k-1)}) \notin B, h(Q_N^{(k)}) \in B$. Then individual $k$ can manipulate the outcome at the preference profile $Q_N^{(k-1)}$, since changing the stated preference relation to $R_k$ will give a better result according to $Q_k$. Thus, we have a contradicting proving the claim.

We notice also that if $Q_N$ and $Q'_N$ are two profiles of the above type such that all alternatives in $B$ are ordered in the same way by all individuals, then $h(Q_N) = h(Q'_N)$, since otherwise there would be an instance of manipulation.

*Step 2.* We can now define a coalition $S$ of individuals to be *decisive for $x$ against $y$* if $h(Q_N) \neq y$ for every preference profile $Q_N$ with

$$x \; Q_i \; y \; Q_i \; A \setminus \{x, y\} \text{ for } i \in S,$$
$$y \; Q_i \; x \; Q_i \; A \setminus \{x, y\} \text{ for } i \notin S$$

*Step 3.* Suppose that $S$ is decisive for $x$ against $y$, and let $z \in A\backslash\{x,y\}$. We claim that $S$ is decisive for $x$ over $z$. To show this, consider first a profile $Q_N$ with

$$x \, Q_i \, y \, Q_i \, z \, Q_i \, A\backslash\{x,y,z\} \text{ for } i \in S,$$
$$y \, Q_i \, z \, Q_i \, x \, Q_i \, A\backslash\{x,y,z\} \text{ for } i \notin S, \tag{4}$$

We have that $h(Q_N) \in \{x,y,z\}$ by Step 1, but $h(Q_N) \neq z$, since $y \, Q_i \, z$ for all $i$. If $h(Q_N) = y$, then there is a sequence $Q_N^{(0)}, \ldots, Q_N^{(n)}$ obtained by successively changing $Q_N$ to a profile with $\{x,y\}$ above all other alternatives, and letting $k$ be the step at which outcome changes from $y$ to $x$, we get that $h$ is manipulable at $Q_N^{(n)}$ by individual $k-1$.

We conclude that $h(Q_N) = x$, where $Q_N$ is the profile in (4). To see that $h(R_N) = x$ for the profile obtained from $Q_N$ by successively moving $y$ below $x$ and $z$, we argue as before, using that otherwise there would be an instance of manipulation. Thus, $S$ is decisive also for $x$ against $z$.

*Step 4.* We show now that $S$ is decisive for $y$ against $z$. Consider a profile $Q_N$ with

$$y \, Q_i \, x \, Q_i \, z \, Q_i \, A\backslash\{x,y,z\} \text{ for } i \in S,$$
$$z \, Q_i \, y \, Q_i \, x \, Q_i \, A\backslash\{x,y,z\} \text{ for } i \notin S, \tag{5}$$

We have that $h(Q_N) \in \{y,z\}$ by Step 1. If $h(Q_N) = z$, then successively moving $y$ below $x$ and $z$ would give an instance of manipulation, since $S$ is decisive for $x$ against $z$. Consequently, $h(Q_N) \neq z$, and arguing as above we obtain that $S$ is decisive also for $y$ against $z$.

*Step 5.* By the previous steps, we have that if $S$ is decisive for $x$ against $y$, then $S$ is decisive for any $a \in A$ against any $b \in B$. Therefore, we say in the following that $S$ is decisive, and we claim that the decisive coalitions constitute an ultrafilter on $S$, so that (a) $S_1$ decisive and $S_1 \subset S_2$ implies $S_2$ decisive, (b) $S_1, S_2$ decisive implies $S_1 \cap S_2$ decisive, and (c) for each $S \subset N$, either $S$ or $N\backslash S$ is decisive. To show (a) consider a pofile $Q_N$ with

$$x \, Q_i \, y \, Q_i \, A\backslash\{x,y\} \text{ for } i \in S_2,$$
$$y \, Q_i \, x \, Q_i \, A\backslash\{x,y\} \text{ for } i \notin S_2.$$

If $h(Q_N) = y$, then successive reversals of the order of $x$ and $y$ for individuals in $S_2\backslash S_1$ would produce an instance of manipulation, so $h(Q_N) = x$ and $S_2$ is indeed decisive. To show (b), we consider the profile $Q_N$ with

$$x \, Q_i \, y \, Q_i \, z \, Q_i \, A\backslash\{x,y,z\} \text{ for } i \in S_1 \cap S_2,$$
$$y \, Q_i \, z \, Q_i \, x \, Q_i \, A\backslash\{x,y,z\} \text{ for } i \notin S_1\backslash S_2,$$
$$z \, Q_i \, x \, Q_i \, y \, Q_i \, A\backslash\{x,y,z\} \text{ for } i \notin S_2\backslash S_1$$

Since $S_1$ is decisive for $y$ over $z$, we cannot have $h(Q_N) = z$ (if $h(Q_N) = z$, successively moving $x$ below $y$ and $z$ would yield a case of manipulation), and since $S_2$ is decisive for $x$ over $y$, we cannot have $h(Q_N) = y$, consequently $h(Q_N) = x$. Moving down

$y$ successively, we get that $S_1 \cap S_2$ is decisive for $x$ against $z$, and by the previous steps, $S_1 \cap S_2$ is decisive, which gives us (b). Finally, (c) is an easy consequence of the definition of decisiveness.

*Step 6.* Since the decisive coalitions form an ultrafilter on the finite set $N$ of individuals, there is a unique individual $i$ such that $\{i\}$ is decisive. Let $Q_N$ be arbitrary, and suppose that

$$x\ Q_i\ y\ Q_i\ A \backslash \{x, y\}. \tag{6}$$

If also

$$\{x, y\}\ Q_j\ A \backslash \{x, y\},\ j \neq i, \tag{7}$$

then $h(Q_N) = x$. If $h(Q_N) \neq x$, then by successive change from a profile satisfying (6) and (7) will give rise to an instance of manipulation. We conclude that $h$ is dictatorial. □

The Gibbard-Satterthwaite theorem is closely related to the famous *Arrow impossibility theorem* [Arrow, 1963], which pertains to mappings from $Q(A)^N \to Q(A)$, aggregating preference profiles into preference relations which might be considered as society's preferences, and stating that if aggregation should have certain simple and reasonable properties, then it must be dictatorial, taking a particular agent's preferences as society's.

## 2  Game forms and effectivity functions

Starting from a game form, one obtains a normal form game whenever the preferences over outcomes of all the players are specified. Turning to cooperative solutions, which are studied mainly in the context of the characteristic functions of the game, it might be useful to have a basic form of the cooperative game which does not yet involve the preferences of the players. This is where the *effectivity function* comes in.

We use the notation $\mathcal{P}(D)$ for the set of all subsets of a set $D$, and we let $\mathcal{P}^2(D) = \mathcal{P}(\mathcal{P}(D))$.

DEFINITION 1 *An effectivity function is a map* $E : \mathcal{P}(N) \to \mathcal{P}^2(A)$ *which satisfies the following conditions:*

> (i)    $E(N) = \mathcal{P}(A)$,
> (ii)   $E(\emptyset) = \emptyset$,
> (iii)  *for all* $C \in P(N)$, $A \in E(C)$,
> (iv)   *for all* $C \in P(N)$, $\emptyset \notin E(C)$.

Effectivity functions may be considered as a formalization of *constitutions*, specifying for each coalition $C$ the subsets $B$ of alternatives such that the coalition has a right to restrict society's choice to the set $B$.

From a given game form $G = (N, (S_i)_{i \in N}, \pi)$, one can derive an effectivity function in two different ways (corresponding to the two basic methods of deriving a characteristic function from a normal form game): The $\alpha$-effectivity function $E_\alpha^G$ is defined by

$$E_\alpha^G(C) = \{B \in \mathcal{P}(A) \mid \exists t_C \in S_C \; \forall s_{N \setminus C} : \pi(t_S, s_{N \setminus C}) \in B\},$$

$E_\alpha^G(\emptyset) = \emptyset$, so that $E_\alpha^G(S)$ contains all the subsets $B$ of $A$ for which $C$ can guarantee that outcome will be in $B$ no matter what the complementary coalition will choose, and the $\beta$-effectivity function is defined by

$$E_\beta^G(C) = \{B \in \mathcal{P}(A) \mid \forall s_{N \setminus S} \; \exists t_C \in S_C : \pi(t_S, s_{N \setminus C}) \in B\},$$

$E_\beta^G(\emptyset) = \emptyset$, so that for $B \in E_\beta^G(C)$, the coalition $C$ has a response to every choice by the complement such that outcome is in $B$.

The cooperative solution concepts discussed in the previous chapters have a formulation in terms of effectivity functions. We restrict attention to the core, which as usual is defined using the notion of domination: Let $E : \mathcal{P}(N) \to \mathcal{P}^2(A)$ be an effectivity function and $R_N \in \mathcal{L}(A)^N$ a preference profile, then an alternative $x \in A$ is *dominated* at $R_N$ (by $B \subset A$ via the coalition $C$) if

$$x \notin B, B \in E(C), y \, R_i \, x, \text{ all } i \in C.$$

The core of $E$ and $R_N$, written $\text{Core}(E, R_N)$, is the set of alternatives which are not dominated at $R_N$. The effectivity function $E$ is *stable* if $\text{Core}(E, R_N) \neq \emptyset$ for all $R_S \in \mathcal{L}^N(A)$.

There is a close connection between cores of effectivity functions and implementation in strong Nash equilibrium.

THEOREM 2 *Let $H : \mathcal{L}(A)^N \to A$ be a social choice correspondence which is implementable in strong Nash equilibrium with game form $G$, so that for each $R_N \in \mathcal{L}(A)^N$, the set of strong Nash equilibria of the game $(G, R_N)$ equals $H(R_N)$. Then $E_\alpha^G = E_\beta^G = E$, and $E$ is stable and maximal in the sense that $B \notin E(C) \Rightarrow A \setminus B \in E(N \setminus C)$.*

PROOF: Suppose that $s = (s_1, \ldots, s_n)$ is a strong Nash equilibrium of $(G, R_N)$ with outcome $\pi(s) = x$. If $x \notin \text{Core}(E_\beta^G, R_N)$, then there must be a coalition $C$ such that the set $\{y \mid y \, R_i \, x\}$ contains a member of $E_\beta^G(C)$, meaning that $C$ has a response to $s_{N \setminus C}$ giving a preferred outcome, contradicting that $s$ is a strong Nash equilibrium. Since $E_\alpha^G(C) \subset E_\beta^G(C)$ for all $C$, we must have $\text{Core}(E_\beta^G, R_N) \subset \text{Core}(E_\alpha^G, R_N)$ for all $R_N$, in particular, $E_\alpha^G$ is stable.

Suppose next that $B \notin E_\alpha^G(C)$ for some coalition $C$. Then for each strategy $s_C$ chosen by $C$, $N \setminus C$ has a strategy $t_{N \setminus C}$ for which $\pi(s_C, t_{N \setminus C}) \notin B$, meaning that $A \setminus B \in E_\beta^G(N \setminus C)$. Conversely, if $A \setminus B \in E_\beta^G(N \setminus C)$, consider a profile $Q_N$ such that

$$x \, Q_i \, y, \quad x \in B, y \notin B, \quad i \in C,$$
$$y \, Q_i \, x, \quad x \in B, y \notin B, \quad i \notin C.$$

Then the game $(G, Q_N)$ has no strong equilibrium with outcome in $B$ (any such strategy can be met with a response from $N \backslash C$ forcing outcome to be outside $B$), and since $(G, Q_N)$ has a strong Nash equilibrium, the resulting outcome must be in $A \backslash B$, from which we conclude that $B \notin E_\beta^G(C)$. It follows that $E_\alpha^G = E_\beta^G = E$, and the remaining properties follow directly. $\qquad \square$

So far we have that if the effectivity function $E$ is derived from a game form which implements a social choice correspondence in strong Nash equilibria, then the core correspondence $H(R_N) \subseteq \text{Core}(E, R_N)$ for all $R_N \in \mathcal{L}(A)^N$. The theorem has a converse. We state and prove a somewhat weaker version, the more general result is in Moulin and Peleg [1982], cf. Abdou and Keiding [1991].

THEOREM 3 *Let* $E : \mathcal{P}(N) \to \mathcal{P}^2(A)$ *be an effectivity function which is stable and maximal. Then the social choice correspondence* $\text{Core}(E, \cdot)$ *is implementable in strong Nash equilibrium.*

Proof: Define the implementing game form $G = (N, (S_i)_{i \in N}, \pi)$ as follows: For each $i$, the strategy set is

$$S_i = \text{Graph}\,[\text{Core}(E, \cdot)] \times \mathcal{L}(A) \times \mathbf{N},$$

where Graph $[\text{Core}(E, \cdot)]$ is the set of all pairs $(R_n, x)$ with $x \in \text{Core}(E, R_N))$. Thus, each individual proposes a pair $(Q_N^i, x_i) \in C$, a linear order $T^i$, and a number $t^i$. The outcome function $\pi$ will be defined in several steps:

Let the strategy array $s$ be given by $s_i = ((Q_N^i, x_i), T^i, t^i)$ for each $i \in N$, and define the partition $\Pi[s]$ induced by $s$ as the coarsest partition $\{C_1, \ldots, C_k\}$ such that $(Q_N^i, x_i) = (Q_N^j, x_j)$ for $i, j \in S_h, h = 1, \ldots, k$. For each $h$, let

$$D_h(s) = \{y \in A | y \neq x_h, y \, Q_i^h \, x_h, i \in S' \cup (N \backslash S_h), S' \subset S_h\}.$$

Since $x_h \in \text{Core}(E, Q_N^h)$ we have that $D_h(s) \notin E(N \backslash C_h)$, and by maximality of $E$, $B_h(s) = A \backslash D_h(s) \in E(S_h)$.

Now we define

$$\pi(s) = \max(T^{i_0} | \cap_{h=1}^k B_h),$$

where $i_0$ is the individual with lowest index such that $t^{i_0} = \max_{i \in N} t^i$; this individual chooses the maximal element in $\cap_{h=1}^k B_h$ for the stated linear order $T^{i_0}$.

Let $R_N \in \mathcal{L}(A)^N$ be arbitrary, and let $x \in \text{Core}(E, R_N)$. We show that there is a strong Nash equilibrium $s$ in $(G, R^N)$ with $\pi(s) = x$.

Let

$$s_i = ((R_N, x), T^i, 1),$$

where $T^i \in \mathcal{L}(A)$ is a linear order with $x \, T^i \, A \backslash \{x\}$. Clearly $\pi(s) = x$. Suppose that there is a coalition $C$ and a $C$-strategy $t_C$ such that $\pi(t_C, s_{N \backslash C}) = y$ and $y \, R_i \, x$ for all $i \in C$. Let $\Pi[(t_C, s_{N \backslash S}] = \{C_1, \ldots, C_k\}$, then there is at least one of the sets $C_1, \ldots, C_k$, say $C_1$, which contains $N \backslash S$. By our construction, $y$ belongs to $D_1(t_C, s_{N \backslash C})$, consequently

$y \notin B_1(t_S, s_{N \setminus S})$, and we conclude that $\pi(t_C, s_{N \setminus C}) \neq y$, a contradiction, showing that $s$ is strong Nash equilibrium. □

As a by-product of the considerations above, where the questions of implementation were at the forefront, we have that every stable and maximal effectivity function $E$ has a *representation*, in the sense that there is a game form $G = (N, (S_i)_{i \in N}, \pi)$, such that $E = E_\alpha^G$. Remembering that effectivity functions may be viewed as *constitutions*, specifications of the choice possibilities of each coalition of individuals, our result pertains to the *consistency problem*. For a given constitution, defined by an effectivity function, the problem is to find out whether there exists a choice procedure such that the constitution is effective, in the sense that the subsets of alternatives to which a coalition can restrict the outcome of the choice procedure to $B$ coincides with the family prescribed by the effectivity function.

For the case studied here, where $A$ is a finite set, the question of existence of a representation has a simple solution (see Problem **16.3**). For a generalization to infinite spaces of alternatives, we let $A$ be a compact metric space. By $\mathcal{K}(A)$ we denote the set of all compact subsets of $A$. Now an effectivity function is a map $E : P(N) \to P(\mathcal{K}(A))$ which satisfies the boundary conditions

  (i)    $E(N) = \mathcal{K}(A)$,
  (ii)   $E(\emptyset) = \emptyset$,
  (iii)  for all $S \in P(N)$, $A \in E(S)$,
  (iv)   for all $S \in P(N)$, $\emptyset \notin E(S)$.

A game form $G = (N, (S_i)_{i \in N}, \pi; A)$ is a *representation* of the effectivity function $E : P(N) \to P(\mathcal{K}(A))$ if $E = E_\alpha^G$. However, in the present context it seems natural to consider representations with more structure: A representation $G = (N, (S_i)_{i \in N}, \pi; A)$ of an effectivity function $E$ is *continuous* if the strategy sets $S_i$ are compact metric spaces for $i = 1, \ldots, n$, and $\pi : S_1 \times \cdots \times S_n \to A$ is continuous. The existence problem for continuous representations is far from trivial. There are very simple effectivity functions which have no representations; one such case is given as Example 16.1.

## 3   Solvability of game forms

In the present section, we shall be concerned with the problem of solvability of game forms. A game form is *solvable* for a given type of game-theoretic solution if for each assignment of preferences over outcomes to players, the resulting game has at least one solution. Thus, investigating solvability conditions amounts to finding properties of the underlying game form, which describe the basic rules of the game, such that there are solutions no matter how the players assess the outcomes.

We shall restrict our treatment to a particular type of solution, namely strong Nash equilibria, which are however quite important in the context of implementation and representation theory, as seen in the previous section. In addition, we shall work here only with game forms where all strategy sets are finite, so that in particular we are considering solutions in pure strategies.

We need an extension of the concept of an effectivity function, allowing us to deal with situations where several coalitions may be effective for particular subsets

---

**Example 16.1. An effectivity function which has no continuous representation.**
Let $A$ be the set $\{x = (x_1, x_2) \in \mathbb{R}_+^2 \mid x_1 + x_2 \leq 1\}$, and let the effectivity function $E : P(N) \to P(\mathcal{K}(A))$ be such that line segments $[(0, x_2), (1, 0)]$, for $x_2 \in [0, 1]$, belong to and are minimal for inclusion in $E(\{1\})$, and similarly all line segments $[(x_1, 0), (0, 1)]$, for $x_1 \in [0, 1]$, are minimal for inclusion in $E(\{2\})$. We show that $E$ has no continuous representation.

Suppose, on the contrary, that $G = (\{1, 2\}, S_1, S_2, \pi; A)$ is a continuous representation of $E$. Then for each minimal (for inclusion) element $B_1$ of $E(\{1\})$, there is $\sigma^1 \in \Sigma^1$ such that $\pi(\sigma^1, \Sigma^2) = B_1$. In particular, there is a sequence $(\sigma_n^1)_{n=1}^\infty$ such that

$$\pi(\sigma_n^1, \Sigma^2) = \left[\left(0, 1 - \frac{1}{n}\right), (1, 0)\right],$$

and by compactness, we may assume that $(\sigma_n^1)_{n=1}^\infty$ converges to some $\sigma_0^1 \in \Sigma^1$ with $\pi(\sigma_0^1, \Sigma^2) = [(0, 1), (1, 0)]$.

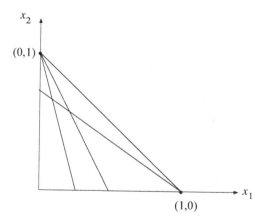

Next, for each $n \in \mathbb{N}$ choose $\sigma_n^2 \in \Sigma^2$ such that

$$\pi(\Sigma^1, \sigma_n^2) = \left[\left(1 - \frac{1}{n}, 0\right), (0, 1)\right].$$

Then $\pi(\sigma_0^1, \sigma_n^2) \in [(0, 1), (1, 0)] \cap \left[\left(1 - \frac{1}{n}, 0\right), (0, 1)\right]$, i.e. $\pi(\sigma_0^1, \sigma_n^2) = (0, 1)$, and by compactness of $\Sigma^2$ we may assume that $(\sigma_n^2)_{n=1}^\infty$ converges to some $\sigma_0^2$ such that $\pi(\sigma_0^1, \sigma_0^2) = (0, 1)$.

**Example 16.1, continued.** Now, by (joint) continuity of $\pi$ for any $\varepsilon > 0$ there is a neighborhood $U$ of $(\sigma_0^1, \sigma_0^2) \in \Sigma^1 \times \Sigma^2$ such that $\pi(\sigma^1, \sigma^2) \in B_\varepsilon((0,1))$ for all $(\sigma^1, \sigma^2) \in U$. However, choosing $n_1$ and $n_2$ sufficiently large with $n_1 > Kn_2$ for large enough $K$, we may achieve that $(\sigma_{n_1}^1, \sigma_{n_2}^2) \in U$ but

$$\pi(\sigma_{n_1}^1, \sigma_{n_2}^2) \in \left[\left(0, 1 - \frac{1}{n_1}\right), (1,0)\right] \cap \left[\left(1 - \frac{1}{n_2}, 0\right), (0,1)\right]$$

is as close to $(1,0)$ as desired, a contradiction. We conclude that $E$ has no continuous representation.

---

*simultaneously*, i.e. at a given profile of preferences. As usually, $N = \{1, \ldots, n\}$ is a set of players, and $X$ is a set of outcomes.

DEFINITION 2 *An effectivity pattern on* $(N, X)$ *is an array* $(D_1, \ldots, D_n)$ *of subsets of* $X$, *not all empty. An effectivity structure over* $(N, X)$ *is a family* $\mathcal{E} = (E_x)_{x \in X}$ *of sets of effectivity patterns* $X$ *satisfying the monotonicity condition*

$$[x \in X, (D_1, \ldots, D_n) \in E_x, D_i \subseteq D'] \Rightarrow (D_1', \ldots, D_n') \in E_x. \tag{8}$$

The effectivity structure may be considered as an elaboration of the concept of an effectivity function, allowing for the possibility that several coalitions may exert influence on the outcome at the same time. If $E$ is an effectivity function, and $B \in E(S)$, then $(S, B)$ gives rise to an effectivity patterns $(D_1, \ldots, D_n)$ with $D_i = B$ for $i \in S$ and $D_i \in X$ otherwise. In general, the sets $D_i$ of an effectivity pattern represents the power of an individual $i$, in the sense that if the coalition $S$ forms, then it can assure that outcome will be in $\cap_{i \in S} D_i$.

A game form $G = (N, (S_i)_{i \in N}, \pi)$ gives rise to an effectivity structure $\mathcal{E}^G = (E_x^G)_{x \in X}$, namely by

$$E_x^G = \{(D_1, \ldots, D_n) \mid \forall s \in \pi^{-1}(x), \exists T \in C, s_T' \in S_T : \pi(s_T', s_{N \setminus T}) \in \cap_{i \in T} D_i\}. \tag{9}$$

The effectivity structure $\mathcal{E}^G$ is said to be *associated* with $G$. The construction resembles that of the $\beta$-effectivity function considered previously: For all strategy arrays giving rise to the outcome $x$, the coalition $T$ has a reply which makes the outcome belong to the intersection of the $D_i$ for $i$ a member of $T$.

If $R_N$ is a profile of linear orders over $X$, then an outcome $x$ is *dominated* in $\mathcal{E}$ at $R^N$ if the set

$$(\overline{R}_1(x), \ldots, \overline{R}_n(x)) \in E_x, \tag{10}$$

where $\overline{R}_i(x) = \{x' \mid x' \, \overline{R}_i \, x\}$ for $i \in N$ (here $\overline{R}$ is the strict preference relation defined from $R$), and $x$ is undominated otherwise. The notions introduced so far are connected as shown by the following result.

THEOREM 4 *Let* $G = (N, (S_i)_{i \in N}, \pi)$ *be a game form with associated effectivity structure* $\mathcal{E}^G$, *and let* $R_N \in Q(X)^N$ *be a preference profile. Then the following are equivalent:*

  (i) *x is a strong Nash equilibrium outcome of the game* $(G, R_N)$,
  (ii) *x is undominated in* $\mathcal{E}^G$.

PROOF. (ii)$\Rightarrow$(i): If $x$ is not a strong Nash equilibrium outcome for the game $(G, R_N)$, then for all $s \in \times_{i \in N} S_i$ there is some coalition $T$ which can improve, that is $T$ has a strategy $s'_T$ such that $\pi(s'_T, s_{N \setminus S}) \in \overline{R}_i(x)$ for all $i \in T$, or equivalently,

$$\pi(s'_T, s_{N \setminus T}) \in \cap_{i \in T} \overline{R}_i(x),$$

which by (9) means that $(\overline{R}_1(x), \dots, \overline{R}_n(x)) \in E^G_x$, so that $x$ is dominated in $\mathcal{E}^G$.

  (i)$\Rightarrow$(ii): If $x$ is dominated in $\mathcal{E}^G$, then by (10) $(\overline{R}_1(x), \dots, \overline{R}_n(x))$ is an effectivity pattern in $E^G_x$, and using the definition of $E^G_x$, we get that for each strategy array $s$ with $\pi(s) = x$ there is a coalition which has an improvement of $s$, so that $x$ cannot be a strong Nash equilibrium outcome for $(G, R^N)$. $\square$

From the result of Theorem 4 we obtain that a game form is solvable if the associated effectivity structure has undominated outcomes at every profile. An effectivity structure with this property is said to be *stable,* and we look for conditions on effectivity structures which guarantee their stability. One such condition is *acyclicity* which is defined below.

DEFINITION 3 *Let* $\mathcal{E}$ *be an effectivity structure over* $(N, X)$. *A cycle in* $\mathcal{E}$ *is a family* $((C^1, \mathcal{B}^1), \dots, (C^r, \mathcal{B}^r))$, *where for* $k = 1, \dots, r$, $S^k \subseteq N$ *is a coalition,* $C^k$ *is a nonempty subset of X, and* $\mathcal{B}^k$ *is an effectivity pattern in* $E_x$ *for all* $x \in C^k$, *such that*

  (i) $\{C^1, \dots, C^r\}$ *is a partition of X,*
  (ii) *for* $i \in N$ *and any nonempty subset J of* $\{1, \dots, r\}$, *there is* $k \in J$ *such that* $C^k \cap D^k_i = \emptyset$ *for all* $j \in J$.

*The effectivity structure* $\mathcal{E}$ *is acyclic if there are no cycles in* $\mathcal{E}$.

Intuitively, the cycle is an arrangement of domination patterns such that (i) every outcome is dominated, and (ii) this pattern of domination is compatible with an assignment of quasi-orders to the individuals. The exact formulation is given by the next theorem.

THEOREM 5 *Let* $\mathcal{E}$ *be an effectivity structure over* $(N, X)$. *Then* $\mathcal{E}$ *is stable if and only if it is acyclic.*

PROOF: Let $\mathcal{E}$ be an effectivity function which is not stable; we show that $\mathcal{E}$ has a cycle. Then there is a profile $R_N \in Q(X)^N$ such that for each $x \in X$, $\mathcal{B}_x = (\overline{R}_1(x), \dots, \overline{R}_n(x))$ belongs to $E_x$. We claim that the family $(\{x\}, \mathcal{B}_x)_{x \in X}$ is a cycle in $\mathcal{E}$. Indeed, the property (i) in Definition 3 is immediate, and to show that (ii) holds, choose an arbitrary $i \in N$ and a subset $J$ of $X$. Since $R_i$ is a quasiorder, there is $x \in J$ such that $x' R_i x$ for all $x' \in J$. But then $\{x\} \cap \overline{R}_i(x') = \emptyset$ for all $x' \in J$, and property (ii) is satisfied.

Assume next that $\mathcal{E}$ is stable but has a cycle $((C^1, \mathcal{B}^1), \ldots, (C^r, \mathcal{B}^r))$. We construct a profile $R_N \in Q(X)^N$ such that every outcome in $X$ is dominated, obtaining a contradiction.

Let $i \in N$ be arbitrary. By property (ii) of Definition 3, there is $k_1 \in \{1, \ldots, r\}$ such that $C^{k_1^i} \cap D_i^j = \emptyset$ for $j = 1, \ldots, r$. By the same reasoning, there is $k_2^i \in J_1 = \{1, \ldots, r\} \setminus \{k_1^i\}$ with $C^{k_2^i} \cap D_i^j = \emptyset$ for $j \in J_1$, and in general, for $t = 2, \ldots, r$, there is $k_t^i \in J_{t-1} = \{1, \ldots, r\} \setminus \{k_1^i, \ldots, k_{t-1}^i\}$ with $C^{k_t^i} \cap D_i^j = \emptyset$ for all $j \in J_{t-1}$. Let $R_i \in Q(A)^N$ be such that $x R_i x'$ whenever $x \in C^{k_t^i}$, $x' \in C^{k_{t'}^i}$. We then have that for $x \in X$ arbitrary, if $t \in \{1, \ldots, r\}$ is such that $x \in C^{k_t^i}$, then $C^{k_t^i} \cap D_i^{k_{t'}^i} = \emptyset$ for $t' < t$. It follows that $D_i^{k_t^i} \subseteq X \setminus \cup_{t' < t} C^{k_{t'}^i}$. From our construction of $R_i$, we have that $X \setminus \cup_{t' < t} C^{k_{t'}^i} = \overline{R}_i(x)$.

Let $R_N = (R_i)_{i \in N}$ be the profile consisting of the quasiorders $R_i$ constructed above. Using the monotonicity condition (ii) of Definition 2, we get that $(\overline{R}_1(x), \ldots, \overline{R}_n(x))$ belongs to $E_x$ for each $x \in X$, so that each element of $X$ is dominated in $\mathcal{E}$, contradicting the stability of $\mathcal{E}$.                                                                     $\square$

## 4   Stable matching

The study of *matchings* is a traditional field of applied game theory, going back to the initial contribution by Gale and Shapley [1962] and having received much attention over the years, cf. the survey in Roth and Sotomayor [1992].

**4.1. The marriage problem and its extensions.** The basic problem in matching is illustrated by its simplest version, that of the *marriage problem*. There are given sets $M$ and $W$ of males and females. A *matching* for $(M, W)$ is a bijection $\mu$ from $M \cup W$ to itself, such that $\mu^2(i) = \mu(\mu(i))$ for each $i \in M \cup W$, and having the property that $\mu(m) \in W \cup \{m\}$, $\mu(w) = M \cup \{w\}$ for all $m \in M$, $w \in M$. Thus, the matching assigns to each male either himself (meaning that $m$ remains unmarried) or a female, and similarly each woman either remains unmarried or is matched to a male. The condition $\mu^2(i) = i$ assures that the matching is consistent. Let $\mathcal{M}$ be the set of all matchings for $(M, W)$.

For $m \in M$, let $\mathcal{P}_m$ be the set of linear orders on $W \cup \{m\}$, and for $w \in W$, $\mathcal{P}_w$ is the set of linear orders on $M \cup \{w\}$. Clearly, matchings $\mu \in \mathcal{M}$ can be ordered by each individual $i \in M \cup W$ using the preference relation $P_i \in \mathcal{P}_i$ (so that $x$ is indifferent w.r.t. the aspects of the matching which does not involve $x$). If each of the preference relations $P_i$ is represented by a utility function $u_i$, we obtain a cooperative NTU game $(M \cup W, \mathcal{S}, V^{\mathcal{M}})$, where $\mathcal{S}$ is a coalition structure on $M \cup W$, and

$$V^{\mathcal{M}}(S) = \{(z_i)_{i \in S} \mid \exists \mu \in \mathcal{M} : z_i \leq u_i(\mu(i))\}.$$

The coalition structure $\mathcal{S}$ will be restricted to contain singletons and pairs $(m, w) \in M \times W$. A matching $\mu \in \mathcal{M}$ in the marriage problem is *stable* if it belongs to the core of $(M \cup W, \mathcal{S}, V^M)$. This means that no individual can improve (by getting unmarried) and no pair $(m, w)$ can join in a way that

$$u_m(w) > u_m(\mu(m))$$
$$u_w(m) > u_w(\mu(w))$$

THEOREM 6 *The core of the marriage game (and thus the set of stable matchings) is nonempty.*

To prove the theorem, we exhibit a particular method for finding a matching, the *deferred acceptance* algorithm proposed by Gale and Shapley [1962]. This method will work also for more complicated matching problems, as we shall see below.

The deferred acceptance algorithm goes as follows: Choose one of the subsets $M$ or $W$, say $M$. At stage 1, each $m \in M$ proposes to its most preferred element of $W \cup \{m\}$, and each $w \in W$, who has received a proposal, preliminarily accepts the proposal from the most preferred $m \in M$ (if better than $w$) and rejects the other proposals. In the $k$th stage, each $m \in M$ who was rejected at the previous stage renews the proposal, now selecting the most preferred element from those that have not yet rejected him and each $w$ chooses the best from all proposals received so far and rejects the rest. The algorithm stops if there are no rejections.

PROOF OF THEOREM 6: The deferred acceptance algorithm as outlined above must stop after finitely many steps: Indeed, let $W_m^{(1)}$ be the set of elements in $W \cup \{m\}$ which are as good as $m$ according to $P_m$, and for $k > 1$, let $W^{(k)}$ be the subset of $W_m^{(1)}$ obtained by removing those of $m$'s previous proposals that have been rejected. Clearly, $W_m^{(k)} \subset W_m^{(k-1)}$ for each $k$, and $\sum_{m \in M} |W_m^{(k)}|$ decreases at each step, so the algorithm must terminate after at most $\sum_{m \in M} |W_m^{(1)}|$ steps.

Let $\mu$ be a matching obtained by the deferred acceptance algorithm. Then no singleton coalition can improve: Males never propose a $w$ with $m \, R_m w$, and females reject proposals from $m$ if $w \, P_w \, m$. Suppose that there is a pair $(m, w) \in M \times W$ such that $w \, P_m \, \mu(m)$, $m \, P_w \, \mu(w)$. Then $w$ must have received and rejected a proposal from $m$ at some stage, but since $w$ can choose among all previously received proposals at any stage, she would not have accepted a proposal inferior to $m$, and we have a contradiction, showing that $\mu$ belongs to the core. □

Since the deferred acceptance algorithm started with a choice of the proposing side (either $M$ or $W$), it follows that there may be several stable matchings. Indeed, it can be shown that the set of stable matchings contains an $M$-optimal matching $\mu_M$ (no other stable matching $\mu \neq \mu_M$ satisfies $\mu(m) \, P_m \, \mu_M(m)$ for all $m \in M$) and a $W$-optimal matching $w_M$ (defined in similar way):

THEOREM 7 *The set of stable matchings for the marriage problem contains an M-optimal matching.*

PROOF: We show that the matching $\mu^M$ produced by the deferred acceptance algorithm with $M$ as proposers is $M$-optimal. Define $w \in W$ as *achievable* for $m$ if there is a stable matching such that $\mu(m) = w$. Suppose that in the first step, $w$ receives proposals, rejects $m$ and accepts $m'$. If $w$ is achievable for $m$, there must be some stable matching $\mu$ such that $w = \mu(m)$. Let $w' = \mu(m')$; then $m'$ prefers $w$ to $w'$ since he proposed to $w$, consequently the pair $(m', w)$ can improve the matching $\mu$, contradicting stability of $\mu$. We conclude that $w$ is not achievable for $m$.

At step $k$, suppose that $w$ has not yet rejected any proposals but now rejects $m$ and chooses $m'$. Arguing as before, we again obtain that $w$ cannot be achievable for $m$.

It now follows that there can be no stable matching $\mu$ with $\mu(m)$ as good as $\mu^M(m)$ for each $m$ and better for at least one $m' \in M$, since in that case then $\mu(m')$ would have been proposed by $m'$ with rejection as result, so that $\mu(m')$ would not be achievable for $m'$, so that $\mu$ cannot be stable.                                           □

The marriage model can be extended so as to deal with situations of many-to-one matching. The classical case is the *college admissions* model. Here we have a set $C$ of colleges and a set $S$ of students. Each college $C \in C$ has a quota $q_C$, which is the maximal number of students that can be enrolled, but there may be vacancies. A matching is a map $\mu$ which to each $s \in S$ assigns an element of $C \cup \{s\}$ (where $\mu(s) = s$ means that student $s$ is not admitted to a college), and to each $C \in C$ a $q_C$-tuple $(s_1, s_2, \ldots, s_k, C, \ldots, C)$, meaning that students $s_1, \ldots, s_k$ with $k \le q_C$ are enrolled, and remaining $q_C - k$ places are vacant. The matching $\mu$ must satisfy the condition

$$\mu(s) = C \Leftrightarrow s \in \mu(C).$$

As before, it is assumed that colleges have preferences over students and students have preferences over colleges (where, as before, the position of $C$ for a college and $s$ for a student determines the admissible alternatives). The *stable matchings* are those which cannot be improved by any individual agent (college or student) or any college-student pair.

The deferred acceptance algorithm may be adapted to the situation (colleges may send out $q_c$ proposals) to give a stable matching, and the stable matchings will again contain both $C$-optimal and $S$-optimal matchings.

**4.2. Strategic aspects.** So far, we have considered matchings which were stable given the preferences of the individuals, and we have exhibited a particular method for finding such matchings, the deferred acceptance algorithm. In general, such methods can be seen as a game form $G = (M \cup W, (\mathcal{P}_x)_{x \in M \cup W}, \pi)$, where the outcome map $\pi$ assigns a matching to each preference information $(P_x)_{x \in M \cup W}$. $G$ is a *stable matching mechanism* if $\pi((P_x))_{x \in M \cup W}$ is always a stable matching.

The natural question to pose in this situation is whether truth is an equilibrium strategy in the game $(G, (P_x)_{x \in M \cup W})$, and the question is answered to the negative by the following impossibility result due to Roth [1982].

THEOREM 8 *There is no stable matching mechanism such that truth is always dominating strategy for each individual.*

PROOF: It suffices to exhibit an example of a marriage market and preference assignments for each individual such that there is no stable mechanism with truthtelling a dominating strategy. We consider a case with $M = \{m_1, m_2\}$, $W = \{w_1, w_2\}$, and preferences such that

$$w_1 \; P_{m_1} \; w_2 \; P_{m_1} \; m_1 \qquad m_2 \; P_{w_1} \; m_1 \; P_{w_1} \; w_1$$
$$w_2 \; P_{m_2} \; w_1 \; P_{m_2} \; m_2 \qquad m_1 \; P_{w_2} \; m_2 \; P_{w_2} \; w_2.$$

There are two stable matchings, namely $\mu$ defined by $\mu(m_i) = w_i$ for $i = 1, 2$, and $\nu$ given by $\nu(m_i) = w_j$ for $i, j \in \{1, 2\}, i \neq j$. Therefore any stable matching mechanism $\pi$ must result in either $\mu$ or $\nu$ with these preferences. Suppose that the mechanism chooses $\mu$; if $w_2$ changes her stated preference to $Q_{w_2}$ with

$$m_1 \; Q_{w_2} \; w_2 \; Q_{w_2} \; m_2,$$

then $\nu$ is the only stable matching for the preferences $(P_{m_1}, P_{m_2}, P_{w_1}, Q_{w_2})$, so that the mechanism must choose $\nu$ in this situation, giving a result which is better for $w_2$. If the mechanism chooses $\nu$, then $m_2$ can misrepresent preferences and obtain a better result.                                                                           □

That any stable matching mechanism can be manipulated is perhaps not surprising, in particular in view of the general manipulability results earlier in this chapter. It can be shown that the cases where manipulation may occur, that is the set of preference profiles such that truthfulness is not dominating, may not be very large. However, the very occurrence of misrepresentation remains a disturbing element for the construction of stable matching mechanisms.

## 5   Problems

**1.** For $R_N \in Q(A)^N$ a preference profile, define a *Condorcet winner* at $R_N$ as an alternative $a \in A$ such that $|\{i \mid a \; R_i \; b\}| > n/2$ for all $b \in A$.

Check whether the Condorcet winners constitute a social choice function or correspondence. Let $R_N$ be a preference profile for which there is a unique Condorcet winner. Can the social choice rule be manipulated at this profile?

**2.** A social choice function $h : Q(A)^N \to A$ satisfies *Maskin monotonicity* if for all $R_N, R'_N \in Q(A)^N$, if $h(R_N) = a$ and $\{a' \mid a' \; R'_i \; a\} \subseteq \{a' \mid a' \; R_i \; a\}$ for all $i \in N$, then $h(R'_N) = a$. Show that if $h$ can be implemented in Nash equilibria, then $h$ satisfies Maskin monotonicity. (The converse is also true, as shown in Maskin [1999]).

**3.** An effectivity function $E : \mathcal{P}(N) \to \mathcal{P}^2(A)$ is *convex* if for $C_1, C_2 \in \mathcal{P}(N), B_1 \in E(C_1)$, $B_2 \in E(C_2)$, either $B_1 \cup B_2 \in E(C_1 \cap C_2)$ or $B_1 \cap B_2 \in E(C_1 \cup C_2)$.

Show that a convex effectivity function is stable.

**4.** An effectivity function $E : \mathcal{P}(N) \to \mathcal{P}^2(A)$ is *superadditive* if $B_1 \cap B_2 \in E(C_1 \cup C_2)$ for all $C_1, C_2 \in \mathcal{P}(N)$ with $C_1 \cap C_2 \neq \emptyset$ and $B_1 \in E(C_1)$, $B_2 \in E(C_2)$, and *monotonic* if $B \in E(C)$, $B \subseteq B'$, $C \subseteq C'$ implies $B' \in E(C')$.

Show that if $A$ is a finite set, then $E$ is representable if and only if $E$ is superadditive and monotonic.

**5.** Show that the deferred acceptance algorithm leads to stable matchings in the college admission problem.

**6.** Show that two-sidedness matters for the existence of stable matchings: Construct a 'room-mates problem' ($n$ individuals must be paired, and each individual has preferences over the other individuals, including the situation of not being paired with anyone) for which there are no stable matchings.

**7.** Show for the marriage problem that the set of individuals who are single (unmatched) is the same in all stable matchings.

# Bibliography

J. Abdou and H. Keiding. *Effectivity functions in social choice.* Kluwer, Dordrecht, 1991.

G.A. Akerlof. The market for "lemons": Quality uncertainty and the market mechanism. *The Quarterly Journal of Economics,* 84:488–500, 1970.

C.D. Aliprantis and K.C. Border. *Infinite dimensional analysis, A hitchhiker's guide.* Springer, Berlin, 1999.

M. Allais. Le comportement de l'homme rationnel devant le risque: critique des postulats et axiomes de l'ecole Americain. *Econometrica,* 21:503–546, 1953.

K.J. Arrow. *Social choice and individual values.* Wiley, New York, 2nd edition, 1963.

R.J. Aumann. Acceptable points in general cooperative $n$-person games. In A.W. Tucker and R.C. Luce, editors, *Contributions to the theory of games,* volume IV, pages 287–324. Princeton, 1959.

R.J. Aumann. Subjectivity and correlation in randomized strategies. *Journal of Mathematical Economics,* 1:67–96, 1974.

R.J. Aumann. Agreeing to disagree. *The Annals of Statistics,* 4:1236–1239, 1976.

R.J. Aumann. An axiomatization of the non-transferable utility value. *Econometrica,* 53: 599–612, 1985.

R.J. Aumann and M. Maschler. The bargaining set for cooperative games. In M. Dresher, L.S. Shapley, and A.W. Tucker, editors, *Advances in Game Theory,* pages 443–476. Princeton University Press, Princeton, 1964.

R.J. Aumann and M. Maschler. Some thoughts on the minimax principle. *Management Science,* 18:54–63, 1972.

G.S. Becker. Altrusim in family and selfishness in the market place. *Economica,* 48:1–15, 1981.

B.D. Bernheim. Rationalizable strategic behavior. *Econometrica,* 52:1007–1028, 1984.

B.D. Bernheim, M.D. Whinston, and B. Peleg. Coalition-proof nash equilibria i. concepts. *Journal of Economic Theory,* 42:1–12, 1987.

L.J. Billera. Some theorems on the core of an $n$-person game without side payments. *SIAM J. Appl. Math.,* 18(567–579), 1970.

D. Blackwell. An analog of the minimax theorem for vector payoffs. *Pacific Journal of Mathematics,* 6:1–8, 1956.

O.N. Bondareva. The theory of the core in an $n$-person game. *Vestnik LGU,* 13:141–142., 1962.

C.L. Bouton. Nim, a game with a complete mathematical theory. *Annals of Math.,* 1:35–39, 1905.

H. Carlsson and E. van Damme. Global games and equilibrium selection. *Econometrica,* 61: 989–1018, 1993.

J.H. Conway. *On numbers and games.* Academic Press, London, 1976.

A. Cournot. *Recherches sur les principes matheématiques de la theéorie des richesses.* Paris, 1838.

V.P. Crawford and J. Sobel. Strategic information transmission. *Econometrica*, 50:1431–1451, 1982.

M. Davis and M. Maschler. The kernel of a cooperative game. *Naval Research Logistics Quarterly*, 12:223–259, 1965.

J.S. Demski. The general impossibility of normative accounting standards. *The Accounting Review*, 48:718–723, 1973.

P. Dubey. On the uniqueness of the Shapley value. *International Journal of Game Theory*, 4: 131–139, 1975.

J. Engwerda. *LQ dynamic optimization and differential games*. John Wiley and Sons, London, 2005.

L.C. Evans and P.E. Souganidis. Differential games and representation formulas for solutions of hamilton-jacobi equations. *Indiana University Mathematical Journal*, 33:773–797, 1984.

C. Futia. The complexity of economic decision rules. *Journal of Mathematical Economics*, 4: 289–299, 1977.

D. Gale. *The theory of linear economic models*. McGraw-Hill, New York, 1960.

D. Gale and L.S. Shapley. College admission and the stability of marriage. *American Mathematical Monthly*, 69:9–15, 1962.

D. Gale and F.M. Stewart. Infinite games with perfect information. In H.W. Kuhn, editor, *Contribution to the theory of games*, volume 2, pages 245–266. Princeton University Press, Princeton, 1953.

W. Gartner. A brief introduction to combinatorial game theory. University of California San Diego, 2007. URL math.ucsd.edu/~wgarner/research/pdf/brief_intro_cgt.pdf.

A. Gibbard. Manipulation of voting schemes: A general result. *Econometrica*, 41:587–601, 1973.

T. Gilligan and K. Krehbiel. Collective decision making and standing committees: An informational rationale for restrictive amendment procedures. *Journal of Law, Economics and Organization*, 3:287–335, 1987.

J. Greenberg. Cores of convex games without side payment. *Mathematics of Operations Research*, 10:523–525, 1985.

T. Groves and J. Ledyard. Optimal allocation of public goods: a solution to the "free rider" problem. *Econometrica*, 45:783–810, 1977.

P.M. Grundy. Mathematics and games. *Eureka*, 2:6–8, 1939.

V.A. Gurvič. On the normal form of positional games. *Soviet Math. Dokl.*, 25:572–575, 1982.

J.C. Harsanyi. A simplified bargaining model for the $n$-person cooperative game. *International Economic Review*, 4:194–220, 1963.

J.C. Harsanyi. Games with incomplete information played by "Bayesian" players. i. the basic model. *Management Science*, 14:320–334, 1967.

J.C. Harsanyi. Oddness of the number of equilibrium points: A new proof. *International Journal of Game Theory*, 2:235–250, 1973.

J.C. Harsanyi and R. Selten. A generalized Nash bargaining solution for two-person bargaining games with incomplete information. *Management Science*, 18:80–106, 1972.

D. Hart. An axiomatization of Harsanyi's nontransferable utility solution. *Econometrica*, 53: 1295–1313, 1985.

S. Hart. Games in extensive and strategic forms. In R.J. Aumann and S. Hart, editors, *Handbook of Game Theory*, volume 1. Elsevier Science Publishers, Amsterdam, 1992.

S. Hart and A. Mas-Colell. A simple adaptive procedure leading to correlated equilibrium. *Econometrica*, 68(1127–1150), 2000.

S. Hart and D. Schmeidler. Existence of correlated equilibria. *Mathematics of Operations Research*, 14:18–25, 1989.

J. Hillas. On the definition of the strategic stability of equilibria. *Econometrica*, 58:1365–1390, 1990.

E. Kalai and M. Smorodinsky. Other solutions to Nash's bargaining problem. *Econometrica*, 43:513–518, 1975.

M. Kaneko. Some remarks on the folk theorem in game theory. *Mathematical Social Sciences*, 3:281–290, 1982.

H. Keiding and L. Thorlund-Petersen. The core of a cooperative game without side payments. *Journal of Optimization Theory*, 54:273–288, 1987.

B. Knaster, C. Kuratowski, and S. Mazurkiewicz. Ein Beweis des Fixpunktsatzes für *n*-dimensionale Simplexe. *Fundamentae Mathematicae*, 14:132–137, 1929.

E. Kohlberg and J.-F. Mertens. On the strategic stability of equilibria. *Econometrica*, 54: 1003–1038, 1986.

K. Krohn and J. Rhodes. Algebraic theory of machines. i. Prime decomposition theorem for finite. *Transactions of the AMS*, 116:450–464, 1965.

H.W. Kuhn. Extensive games and the problem of information. In H.W. Kuhn and A.W. Tucker, editors, *Contributions to the theory of games*, volume II, pages 193–216. Princeton University Press, Princeton, 1953.

W.F. Lucas. The proof that a game may not have a solution. *Transactions of the AMS*, 137: 219–229, 1969.

D. Luce and H. Raiffa. *Games and Decisions*. Wiley, New York, 1957.

G. Mailath and L. Mester. A positive analysis of bank closure. *Journal of Financial Intermediation*, 3:272–299, 1994.

D.A. Martin. Borel determinacy. *ZAnnals of Mathematics*, 102:363–371, 1975.

L.M. Marx and J.M. Swinkels. Order independence for iterated weak dominance. *Games and Economic Behavior*, 18(219–247), 1997.

M. Maschler. The bargaining set, kernel, and nucleolus. In R.J. Aumann and S. Hart, editors, *Handbook of Game Theory*, volume 1, pages 591–667. Elsevier Science Publishers, Amsterdam, 1992.

M. Maschler and M.A. Perles. The superadditive solution for the Nash bargaining game. *International Journal of Game Theory*, 10:163–193, 1981.

E. Maskin. Nash equilibrium and welfare optimality. *The Review of Economic Studies*, 66: 23–38, 1999.

J.-F. Mertens and S. Zamir. Formulation of Bayesian analysis for games with incomplete information. *International Journal of Game Theory*, 14:1–29, 1985.

S. Morris and H. Shin. Global games: theory and applications. In M. Dewatripont, L. Hansen, and S. Turnovsky, editors, *Advances in Economics and Econometrics: Theory and Applications, Eighth World Congress*, pages 56–114. Cambridge University Press, 2003.

H. Moulin. Le choix social utilitariste. Technical report, Ecole Polytechnique DP, 1983.

H. Moulin and B. Peleg. Cores of effectivity functions and implementation theory. *Journal of Mathematical Economics*, 10:115–145, 1982.

R. Myerson. Optimal auction design. *Mathematics of Operations Research*, 6:58–73, 1981.

R. Myerson. Two-person bargaining problems with incomplete information. *Econometrica*, 52:461–487, 1984.

R.B. Myerson. Refinements of the Nash equilibrium concept. *International Journal of Game Theory*, 7:73–80, 1978.

M. Nakayama. A note on a generalization of the nucleolus to games without sidepayments. *International Journal of Game Theory*, 12:115–122, 1983.

J. Nash. Two-person cooperative games. *Econometrica*, 21:129–140, 1953.

D.G. Pearce. Rationalizable strategic behavior and the problem of perfection. *Econometrica*, 52:1029–1050, 1984.

B. Peleg and P. Südholter. *Introduction to the theory of cooperative games*. Springer, Berlin, 2nd edition, 2007.

A. Predtetchinski and P.J.-J. Herings. A necessary and sufficient condition for non-emptiness of the core of a non-transferable utility game. *Journal of Economic Theory*, 116:84–92, 2004.

P.J. Reny and M.H. Wooders. The partnered core of a game without side payments. *Journal of Economic Theory*, 70:298–311, 1996.

J.C. Rochet. Selection of an unique equilibrium value for extensive games with perfect information. Discussion paper, Ceremade, Université Paris IX, 1980.

R.T. Rockafellar. *Convex analysis*. Princeton University Press, Princeton, New Jersey, 1970.

A.E. Roth. Values for games without sidepayments: Some difficulties with current concepts. *Econometrica*, 48:457–466, 1980.

A.E. Roth. The economics of matching: Stability and incentives. *Mathematics of Operations Research*, 7:617–628, 1982.

A.E. Roth and M. Sotomayor. Two-sided matching. In R.J. Aumann and D. Hart, editors, *Handbook of Game Theory*, volume 1, pages 485–541. Elsevier Science Publishers, Amsterdam, 1992.

A. Rubinstein. Equilibrium in supergames with the overtaking criterion. *Journal of Economic Theory*, 21:1–9, 1979.

A. Rubinstein. Perfect equilibrium in a bargaining model. *Econometrica*, 50:97–109, 1982.

A. Rubinstein. Finite automata play the repeated prisoner's dilemma. *Journal of Economic Theory*, 39:83–96, 1986.

M. Satterthwaite. Strategy-proofness and Arrow's conditions: Existence and correspondence theorems for voting preferences and social welfare functions. *Journal of Economic Theory*, 10:187–217, 1975.

L.J. Savage. *The foundations of statistics*. Wiley, New York, 1954.

D. Schleicher and M. Stoll. An introduction to Conway's games and numbers. *Mosc. Math. J.*, 6:359–388, 2006.

D. Schmeidler. The nucleolus of a characteristic function game. *SIAM J. Appl. Math.*, 17: 1163–1170, 1969.

R. Selten. Spieltheoretische Behandlung eines Oligopolmodells mit Nachfrageträgheit. *Zeitschrift für die gesamte Staatswissenschaft*, 121:301–324, 667–689, 1965.

L.S. Shapley. A solution with an arbitrary closed component. In A.W. Tucker and R.D. Luce, editors, *Contributions to the theory of games IV*, pages 87–93. Princeton University Press, Princeton, 1959.

L.S. Shapley. Utility comparison and the theory of games. In *La Décision*, pages 251–263. Editions du CNRS, 1969.

M. Shubik. *Game theory in the social sciences: Concepts and solutions*. The MIT Press, Cambridge, Massachusetts, 1984.

A.I. Sobolev. A characterization by functional equations of optimality in cooperative games (in Russian). *Mat. metody v sots. naukach*, 6:94–151, 1975.

A.I. Sobolev and S.L. Pechorsky. *Optimal distribution in socialeconomic problems and cooperative games (in Russian)*. Nauka, Leningrad, 1983.

S. Sorin. Repeated games with complete information. In R.J. Aumann and S. Hart, editors, *Handbook of Game Theory*, volume 1, chapter 4, pages 71–107. Elsevier, Amsterdam, 1992.

R.P. Sprague. Über mathematische Kampfspiele. *Tohoku Mathematical Journal*, 41:438–444, 1936.

S. Tadelis. *Game theory: an introduction*. Princeton University Press, Princeton, 2013.

S. Tijs. Bounds for the core and the $\tau$-value. In O. Moeschlin and D. Pallaschke, editors,

*Game theory and mathematical economics*, pages 123–132. North-Holland, Amsterdam, 1981.

J. Tirole. *The theory of industrial organization*. The MIT Press, Cambridge, Massachusetts, 1988.

D. Vermeulen and M. Jansen. The reduced form of a game. *European Journal of Operational Research*, 106:204–211, 1998.

J. von Neumann and O. Morgenstern. *Theory of games and economic behavior*. Princeton University Press, Princeton, 1944.

E.C. Zeeman. Dynamics of the evolution of animal conflicts. *Journal of Theoretical Biology*, 89:249–270, 1981.

E. Zermelo. Über eine Anwendung der Mengenlehre auf der Theorie des Schachspiels. In *Proceedings of the Fifth Ingternational Congress on Mathematics*, Cambridge, 1913. Cambridge University Press.

# Index

Printed in the United States
By Bookmasters